Algebra, Geometry, and Software Systems

T0181096

Springer
Berlin
Heidelberg
New York
Hong Kong
London
Milan
Paris
Tokyo

Michael Joswig · Nobuki Takayama (Eds.)

Algebra, Geometry, and Software Systems

With 68 Figures

 Springer

Editors

Michael Joswig

Technische Universität Berlin
Institut für Mathematik, MA 6-2
Straße des 17. Juni 136
10623 Berlin
Germany
e-mail: joswig@math.tu-berlin.de

Nobuki Takayama

Kobe University
Department of Mathematics
Rokko, Kobe 657-8501
Japan
e-mail: takayama@math.kobe-u.ac.jp

Cataloging-in-Publication Data applied for

A catalog record for this book is available from the Library of Congress.

Bibliographic information published by Die Deutsche Bibliothek
Die Deutsche Bibliothek lists this publication in the Deutsche Nationalbibliografie;
detailed bibliographic data is available in the Internet at http://dnb.ddb.de

Mathematics Subject Classification (2000): M14042 Mathematical Software

ISBN 978-3-642-05539-3

Springer-Verlag Berlin Heidelberg New York
a member of BertelsmannSpringer Science+Business Media GmbH

http://www.springer.de

© Springer-Verlag Berlin Heidelberg 2010
Printed in Germany

Cover production: Erich Kirchner, Heidelberg

Printed on acid-free paper   46/3142db - 5 4 3 2 1 0 –

Preface

This book is about computer mathematics and mathematical software. Within its range of topics it tries to address experts as well as graduate level students. To this end this volume comprises research papers and surveys with an abundance of explicit examples.

There are several angles from which this book can be read. Firstly, as far as the mathematical content is concerned, many articles in this book deal with topics on the border between algebra and geometry, e.g., elimination theory. There are few exceptions to this rule, and these chapters can be considered as being supplementary material to the main thread.

Since we are particularly interested in the algorithmic aspects, and beyond that, computational results, the second focus is on software. Next to the mathematics almost every paper gives an introduction to some software system developed by the authors. Several articles contain comparisons of the software presented to other systems with similar functionality. This method shall enable the reader to gain an overview of available mathematical software in algebra and geometry. Moreover, a few papers discuss about designs and internals of software systems. This is a new direction that we follow in this book; there does not seem to be a forum for developers of mathematical software to discuss such topics.

From our own experience we are led to say that developing software on the border of two rich mathematical subjects — like algebra and geometry — is both particularly challenging and particularly rewarding. The benefit is the wide range of possible applications, and from relying on results in several fields together emerges the challenge to combine many different algorithms. It then becomes indispensable to take into account the technological aspects of mathematical software. This leads us to the third topic dealt with in this book: The economic imperative requires to combine existing software into meta-systems of some sort. For several of the systems and packages presented this particular consideration was pivotal already in the early design phase. Additionally, other articles give indications of how to overcome the barriers between existing software systems.

The ordering of the articles suggest one possible sequence of reading. Although the papers are mostly self contained the non-expert will find that the papers more at the beginning of the book contain large introductions to certain topics in discrete geometry. These may form a helpful background in particular for the following sequence of articles on elimination theory. The book then touches upon further topics in computer algebra and symbolic computation. This is followed by articles on algebraic surfaces and geometric theorem proving. Finally, the book closes with two papers devoted to communication issues within and between mathematical software systems.

The idea to compile this book grew out of the workshop "Integration of Algebra and Geometry Software Systems" held at Schloss Dagstuhl in October 2001. In spite of the fact that there is a very large intersection with the workshop's participants and the authors, this is not a mere proceedings volume. It is our hope that the significant effort of the referees and the numerous discussions with the authors helped to conglomerate the individual chapters into one coherent work.

We started the Dagstuhl workshop with the conviction that there is a need for a joint platform for mathematicians who develop software to do research in mathematics. At the same time we felt that our enterprise could fail badly: Since we tried to bring together many people from diverse areas it was not clear in the beginning whether there could be a sufficient common ground for useful discussions. To our relief the participants debated passionately until late in the evening. We hope that at least some of the spirit of this week in Dagstuhl becomes felt from the pages of this book.

We thank the participants of the workshop, the contributors to this book, and also the referees of the individual articles. Oliver Labs suggested the surface* on the front cover, and he also provided the image, for which we are thankful. Further we gratefully acknowledge the wonderful atmosphere of Schloss Dagstuhl and the cordial reception by Rainer Wilhelm and his staff. Special thanks go to Christoph Eyrich who helped out here and there with his typographic experience and his abundant LaTeX knowledge. Finally we are indebted to Martin Peters at Springer whose enthusiasm made this book possible.

Berlin, Kobe,   *Michael Joswig*
November 2002   *Nobuki Takayama*

---

* The algebraic surface $C_8 \subset \mathbf{P}^3$ on the cover belongs to a series similar to the one introduced by Chmutov, see page 419 in the monograph by V.I. Arnold, S.M. Gusein-Zade, and A.N. Varchenko: Singularities of Differential Maps, Vol. II, Birkhäuser 1998. The surface C_8 has exactly 144 real singularities of type A_1 and is given by the affine equation $\sum_{i=1}^{3} (T_4(x_i))^2 = 1$, where $T_4(x_i) = 8x_i^4 - 8x_i^2 + 1$ denotes the Chebyshev polynomial of degree 4 with critical values ± 1. For more information see http://www.OliverLabs.net/series.

Contents

Contents

Beneath-and-Beyond Revisited

Michael Joswig*

Technische Universität Berlin, Institut für Mathematik, MA 6–2
Straße des 17. Juni 136, 10623 Berlin, Germany
joswig@math.tu-berlin.de, http://www.math.tu-berlin.de/~joswig

Abstract. It is shown how the Beneath-and-Beyond algorithm can be used to yield another proof of the equivalence of V- and H-representations of convex polytopes. In this sense this paper serves as the sketch of an introduction to polytope theory with a focus on algorithmic aspects. Moreover, computational results are presented to compare Beneath-and-Beyond to other convex hull implementations.

1 Introduction

One of the most prominent algorithmic problems in computational geometry is the convex hull problem: Given a finite set of points $S \in \mathbb{R}^d$, compute the facets (for instance, in terms of their normal vectors) of the convex hull conv S. To the non-expert it may come as a surprise that fundamental questions concerning the complexity of this problem are still unsettled. Numerous methods have been invented through the last three decades. However, the performance of each known algorithm heavily depends on specific combinatorial properties of the input polytope conv S. Our current knowledge can largely be summarized by three statements: For many algorithms there is a class of polytopes for which the given algorithm performs well. For all known algorithms there is a class of polytopes for which the given algorithm performs badly. There are classes of polytopes for which all known algorithms perform badly.

For a comprehensive survey on convex hull algorithms we refer the reader to the paper [3] by Avis, Bremner, and Seidel.

The Beneath-and-Beyond algorithm is among the most natural methods for convex hull computation. It is treated thoroughly in many text books, see, for instance, Grünbaum [20, 5.2] and Edelsbrunner [10, Section 8.4]. The purpose of this paper is twofold: Firstly, we want to sketch how this particular algorithm can be used to give yet another proof of the "Main theorem for polytopes", see Ziegler [31, Theorem 1.1], which says that the convex hull of finitely many points is exactly the same as a bounded intersection of finitely many affine halfspaces. Previously known proofs are based on Fourier-Motzkin elimination, as in Grötschel [19, Kapitel 3] and Ziegler [31], on the

* Supported by Deutsche Forschungsgemeinschaft, Sonderforschungsbereich 288 "Differentialgeometrie und Quantenphysik" and Forschungszentrum "Mathematik für Schlüsseltechnologien."

simplex method for linear optimization, see Schrijver [29, Chapter 7], or non-algorithmic, as in Grünbaum [20, 3.1.1] or Matoušek [26, Theorem 5.2.2]. Our strategy for a proof of the "Main theorem for polytopes" via the Beneath-and-Beyond algorithm could be phrased as: Prove everything directly for simplices and then inductively use triangulations to extend the results to arbitrary polytopes. Secondly, we give a brief description of the implementation of the Beneath-and-Beyond algorithm in `polymake` [14–16]. The paper closes with a survey of computational results of Beneath-and-Beyond in comparison with Fourier-Motzkin elimination (or, dually, the double description method), implemented by Fukuda [12], and reverse search, implemented by Avis [2]. This complements the computational results in [3].

I am indebted to Thilo Schröder who helped to obtain the timing data in Section 5. Thanks to Volker Kaibel, Marc E. Pfetsch, and Günter M. Ziegler for their comments and many helpful discussions. And, finally, thanks to Ewgenij Gawrilow for co-authoring `polymake` and his unfailing technical advice. The polytope images were produced with `JavaView` [28] and `polymake`.

2   Definitions, an Algorithm, and a Classical Theorem

A subset S of the Euclidean space $\mathbb{R}^d$ is *convex* if for any two points $x, y \in S$ the line segment $[x, y] = \{\lambda x + (1 - \lambda)y \mid 0 \leq \lambda \leq 1\}$ between x and y is contained in S. We define the *convex hull* of S, which is denoted by $\mathrm{conv}(S)$, as the smallest convex set containing S. It is easy to see that, equivalently,

$$
\mathrm{conv}(S) = \left\{ \sum_{i=1}^{n} \lambda_i x_i \;\middle|\; x_i \in S,\; \lambda_i \geq 0,\; \sum_{j=1}^{n} \lambda_i = 1 \right\}. \tag{1}
$$

For the purpose of this paper the key objects are *(convex) polytopes*, that is, the convex hulls of finite point sets in $\mathbb{R}^d$.

An affine hyperplane H defines two (closed) affine halfspaces H^+ and H^- with $H^+ \cup H^- = \mathbb{R}^d$ and $H^+ \cap H^- = H$. Let $S \subset \mathbb{R}^d$ be any set. A hyperplane H with $(H^+ \setminus H) \cap S \neq \emptyset$ and $(H^- \setminus H) \cap S \neq \emptyset$ is said to *separate* S. A hyperplane H *supports* the set S if H intersects S non-trivially and if, moreover, either $S \subset H^+$ or $S \subset H^-$. We will always assume that $S \subset H^+$, that is, if H supports S we consider H to be *positively oriented* toward S.

The *dimension* of any set $S \subset \mathbb{R}^d$ is defined as the dimension of its affine span $\mathrm{aff}(S)$; it is denoted by $\dim S$.

Throughout the following let $X \subset \mathbb{R}^d$ be finite. A *proper face* of the polytope $P = \mathrm{conv}(X)$ is the intersection of P with a supporting hyperplane which does not contain P. Note that this last condition is superfluous in the case where $\dim P = d$, that is, $\mathrm{aff}(P) = \mathbb{R}^d$. Occasionally, the empty set and the set P are also considered as (non-proper) faces of P. Some types of faces of a polytope will play particularly important roles in our investigations: the faces of dimensions 0, 1, $d - 2$, and $d - 1$, are called *vertices, edges, ridges,* and *facets*, respectively.

A point $x \in S \subseteq \mathbb{R}^d$ is a *relatively interior point* of S if there is a small $\dim S$-dimensional ball around x which is contained in S. The *boundary* of S is the complement of the relative interior of S in S; it is denoted by $\mathrm{bnd}(S)$.

For each facet F of a polytope P one can choose a *facet defining* affine halfspace F^+ whose boundary hyperplane supports P in F. A *complete* set of facet defining halfspaces contains exactly one such halfspace for each facet. If $\dim P = d$ then this choice is unique.

In the following we abbreviate "d-dimensional polytope" and "k-dimensional face" by "d-polytope" and "k-face", respectively. The number of k-faces of a polytope P is denoted as $f_k(P)$. The sequence $(f_0(P), \ldots, f_{d-1}(P))$ is called the f-vector of the d-polytope P.

Consider $n+1$ affinely independent points $x_1, \ldots, x_{n+1} \in \mathbb{R}^d$. Their convex hull is an *n-simplex*. A direct computation yields the following result.

Proposition 2.1. *The k-faces of any n-simplex $\Delta_n = \mathrm{conv}(x_1, \ldots, x_{n+1})$ are precisely the convex hulls of all $k+1$ element subsets of $\{x_1, \ldots, x_{n+1}\}$. In particular, the proper k-faces are k-simplices and $f_k(\Delta_n) = \binom{n+1}{k+1}$. Moreover, each n-simplex in $\mathbb{R}^n$ is the intersection of its $n+1$ facet defining halfspaces, and the boundary of a simplex is the union of its facets.*

Observe that
$$\left\{ \lambda \in \mathbb{R}^n \mid \lambda_i \geq 0, \sum \lambda_i = 1 \right\}$$
is an $(n-1)$-dimensional simplex. Hence the Equation (1) implies that each polytope is the linear projection of a high-dimensional simplex.

A *(geometric) simplicial complex* is a finite collection $\mathcal{T}$ of simplices in $\mathbb{R}^d$ with the following two properties:

a. Each proper face of a simplex in $\mathcal{T}$ is also contained in $\mathcal{T}$.
b. The intersection of any two simplices $\Delta, \Delta' \in \mathcal{T}$ is a (possibly empty) face of both, Δ and Δ'.

A simplicial complex $\mathcal{T}$ is a *triangulation* of a set $S \subset \mathbb{R}^d$ if the union of all simplices in $\mathcal{T}$ is S. The k-dimensional elements of a simplicial complex $\mathcal{T}$ again are called k-*faces* of $\mathcal{T}$. A trivial example: Any simplex (together with its collection of faces) is a triangulation of itself.

Throughout the following we will assume that simplicial complexes (and triangulations) are *pure*, that is, all (with respect to inclusion) maximal faces have the same dimension.

Usually, it is more convenient to assume that a given polytope $P = \mathrm{conv}(X)$ affinely spans its ambient space $\mathbb{R}^d$. This is justified — also in an algorithmic setting — due to the following reasoning. By performing Gaussian elimination we can determine the dimension $\dim P = \dim X$, and we can even select an affine basis of the span of X. Moreover, for instance by omitting redundant coordinates, we can project X affinely isomorphic to a linear subspace L of $\mathbb{R}^d$ with $\dim L = \dim X$. A triangulation of the projection of P can directly be lifted back to $P \subset \mathbb{R}^d$. Similarly for facet defining halfspaces.

Lemma 2.2. *Let T be any triangulation of a d-polytope $P \subset \mathbb{R}^d$. Then the following holds.*

a. *For each face F of P the set $T(F) = \{\Delta \in T \mid \Delta \subseteq F\}$ is a triangulation of F.*
b. *For each $(d-1)$-face Δ of T contained in the boundary of P there is a unique facet of P which contains Δ.*

Proof. Let H be a hyperplane which supports P such that the intersection $H \cap P$ is the face F. We have to show that for each point $x \in F$ there is a face $\Delta \in T$ with $x \in \Delta$ and $\Delta \subseteq F$. For each face of T the intersection with H is again a face of T. But, T covers P, that is, some face of T contains x. Its intersection with H is the desired face Δ. This proves the first statement.

Now let $\Delta \in T$ be a $(d-1)$-face in $\mathrm{bnd}(P)$. Since T is pure, there exists a d-face $\Delta' \in T$ with the property that Δ is a facet of the d-simplex Δ'. Therefore, there is a hyperplane H with $H^+ \supset \Delta'$ and $H \cap \Delta' = \Delta$. Choose a point y in the relative interior of Δ. Suppose that H separates P, that is, there is a point $x \in H^- \cap P$. Observe that $\Delta'' = \mathrm{conv}(\Delta, x)$ is a d-simplex, which is contained in P since P is convex. Now $\Delta' \cup \Delta''$ is a d-dimensional ball which contains y in its interior. This contradicts $y \in \mathrm{bnd}(P)$, and thus H defines a facet of P.

A direct consequence of the preceding lemma is the correctness of the Algorithm A below, which computes the complete set of facet defining halfspaces of a polytope from a given triangulation.

Algorithm A: Extracting the facets of a polytope from a triangulation.

> **Data**   : triangulation T of $P = \mathrm{conv}(X)$
> **Result** : complete set of facet defining halfspaces of P
>
> $\mathcal{F} \leftarrow \emptyset$
> **foreach** $(d-1)$-*face* $\Delta \in T$ **do**
>   **if** $\mathrm{aff}(\Delta)$ *does not separate* X **then**
> $\mathcal{F} \leftarrow \mathcal{F} \cup \{\mathrm{aff}(\Delta)^+\}$
> **return** $\mathcal{F}$

Now we are ready to prove the main result of this section: Each polytope is a bounded intersection of finitely many affine halfspaces. Observe that the actual statement of the theorem is, in fact, much stronger. This is necessary to allow for an easy inductive proof.

Theorem 2.3. *Let $P = \text{conv}(X)$ be a polytope and $\mathcal{F}$ a complete set of facet defining affine halfspaces. Then the following holds.*

   a. There is a triangulation $\mathcal{T}$ of P such that the vertices of $\mathcal{T}$ are precisely the points in X.

   b. The polytope P is the intersection of $\text{aff}(P)$ with the intersection of all halfspaces in $\mathcal{F}$.

   c. The boundary of P is the union of its facets.

Proof. We give a constructive proof. More precisely, starting from the finite set X, we construct $\mathcal{T}$ and $\mathcal{F}$ with the desired properties. As pointed out above we can assume that P is full-dimensional.

For the rest of the proof we fix an arbitrary ordering of the set $X = \{x_1, \ldots, x_n\}$ such that the first $d+1$ points $x_1, \ldots, x_{d+1}$ are affinely independent. This ordering gives us a sequence of d-polytopes $P_k = \text{conv}\{x_1, \ldots, x_k\}$ for $k \in \{d+1, \ldots, n\}$. We have $P_{k+1} = \text{conv}(P_k, x_{k+1})$ and we proceed by induction on k.

Since $x_1, \ldots, x_{d+1}$ are affinely independent, their convex hull P_{d+1} is a simplex. The vertices $x_1, \ldots, x_{d+1}$ of P_{d+1} clearly are the vertices of the trivial triangulation of P_{d+1} by itself. The facet defining halfspaces $\mathcal{F}_{d+1}$ of P_{d+1} are given by Proposition 2.1. And we have that $\bigcap \mathcal{F}_{d+1} = P_{d+1}$ and P_{d+1} is the union of its facets.

For the inductive step suppose that we have a triangulation $\mathcal{T}_k$ and the facet defining halfspaces $\mathcal{F}_k$ of P_k. We construct a triangulation $\mathcal{T}_{k+1}$ by using Algorithm B below, where we let $P = P_k$ and $x = x_{k+1}$. It is clear that this yields a simplicial complex, but we have to prove that $\bigcup \mathcal{T}_{k+1}$ is P_{k+1}. Without loss of generality $x_{k+1} \notin P_k$, otherwise $P_{k+1} = P_k$ and the algorithm does not change the triangulation.

The induction hypothesis provides us with a facet F of P_k which is violated by x_{k+1}. By Lemma 2.2a we have a $(d-1)$-face Δ of $\mathcal{T}_k$ which is contained in F, and $\text{conv}(\Delta, x_{k+1}) \in \mathcal{T}_{k+1}$. Further, consider some point $x \in P_k$. By compactness, the line segment $[x, x_k]$ meets the boundary of P_k in a point. Due to compactness the line segment $[x, x_{k+1}]$ meets the boundary in a (not necessarily unique) point y. Again by induction, $\text{bnd}(P_k)$ is the union of its facets, so y is contained in some facet of P_k and therefore also in some $(d-1)$-face of $\mathcal{T}_k$. We obtain

$$P_{k+1} = \bigcup_{x \in P_k} \text{conv}\{x, x_{k+1}\}$$

and hence $\mathcal{T}_{k+1}$ is a triangulation of P_{k+1}.

Now the Algorithm A can be used to determine the complete set $\mathcal{F}_{k+1}$ of facet defining halfspaces of P_{k+1}. It is clear that $P_{k+1} \subseteq \bigcap \mathcal{F}_{k+1}$. For the reverse inclusion suppose that z is a point in $\bigcap \mathcal{F}_{k+1} \setminus P_{k+1}$. By repeating the same argument as before we obtain a $(d-1)$-face $\Delta \in \mathcal{T}_{k+1}$ in $\text{bnd}(P_{k+1})$

such that $\mathrm{aff}(\Delta)$ is a hyperplane which intersects P_{k+1} in a facet and which separates z from P_{k+1}. This yields the desired contradiction.

Making use of both parts of Lemma 2.2 allows to conclude that $\mathrm{bnd}(P_{k+1})$ is the union of the facets of P_{k+1}. The theorem is proved.

Algorithm B: Extending a triangulation.

> **Data**   : triangulation $\mathcal{T}$ of d-polytope P, facet defining halfspaces $\mathcal{F}$
> of P, point x
> **Result** : triangulation of $P' = \mathrm{conv}(P, x)$
>
> $\mathcal{T}' \leftarrow \mathcal{T}$
> **foreach** $F^+ \in \mathcal{F}$ **do**
> $\quad$ **if** $x \notin F^+$ **then**
> $\quad\quad$ **foreach** $(d-1)$-*face* $\Delta \in \mathcal{T}$ *with* $\Delta \subset \mathrm{bnd}(F^+)$ **do**
> $\quad\quad\quad$ $\mathcal{T}' \leftarrow \mathcal{T}' \cup \{\text{faces of } \mathrm{conv}(\Delta, x)\}$
>
> **return** $\mathcal{T}'$

As a direct application we obtain Caratheodory's Theorem:

Corollary 2.4. *Let $X \subset \mathbb{R}^d$ be a (full-dimensional) finite set of points and $p \in \mathrm{conv}(X)$. Then there is an affinely independent subset $X' = \{x_0, \ldots, x_d\} \subseteq X$ with $p \in \mathrm{conv}(X')$.*

Proof. By Theorem 2.3 the polytope $\mathrm{conv}(X)$ admits a triangulation $\mathcal{T}$ with vertex set X. For X' choose the set of vertices of any d-simplex in $\mathcal{T}$ which contains p.

This is where we end our algorithmically inspired introduction to polytope theory. Of course, this is not the end of the story. The next step, which then paves the way to the rest of the theory, would be to prove the converse of Theorem 2.3: Each bounded intersection of finitely many affine halfspaces is a polytope. One classical way of proving this is via the Separation Theorem, see Matoušek [26, Theorem 1.2.4], and duality, see [26, 5.1]. Then Theorem 2.3 itself can be applied to prove its converse. It is interesting that it is also possible to reverse the order in which one proves these results.

3   Sizes of Triangulations and Algorithm Complexity

One of the major open questions in computational geometry is whether there is a *polynomial total time* convex hull algorithm, that is, an algorithm whose running time is bounded by a polynomial in the number of vertices *and* facets. It is unreasonable to hope for an algorithm whose running time depends polynomially on the input size only, since there are families of polytopes whose numbers of facets grow exponentially with the number of vertices. For instance, take the d-dimensional cross polytopes

$$\mathrm{conv}\{\pm e_1, \ldots, \pm e_d\}$$

with $2d$ vertices and 2^d facets.

In the following we relate the complexity of Beneath-and-Beyond to the size of the triangulations produced. Then we review some explicit constructions of polytopes with large triangulations.

3.1 Complexity Analysis

We want to examine how the Beneath-and-Beyond method fits into the picture: What is its complexity? Rather than delving into technicalities we want to exhibit the geometric core of this question. The very coarse and schematic description of the algorithm in steps A and B overestimates the costs of the Beneath-and-Beyond method. For hints to a more practical approach see the next section.

We begin with Algorithm A, which extracts the facets of a d-polytope P from a triangulation $\mathcal{T}$. We call the number of d-faces of $\mathcal{T}$ the *size* of $\mathcal{T}$. If $t = \mathrm{size}(\mathcal{T})$ then $\mathcal{T}$ has at most $(d+1)t$ faces of dimension $d-1$. A facet normal vector can be computed from d affinely independent points contained in the facet by Gaussian elimination, which requires $O(d^3)$ steps. The overall complexity of Algorithm A is then bounded by $O(d^4 nt)$, where n is the number of vertices of P. Now consider Algorithm B: The input is a d-polytope P_k with m_k facets and n_k vertices, a triangulation $\mathcal{T}_k$ with t_k faces of dimension d, and one extra point x_{k+1}. The desired output is a triangulation of $P_{k+1} = \mathrm{conv}(P_k, x_{k+1})$, which can be computed by evaluating $O(dm_k t_k)$ scalar products since each d-simplex of $\mathcal{T}_k$ contains exactly $d+1$ simplices of dimension $d-1$; this gives a total of $O(d^2 m_k t_k)$ arithmetic operations. To clarify the exposition, throughout the following we assume constant costs for each arithmetic operation.[1] Summing up we obtain an upper bound for the complexity of a single Beneath-and-Beyond step.

Lemma 3.1. *The vertices, the facets and the triangulation $\mathcal{T}_{k+1}$ of P_{k+1} can be computed from the triangulation $\mathcal{T}_k$ of P_k in $O(d^4 \max(m_k, n_k) t_k)$ steps.*

We have $t_k \le t_{k+1}$ and $m_k \le (d+1)t_k$. Setting $m = m_n$ and $t = t_n$ we can sum up for all $k \in \{d+1, \ldots, n\}$ to obtain the following result.

---

[1] This severe restriction does not give the complete picture: Goodman, Pollack, and Sturmfels [17] prove that even the coding length of simplicial polytopes in terms of their vertex coordinates is not polynomially bounded.

Proposition 3.2. *The overall complexity of the Beneath-and-Beyond algorithm is bounded by $O(d^5 n t^2)$.*

Of course, an implementation of the algorithm as sketched in Algorithms A and B is far from optimal, see Section 4 below for a few more details. Moreover, the analysis given is very coarse and could be sharpened easily[2]. However, for the purpose intended it is good enough: Since the size of the final triangulation clearly is a lower bound, the Beneath-and-Beyond algorithm runs in polynomial total time if and only if the size of the triangulation constructed is bounded by a polynomial in m and n.

And this could be the end of the story, because it is known that products of simplices form a family of polytopes where the size of *any* triangulation is super-polynomial in the number of vertices and facets, see Haiman [21] or Avis, Bremner, and Seidel [3, Lemma 3]. We conclude that Beneath-and-Beyond has a worst case super-exponential total running-time.

It is a consequence of the Upper Bound Theorem, see Ziegler [31, Section 8.4], that the parameter t is bounded by $O(n^{\lfloor d/2 \rfloor})$. This bound is actually attained, for instance, by the *cyclic polytopes*, see [31, Example 0.6], which arise as convex hulls of finitely many points on the *moment curve*

$$t \mapsto (t, t^2, t^3, \dots, t^d).$$

If the input points are sorted it is possible to avoid looking at all the facets of P_k in Algorithm B. In this way Algorithm B can be replaced by a method which takes time which is proportional to the number of facets of P_k which are not facets of P_{k+1}. Amortized analysis then shows that the size t of the final triangulation T_n only enters linearly into the cost function. In particular, for fixed dimension d, one obtains an $O(n \log n + n^{\lfloor (d+1)/2 \rfloor})$ algorithm, see Edelsbrunner [10, Section 8.4.5].

There is a fairly general result due to Bremner [6] who proves that each incremental convex hull algorithm has a worst-case super-polynomial total running-time, where *incremental* means that the algorithm has to compute the convex hulls of all intermediate polytopes P_k. In particular, this also proves that Beneath-and-Beyond is not a polynomial total time algorithm.

3.2 Polytopes with Large Triangulations

In spite of the fact that there is no polynomial total time convex hull algorithm known, some of the known algorithms have a polynomially bounded running-time on special classes of polytopes. In particular, the reverse search method by Avis and Fukuda [4] runs in $O(dmn)$ time on simplicial polytopes. Each simplicial polytope has a small triangulation: Choose a vertex v and

---

[2] Using the dual graphs of the polytopes P_k in an essential way, it is possible to establish an $O(d^4 n t)$ algorithm.

cone (with apex v) over all the facets not passing through v. Such a triangulation is extremely small, it is clearly of size $O(m)$. Since the size of a
triangulation is the decisive factor for Beneath-and-Beyond's time complexity, this raises the question whether Beneath-and-Beyond could at least have
a polynomial total running-time on simplicial polytopes. We note that the
polytopes in Bremner's construction [6] are products of simplicial polytopes,
that is, they are neither simplicial nor simple.

A *placing triangulation* of a polytope P is a triangulation produced by the
Beneath-and-Beyond algorithm for some ordering of the vertices. For a given
ordering, the size of the corresponding placing triangulation is related to the
question of how many facets of the intermediate polytope P_k are violated by
the next vertex v_{k+1}. A placing triangulation with respect to a given vertex
order is the same as a pushing triangulation with respect to the reverse order,
see Lee [25]. In particular, placing triangulations are lexicographic and thus
regular.

In the dual setting, that is with the roles of vertices and facets inter-
changed, Avis, Bremner, and Seidel [3] introduced the concept of *dwarfing*:
Informally speaking, a separating hyperplane H is called *dwarfing facet* for a
polytope P if all facets of P are also facets of $P \cap H^+$, and if very many vertices of P are not vertices of $P \cap H^+$. Equivalently, the star of the vertex v_H
dual to H in any placing triangulation of the polar polytope $(P \cap H^+)^*$,
where v_H comes last, is large.

The regular d-dimensional cube $C_d = [0,1]^d$ has 2^d vertices and $2d$ facets.
Consider the affine halfspace $H_d^+ = \{x \in \mathbb{R}^d \mid \sum x_i \le 3/2\}$. The boundary
hyperplane $H_d = \partial H_d^+$ separates the cube C_d. Now, as in [3, Theorem 4], call
the simple d-polytope

$$c_d = C_d \cap H_d^+$$

the *dwarfed d-cube*. It has $2d+1$ facets and d^2+1 vertices: The origin 0 and
the d unit vectors $e_1, \ldots, e_d$ are the only vertices of C_d which are contained
in H_d^+; there is one new vertex $e_i + \frac{1}{2}e_j$ for each $i, j \in \{1, \ldots, d\}$ with $i \ne j$.
For illustrations see Figures 1 and 2. By Barnette's Lower Bound Theorem,
see Brønsted [7, §19], d^2+1 is the minimal number of vertices for a simple
d-polytope with $2d+1$ facets. It follows that combinatorially the dwarfed
cubes can be obtained from a simplex by repeated truncation of faces.

Since the cube and the dwarfed cube have $d+1$ vertices in common, there
are $2^d - d - 1$ vertices of C_d which are not vertices of c_d. We conclude the
following result.

Proposition 3.3. *Let $v_1, \ldots, v_{2d+1}$ be an ordering of the vertices of the polar
c_d^* of the dwarfed d-cube c_d such that the last vertex $v' = v_{2d+1}$ corresponds to
the dwarfing facet H_d of c_d. Then the number of d-simplices which contain v'
in the induced placing triangulation of c_d^* equals $2^d - d - 1$.*

The dwarfed cubes form a family of polytopes that are "bad" as input for
the Beneath-and-Beyond algorithm. Since there is only one dwarfed cube per

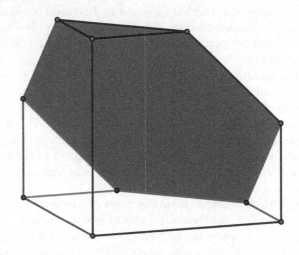

Figure 1. Dwarfed 3-cube c_3 with dwarfing facet marked.

dimension, this does not tell anything about the situation in fixed dimension. However, the same idea can be applied to another (bi-parametric) family of polytopes, whose number of vertices and facets is unbounded even if the dimension is fixed. We sketch the construction from [3], and we omit the proofs.

For $d = 2\delta \geq 4$ and $s \geq 3$ let $G_d(s)$ be the d-dimensional polytope defined by the following list of δs linear inequalities, all of which define facets:

$$y_k \geq 0 \tag{2}$$
$$sx_k - y_k \geq 0 \tag{3}$$
$$(2i+1)x_k + y_k \leq (2i+1)(s+i) - i^2 + s^2 \tag{4}$$
$$(2s-3)x_k + y_k \leq 2s(2s-3), \tag{5}$$

where $i \in \{0, \ldots, s-4\}$, $k \in \{1, \ldots, \delta\}$, and a vector in $\mathbb{R}^d$ is written as $(x_1, y_1, x_2, y_2, \ldots, x_\delta, y_\delta)$. The polytope $G_d(s)$ is the product of δ copies of the s-gon obtained by fixing k in the list of inequalities 2 to 5. This product of s-gons has s^δ vertices. There is a dwarfing halfspace $H_{d,s}^+ = \{x \in \mathbb{R}^d \mid \sum x_i \leq 2s - 1\}$, and the *dwarfed product of s-gons*

$$g_d(s) = G_d(s) \cap H_{d,s}^+$$

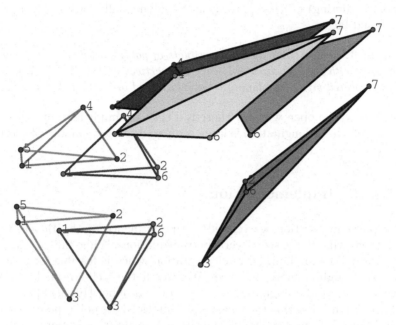

Figure 2. Explosion of the placing triangulation of the polar polytope c_3^* induced by the ordering indicated as vertex labels. Only the four 3-simplices containing the final vertex (numbered 7), which corresponds to the dwarfing facet of c_3, are displayed in solid.

is simple. It has $\delta s + 1$ facets but only $(d-1)(\delta s + 1 - d) + 2$ vertices, which, again, meets Barnette's lower bound; see [3, Theorem 6].

Since $G_d(s)$ and $g_d(s)$ can have at most $(d-1)(\delta s + 1 - d) + 1$ vertices in common, this yields a similar result as for the dwarfed cubes.

Proposition 3.4. *Let $v_1, \ldots, v_{(d-1)(\delta s + 1 - d) + 2}$ be an ordering of the vertices of the polar $g_d(s)^*$ of the dwarfed product of polygons $g_d(s)$ such that the last vertex $v' = v_{(d-1)(\delta s + 1 - d) + 2}$ corresponds to the dwarfing facet $H_{d,s}$ of $g_d(s)$. Then the number of d-simplices which contain v' in the induced placing triangulation of $g_d(s)^*$ is at least $s^\delta - (d-1)(\delta s + 1 - d) - 1$.*

This already shows that the worst case running time of Beneath-and-Beyond is not polynomially bounded in combined the size of the input and the output total time algorithm. Additionally, one can show that $G_d(s)$ and $g_d(s)$ share exactly $\delta(s-2) + 1$ vertices, so the precise number of d-simplices containing v' is $s^\delta - \delta s + 2\delta - 1$.

The closer analysis in [3, Theorem 12] reveals that from Proposition 3.4 it follows that even a typical placing triangulation (that is, with respect to a random ordering) of $g_d(s)^*$ grows super-polynomially with the number of vertices and facets.

Theorem 3.5. *The polar dwarfed products of polygons $g_d(s)^*$ are simplicial d-polytopes with $(d-1)(ds/2+1-d)+2$ vertices and $ds/2+1$ facets, but with an expected size of a placing triangulation of order $\Omega(s^{d/2}/(d+1))$.*

It seems to be open whether there is a class of simplicial polytopes such that *each* placing triangulation is large compared to their number of (vertices and) facets.

4   On the Implementation

In the preceding section we related the performance of the Beneath-and-Beyond algorithm to the size of certain triangulations. While this captures the main ideas of the method, an implementation which is feasible for practical problems is slightly more involved. We sketch the well-known key points below; for a more thorough discussion see Edelsbrunner [10, 8.4.5].

Rather than extracting the facets of the intermediate polytope P_k from the triangulation $\mathcal{T}_k$ from scratch (as Algorithm A suggests) it is more natural to store the facets and their neighborhood structure in terms of the dual graph of P_k. Adding the next vertex v_{k+1} then requires: (i) to find one violated facet, (ii) to perform a breadth-first-search in the dual graph of P_k to find all the other violated facets, (iii) to extend the triangulation $\mathcal{T}_k$ of P_k to the triangulation $\mathcal{T}_{k+1}$ of P_{k+1} by coning over the induced triangulation of the violated facets, (iv) to determine the new facets, that is, the facets of P_{k+1} which are not facets of P_k by examining $\mathcal{T}_{k+1}$.

Edelsbrunner [10, 8.4.5] advocates to sort the input points lexicographically in order to eliminate the time for finding the first violated facet, which then allows to find all the violated facets in linear time (in fixed dimension). Our implementation does not rely on ordered input, but it is programmed in a way such that Edelsbrunner's analysis applies, if the input happens to be ordered. This more flexible algorithm has the advantage that it can possibly benefit from smaller placing triangulations. The best strategy for most cases seems to be to permute the input randomly rather than to sort it.

We omit a discussion of the data structures required in the implementation. Instead we want to to spend a few words on the arithmetic to be used. In a principal way, convex hull computations makes sense over any (computationally feasible) ordered field. Natural choices certainly include the field of rational numbers as well as certain (or even all) algebraic extensions.[3] In the following we focus on the rationals and their extensions by radical expressions.

---

[3] For a more general perspective on the subject see the book of Blum et al. [5].

Examining the Algorithms A and B more closely reveals that computations within the coordinate domain are necessary only to decide whether a given point is contained in a given affine hyperplane and, if not, on which side it lies. That is to say, if we are only interested in combinatorially correct output, it suffices to evaluate signs of determinants without ever knowing their precise values. This approach, sometimes called *robust* or *exact geometric computation*, is taken in several computational geometry libraries including CGAL [8] and LEDA [24]. It has the advantage that it can be extended to radical extensions of ℚ rather easily. This is implemented in LEDA and in the CORE library [30].

If, on the other hand, we do need the exact facet normal vectors then we have to use an exact implementation of the arithmetic functions of our coordinate domain. Today's standard for long integer and rational arithmetic is the GNU Multiprecision Library (GMP) [18]. In fact, this is the arithmetic used in polymake's implementation of Beneath-and-Beyond as well as in the codes of Avis [2] and Fukuda [11]. There are packages in computer algebra and computational number theory which can perform computations in (arbitrary) finite extensions of ℚ, but currently this functionality does not seem to be available as a stand-alone library which encapsulates a given field as a number type to be used in standard C or C++ code.

Even if one agrees to compute with rational coordinates, on a technical level there is still one choice to be made: One can either compute with representations of rational numbers directly, or one can translate everything into integer coordinates, essentially by scaling. While this usually does not make much of a difference for the algorithms, the actual numbers that occur during the computations are different and, in particular, of different sizes. This may affect the performance of an algorithm, but often one does not know in advance which method is superior to the other.

It would be interesting to combine the two techniques, exact geometric computation and exact coordinates, for the Beneath-and-Beyond algorithm in the following way: Use exact geometric computation to produce the combinatorially triangulation, and only in the end compute the facet normals by solving systems of linear equations defined by the $(d-1)$-faces on the boundary. Since in many cases the arithmetic consumes most of the running time it should be possible to save some time this way. The author is not aware of an implementation of any convex hull code taking this approach.

5   Empirical Results

The analysis in Section 3 showed that, from a theoretical point of view, Beneath-and-Beyond seems to be a weak algorithm: It has a super-polynomial total running time even on simplicial polytopes. In this section we display a few computational results, where the performance of polymake's implementation beneath_beyond is compared to two other programs: Fukuda's cdd which implements (dual) Fourier-Motzkin elimination, see Ziegler [31, Sections 1.2 and 1.3] or Fukuda and Prodon [13], and Avis' lrs, which uses reverse search, see Avis and Fukuda [4,1].

The computational results below are intended to complement the corresponding data in the paper by Avis, Bremner, and Seidel [3]. The performance comparison among the three programs is fair in the sense that all programs use the same implementation of exact rational arithmetic, namely from the GNU Multiprecision Library (GMP) [18]. However, beneath_beyond and cdd both use rational coordinates while lrs uses integers only. This could contribute to beneath_beyond's superiority over lrs for the "random spheres", see Figures 5 and 6; but this needs further investigation.

Roughly speaking the input polytopes come in two groups: The first group consists of the intricate polytopes discussed in the previous section. These are the dwarfed cubes, products of simplices, and dwarfed products of polygons. While they are known to be computationally hard for all iterative convex hull algorithms, this study shows that beneath_beyond performs particularly bad. The second class of polytopes investigated are convex hulls of (uniformly distributed) random points on the unit sphere, sometimes called "random spheres". Such polytopes are almost always simplicial (in fact, all the tested ones were simplicial). For the "random spheres" the performance of beneath_beyond is clearly better than that of lrs, which in turn clearly beats cdd.

The "random spheres" were constructed with polymake's client program rand_sphere which produces uniformly distributed random points on the unit sphere with double (i.e., IEEE 64-Bit floating point) coordinates rounded to six digits after the decimal point, which were then transformed into exact rational numbers; the numerators and denominators of such rational numbers typically have 15 to 20 decimal digits each.

Table 1. The timings (in seconds) for the individual runs of cdd for s random points on the unit sphere in $\mathbb{R}^6$, see Figure 6, vary in a rather narrow range. We checked 10 cases for each value of s, produced independently at random.

s	Average	Minimum	Maximum	Standard deviation
100	258.997	247.05	273.58	7.5541
120	427.075	416.67	446.39	8.3978
140	641.947	627.80	666.61	12.0566
160	907.759	895.52	938.18	12.9157
180	1238.253	1202.69	1265.81	18.9353
200	1630.816	1591.89	1670.96	23.6467
220	2078.458	2028.65	2132.95	29.9611
240	2582.256	2495.02	2650.31	41.1209

While in [3] the authors also compared various insertion strategies for the incremental algorithms we did not do that here: cdd was run with the lexmin insertion rule (which is its default behavior). This seems justified in view of the computational results in loc. cit., which showed that lexmin is optimal among the strategies tested in most cases.

A few more remarks on the beneath_beyond timings: As a consequence of the general layout of the polymake system the beneath_beyond client requires point data as input. Therefore, for the dual convex hull problems in the test, namely the dwarfed cubes and the dwarfed products of polygons, the input for beneath_beyond was appropriately translated and dualized. For cdd and lrs the input was not modified in order to allow for better comparison with the results in [3]. Further, in addition to the convex hull beneath_beyond actually computes (and saves) the placing triangulation induced by the respective ordering of the vertices.

All timings are taken on several identical Linux machines with an Athlon XP 1800+ processor (1533.433 MHz clock, 3060.53 bogomips) and 512 MB main memory running RedHat Linux 7.3. The timings were taken via Perl's time() function, and we counted the time in user mode only. The parameter datasize was limited to 400 MB. Since all programs were run through polymake, by calling the clients cdd_ch_client and lrs_ch_client, respectively, there is a certain additional overhead (due to socket communication and data conversion) which should approximately be the same for the three programs. This makes up for the slight difference in shape of some of the curves as compared to the corresponding ones in [3]. In particular, timings below one second are almost impossible to interpret this way. Moreover, the implementations of cdd and lrs tested here are more advanced than the ones tested in loc. cit. In particular, the old versions of cdd and lrs did not use the (considerably faster) GMP arithmetic.

Due to the fact that the timings include some input/output overhead it is impossible to reproduce exact timing data. Therefore we took the average over several runs. Additionally, we randomly permuted the order of the input. While cdd and lrs are almost insensitive to the ordering, this can make a considerable difference for beneath_beyond. The highest variation among the individual timings of beneath_beyond occurred for the dwarfed products of polygons: A more detailed statistical analysis is given in Table 2.

For the "random spheres" we changed our experimental set-up slightly. For each set of parameters $d \in \{3, 4, 5, 6\}$ and $s \in \{100, 120, 140, \ldots, 500\}$ we produced 10 polytopes as the convex hulls of s random points on the unit sphere $\mathbb{S}^{d-1} \subset \mathbb{R}^d$. Each of the three convex hull codes was run once on each polytope (for beneath_beyond we inserted the points in the same order as they were produced). The charts show the average values for each code taken over the 10 samples. The individual timings varied only by a little: The highest deviations occurring for cdd and $d = 6$, see Table 1 on page 14 for more details.

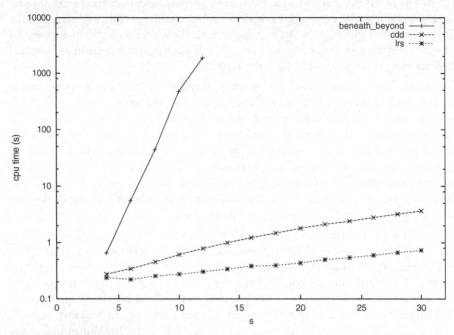

Figure 3. Dwarfed products $g_{10}(s)$ of five s-gons. In this test s is always even. Compare [3, Figure 5]. cdd and lrs: average over 10 runs, beneath_beyond: average over 50 runs. Memory overflow (more than 400 MB required) in beneath_beyond for $s > 12$. For more details of the beneath_beyond timings see Table 2.

Table 2. The timings (in seconds) for the individual runs of beneath_beyond for the dwarfed products of polygons $g_{10}(s)$, see Figure 3, vary rather strongly. 50 runs with a random insertion order were performed.

s	Average	Minimum	Maximum	Standard deviation
4	0.650	0.35	1.05	0.1628
6	5.530	0.78	23.20	4.6607
8	44.091	3.26	226.55	40.7324
10	476.051	10.00	1984.59	452.4177
12	1883.964	239.46	8560.20	1803.4397

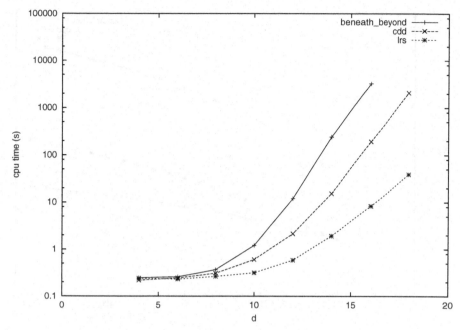

Figure 4. Dwarfed d-cubes c_d, for d even. Compare [3, Figure 4]. `cdd` and `lrs`: average over 10 runs, `beneath_beyond`: average over 50 runs, memory overflow (more than 400 MB required) for $d > 16$.

6   Concluding Remarks

Authors in discrete geometry often talk about "triangulations of finite point sets". They refer to triangulations of the convex hull of these points. For more details see Pfeifle and Rambau [27].

Finite simplicial complexes, as combinatorial abstractions of triangulations, are an indispensible tool in topology. At first sight it may seem accidental that notions from topology appear in questions concerning the complexity of convex hull computations. However, as developed in the paper [22] there is a deeper connection: Up to polynomial equivalence a convex hull computation can be replaced by suitable simplicial homology computations. For details on algorithms and implementations to compute homology see Dumas et al. [9]. The paper [23] by Kaibel and Pfetsch contains more information about the complexity status of the convex hull problem.

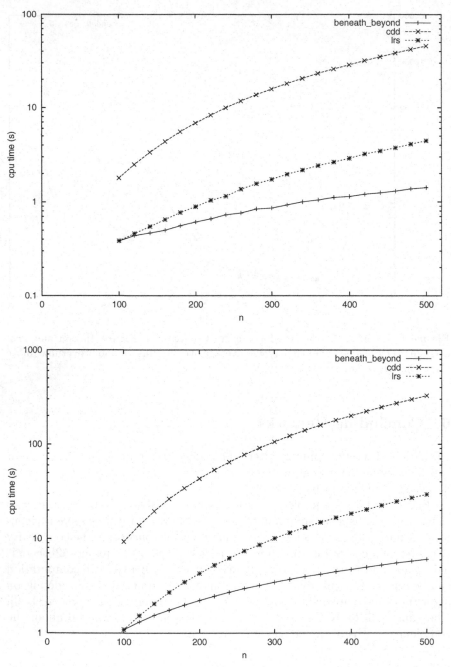

Figure 5. "Random spheres" with n vertices in dimensions 3 (top) and 4 (bottom). Average over 10 polytopes, each program run only once. Note that our timings below 1 second are not very accurate.

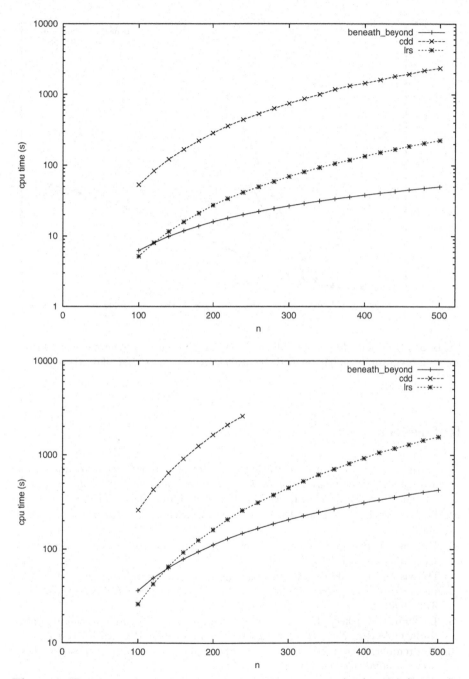

Figure 6. "Random spheres" with n vertices in dimensions 5 (top) and 6 (bottom). Average over 10 polytopes, each program run only once. cdd not tested for input with more than 240 vertices since it takes about three hours per test.

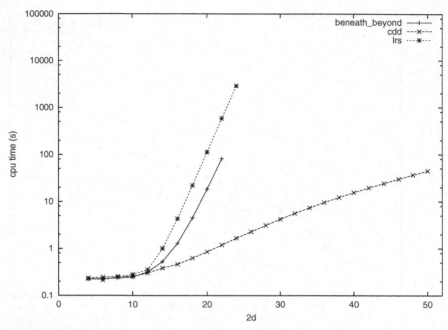

Figure 7. Products of two d-simplices. Compare [3, Figure 2]. Average over 10 runs for each program. `beneath_beyond`: Memory overflow (more than 400 MB required) for $d > 11$.

References

1. D. Avis. A revised implementation of the reverse search vertex enumeration algorithm. In *Polytopes—combinatorics and computation (Oberwolfach, 1997)*, volume 29 of *DMV Sem.*, pages 177–198. Birkhäuser, Basel, 2000.

2. D. Avis. lrs, lrslib, Version 4.1. `http://cgm.cs.mcgill.ca/~avis/C/lrs.html`, 2001.

3. D. Avis, D. Bremner, and R. Seidel. How good are convex hull algorithms? *Comput. Geom.*, 7(5-6):265–301, 1997.

4. D. Avis and K. Fukuda. A pivoting algorithm for convex hulls and vertex enumeration of arrangements and polyhedra. *Discrete Comput. Geom.*, 8(3):295–313, 1992.

5. L. Blum, F. Cucker, M. Shub, and S. Smale. *Complexity and real computation*. Springer-Verlag, New York, 1998. With a foreword by Richard M. Karp.

6. D. Bremner. Incremental convex hull algorithms are not output sensitive. *Discrete Comput. Geom.*, 21(1):57–68, 1999.

7. A. Brøndsted. *An introduction to convex polytopes*, volume 90 of *Graduate Texts in Mathematics*. Springer-Verlag, New York, 1983.

8. CGAL, Version 2.4. `http://www.cgal.org/`, 2002.

9. J.-G. Dumas, F. Heckenbach, D. Saunders, and V. Welker. Computing simplicial homology based on efficient smith normal form algorithms. In *this volume*, pages 177–206.

10. H. Edelsbrunner. *Algorithms in combinatorial geometry.* Springer-Verlag, Berlin, 1987.

11. K. Fukuda. cdd+, Version 0.76a. `http://www.cs.mcgill.ca/~fukuda/soft/cdd_home/cdd.html`, 2001.

12. K. Fukuda. cddlib, Version 0.92a. `http://www.cs.mcgill.ca/~fukuda/soft/cdd_home/cdd.html`, 2001.

13. K. Fukuda and A. Prodon. Double description method revisited. In *Combinatorics and computer science (Brest, 1995)*, volume 1120 of *Lecture Notes in Comput. Sci.*, pages 91–111. Springer, Berlin, 1996.

14. E. Gawrilow and M. Joswig. polymake, version 1.5.1: a software package for analyzing convex polytopes. `http://www.math.tu-berlin.de/diskregeom/polymake`, 1997–2003.

15. E. Gawrilow and M. Joswig. polymake: a framework for analyzing convex polytopes. In G. Kalai and G. M. Ziegler, editors, *Polytopes — Combinatorics and Computation*, pages 43–74. Birkhäuser, 2000.

16. E. Gawrilow and M. Joswig. polymake: an approach to modular software design in computational geometry. In *Proceedings of the 17th Annual Symposium on Computational Geometry*, pages 222–231. ACM, 2001. June 3-5, 2001, Medford, MA.

17. J.E. Goodman, R. Pollack, and B. Sturmfels. The intrinsic spread of a configuration in $\mathbb{R}^d$. *J. Amer. Math. Soc.* 3(3):639–651, 1990.

18. GNU multiprecision library, Version 4.1. `http://www.swox.com/gmp/`, 2002.

19. M. Grötschel. Optimierungsmethoden I. Technical report, Universität Augsburg, 1985. Skriptum zur Vorlesung im WS 1984/85.

20. B. Grünbaum. *Convex polytopes.* Springer, 2003. 2nd edition edited by V. Kaibel, V. Klee, and G.M. Ziegler.

21. M. Haiman. A simple and relatively efficient triangulation of the n-cube. *Discrete Comput. Geom.*, 6(4):287–289, 1991.

22. M. Joswig and G. M. Ziegler. Convex hulls, oracles, and homology. Preprint, 11 pages, `arXiv: math.MG/0301100`.

23. V. Kaibel and M. E. Pfetsch. Some algorithmic problems in polytope theory. In *this volume*, pages 23–47.

24. LEDA, Version 4.3. Algorithmic Solution Software GmbH, `http://www.algorithmic-solutions.com/as_html/products/products.html`.

25. C. W. Lee. Subdivisions and triangulations of polytopes. In J. Goodman and J. O'Rourke, editors, *Handbook of Discrete and Computational Geometry*, pages 271–290. CRC Press, 1997.

26. J. Matoušek. *Lectures on Discrete Geometry.* Springer, 2002.

27. J. Pfeifle and J. Rambau. Computing triangulations using oriented matroids. In *this volume*, pages 49–75.

28. K. Polthier, S. Khadem, E. Preuss, and U. Reitebuch. JavaView, Version 2.21. `http://www.javaview.de`.

29. A. Schrijver. *Theory of linear and integer programming.* Wiley-Interscience Series in Discrete Mathematics. John Wiley & Sons Ltd., Chichester, 1986.

30. C. Yap and Z. Du. Core Library (CORE), Version 1.4. `http://cs.nyu.edu/exact/core/`, 2002.

31. G. M. Ziegler. *Lectures on Polytopes.* Springer, 1998. 2nd ed.

Some Algorithmic Problems
in Polytope Theory

Volker Kaibel* and Marc E. Pfetsch

Technische Universität Berlin, Institut für Mathematik, MA 6-2,
Straße des 17. Juni 136, 10623 Berlin, Germany
{kaibel,pfetsch}@math.tu-berlin.de

Abstract. Many interesting algorithmic problems naturally arise in the theory of convex polytopes. In this article we collect 35 such problems and briefly discuss the current knowledge on their complexity status.

1 Introduction

Convex polyhedra, i.e., the intersections of finitely many closed affine half-spaces in $\mathbb{R}^d$, are important objects in various areas of mathematics and other disciplines. In particular, the compact ones among them (*polytopes*), which equivalently can be defined as the convex hulls of finitely many points in $\mathbb{R}^d$, have been studied since ancient times (e.g., the platonic solids). Polytopes appear as building blocks of more complicated structures, e.g., in (combinatorial) topology, numerical mathematics, or computer aided design. Even in physics polytopes are relevant (e.g., in crystallography or string theory).

Probably the most important reason for the tremendous growth of interest in the theory of convex polyhedra in the second half of the 20th century was the fact that *linear programming* (i.e., optimizing a linear function over the solutions of a system of linear inequalities) became a widespread tool to solve practical problems in industry (and military). Dantzig's Simplex Algorithm, developed in the late 40's, showed that geometric and combinatorial knowledge of polyhedra (as the domains of linear programming problems) is quite helpful for finding and analyzing solution procedures for linear programming problems.

Since the interest in the theory of convex polyhedra to a large extent comes from algorithmic problems, it is not surprising that many algorithmic questions on polyhedra arose in the past. But also inherently, convex polyhedra (in particular: polytopes) give rise to algorithmic questions, because they can be treated as finite objects by definition. This makes it possible to investigate (the smaller ones among) them by computer programs (like the `polymake`-system written by Gawrilow and Joswig, see [26] and [27,28]). Once chosen to exploit this possibility, one immediately finds oneself confronted with many algorithmic challenges.

---

* Supported by the Deutsche Forschungsgemeinschaft, FOR 413/1–1 (Zi 475/3–1).

This paper contains descriptions of 35 algorithmic problems about polyhedra. The goal is to collect for each problem the current knowledge about its computational complexity. Consequently, our treatment is focused on theoretical rather than on practical subjects. We would, however, like to mention that for many of the problems computer codes are available.

Our choice of problems to be included is definitely influenced by personal interest. We have not spent particular efforts to demonstrate for each problem why we consider it to be relevant. It may well be that the reader finds other problems at least as interesting as the ones we discuss. We would be very interested to learn about such problems. The collection of problem descriptions presented in this paper is intended to be maintained as a (hopefully growing) list at http://www.math.tu-berlin.de/~pfetsch/polycomplex/.

Some questions are also interesting for not necessarily bounded polyhedra. It can be tested in polynomial time whether a polyhedron specified by linear inequalities is bounded or not. This can be done by applying Gaussian elimination and solving one linear program.

Roughly, the problems can be divided into two types: problems for which the input are "geometrical" data and problems for which the input is "combinatorial" (see below). Actually, it turned out that it was rather convenient to group the problems we have selected into the five categories "Coordinate Descriptions" (Sect. 2), "Combinatorial Structure" (Sect. 3), "Isomorphism" (Sect. 4), "Optimization" (Sect. 5), and "Realizability" (Sect. 6). Since the boundary complex of a simplicial polytope is a simplicial complex, studying polytopes leads to questions that are concerned with more general (polyhedral) structures: *simplicial complexes*. Therefore, we have added a category "Beyond Polytopes" (Sect. 7), where a few problems concerned with general (abstract) simplicial complexes are collected that are similar to problems on polytopes. We do not consider related areas like oriented matroids.

The problem descriptions proceed along the following scheme. First input and output are specified. Then a summary of the knowledge on the theoretical complexity is given, e.g., it is stated that the complexity is unknown ("Open") or that the problem is $\mathcal{NP}$-hard. This is done for the case where the dimension (usually of the input polytope) is part of the input as well as for the case of fixed dimension; often the (knowledge on the) complexity status differs for the two versions. After that, comments on the problems are given together with references. For each problem we tried to report on the current state of knowledge according to the literature. Unless stated otherwise, all results mentioned without citations are either considered to be "folklore" or "easy to prove." At the end related problems in this paper are listed.

For all notions in the theory of polytopes that we use without explanation we refer to Ziegler's book [65]. Similarly, for the concepts from the theory of computational complexity that play a role here we refer to Garey and Johnson's classical text [24]. Whenever we talk about *polynomial reductions* this refers to polynomial time Turing-reductions. For some of the problems the output can be exponentially large in the input. For these problems the interesting question is whether there is a *polynomial total time* algorithm, i.e., an algorithm whose running time can be bounded by a polynomial in the sizes of the input and the output (in contrast to a *polynomial time* algorithm whose running time would be bounded by a polynomial just in the input size). Note that the notion of "polynomial total time" only makes sense with respect to problems which explicitly require the output to be non-redundant.

A very fundamental result in the theory of convex polyhedra is due to Minkowski [46] and Weyl [64]. For the special case of polytopes (to which we restrict our attention from now on) it can be formulated as follows. Every polytope $P \subset \mathbb{R}^d$ can be specified by an $\mathcal{H}$- or by a $\mathcal{V}$-description. Here, an $\mathcal{H}$-description consists of a finite set of linear inequalities (defining closed affine half-spaces of $\mathbb{R}^d$) such that P is the set of all simultaneous solutions to these inequalities. A $\mathcal{V}$-description consists of a finite set of points in $\mathbb{R}^d$ whose convex hull is P. If any of the two descriptions is rational, then the other one can be chosen to be rational as well. Furthermore, in this case the numbers in the second description can be chosen such that their coding lengths depend only polynomially on the coding lengths of the numbers in the first description (see, e.g., Schrijver [55]). In our context, $\mathcal{H}$- and $\mathcal{V}$-descriptions are usually meant to be rational. By linear programming, each type of description can be made non-redundant in polynomial time (though it is unknown whether this is possible in *strongly* polynomial time, see Problem 24).

One of the basic properties of a polytope is its dimension. If the polytope is given by a $\mathcal{V}$-description, then it can easily be determined by Gaussian elimination (which, carefully done, is a cubic algorithm; see, e.g., [55]). If the polyhedron is specified by an $\mathcal{H}$-description, computing its dimension can be done by linear programming (actually, this is polynomial time equivalent to linear programming).

Furthermore, some of the problems may also be interesting in their polar formulations, i.e., with "the roles of $\mathcal{H}$- and $\mathcal{V}$-descriptions exchanged." Switching to the polar requires to have a relative interior point at hand, which is easy to obtain if a $\mathcal{V}$-description is available, while it needs linear programming if only an $\mathcal{H}$-description is specified.

We will especially be concerned with the combinatorial types of polytopes, i.e., with their *face lattices* (the sets of faces, ordered by inclusion). In particular, some problems will deal with the k-*skeleton* of a polytope, which is the set of its faces of dimensions less than or equal to k, or with its f-*vector*, i.e., the vector $(f_0(P), f_1(P), \ldots, f_d(P))$, where $f_i(P)$ is the number of i-dimensional faces (i-*faces*) of the d-dimensional polytope P (d-*polytope*). Talking of the face lattice $\mathcal{L}_P$ of a polytope P will always refer to the lattice as an abstract object, i.e., to *any* lattice that is isomorphic to the face lattice. In particular, the lattice does not contain any information on coordinates. Similarly, the *vertex-facet incidences* of P are given by any matrix (a_{vf}) with entries from $\{0, 1\}$, whose rows and columns are indexed by the vertices and facets of P, respectively, such that $a_{vf} = 1$ if and only if vertex v is contained in facet f. Note that the vertex-facet incidences of a polytope completely determine its face lattice.

A third important combinatorial structure associated with a polytope P is its (abstract) graph $\mathcal{G}_P$, i.e., any graph that is isomorphic to the graph having the vertices of P as its nodes, where two of them are adjacent if and only if their convex hull is a (one-dimensional) face of P. For simple polytopes, the (abstract) graph determines the entire face lattice as well (see Problem 15). However, for general polytopes this is not true.

Throughout the paper, n refers to the number of vertices or points in the given $\mathcal{V}$-description, respectively, depending on the context. Moreover, m refers to the number of facets or inequalities in the given $\mathcal{H}$-description, respectively, and d refers to the dimension of the polytope or the ambient space, respectively.

Acknowledgment: We thank the referee for many valuable comments and Günter M. Ziegler for carefully reading the manuscript.

2   Coordinate Descriptions

In this section problems are collected whose input are geometrical data, i.e., the $\mathcal{H}$- or $\mathcal{V}$-description of a polytope. Some problems which are also given by geometrical data appear in Sections 4 and 5.

1. VERTEX ENUMERATION

Input: Polytope P in $\mathcal{H}$-description
Output: Non-redundant $\mathcal{V}$-description of P
Status (general): Open; polynomial total time if P is simple or simplicial
Status (fixed dim.): Polynomial time

Let $d = \dim(P)$ and let m be the number of inequalities in the input. It is well known that the number of vertices n can be exponential ($\Omega(m^{\lfloor d/2 \rfloor})$) in the size of the input (e.g., Cartesian products of suitably chosen two-dimensional polytopes and prisms over them).

VERTEX ENUMERATION is strongly polynomially equivalent to Problem 3 (see Avis, Bremner, and Seidel [1]). Since Problem 2 is strongly polynomially equivalent to Problem 3 as well, VERTEX ENUMERATION is also strongly polynomially equivalent to Problem 2.

For fixed d, Chazelle [12] found an $\mathcal{O}(m^{\lfloor d/2 \rfloor})$ polynomial time algorithm, which is optimal by the Upper Bound Theorem of McMullen [43]. There exist algorithms which are faster than Chazelle's algorithm for small n, e.g., an $\mathcal{O}(m \log n + (mn)^{1-1/(\lfloor d/2 \rfloor+1)} \operatorname{polylog} m)$ algorithm of Chan [9].

For general d, the reverse search method of Avis and Fukuda [2] solves the problem for *simple* polyhedra in polynomial total time, using working space (without space for output) bounded polynomially in the input size. An algorithm of Bremner, Fukuda, and Marzetta [8] solves the problem for *simplicial* polytopes. Note that these algorithms need a vertex of P to start from. Provan [52] gives a polynomial total time algorithm for enumerating the vertices of polyhedra arising from networks.

There are many more algorithms known for this problem – none of them is a polynomial total time algorithm for general polytopes. See the overview article of Seidel [57]. Most of these algorithms can be generalized to directly work for unbounded polyhedra, too.

Related problems: 2, 3, 5, 7

2. FACET ENUMERATION

Input: Polytope P in $\mathcal{V}$-description with n points
Output: Non-redundant $\mathcal{H}$-description of P
Status (general): Open; polynomial total time if P is simple or simplicial
Status (fixed dim.): Polynomial time

In [1] it is shown that FACET ENUMERATION is strongly polynomially equivalent to Problem 3 and thus to Problem 1 (see the comments there).

For this problem, one can assume to have an interior point (e.g., the vertex barycenter). FACET ENUMERATION is sometimes called the *convex hull problem*.

Related problems: 1, 3, 5

3. POLYTOPE VERIFICATION

Input: Polytope P given in $\mathcal{H}$-description, polytope Q given in $\mathcal{V}$-description
Output: "Yes" if $P = Q$, "No" otherwise
Status (general): Open; polynomial time if P is simple or simplicial
Status (fixed dim.): Polynomial time

POLYTOPE VERIFICATION is strongly polynomially equivalent to Problem 1 and Problem 2 (see the comments there).

POLYTOPE VERIFICATION is contained in co$\mathcal{NP}$: we can prove $Q \not\subseteq P$ by showing that some vertex of Q violates one of the inequalities describing P. If $Q \subset P$ with $Q \neq P$ then there exists a point p of $P \setminus Q$ with "small" coordinates (e.g., some vertex of P not contained in Q) and a valid inequality for Q, which has "small" coefficients and is violated by p (e.g., an inequality defining a facet of Q that separates p from Q). However, it is unknown whether POLYTOPE VERIFICATION is in $\mathcal{NP}$.

Since it is easy to check whether $Q \subseteq P$, POLYTOPE VERIFICATION is Problem 4 restricted to the case that $Q \subseteq P$.

Related problems: 1, 2, 4

4. Polytope Containment

Input: Polytope P given in $\mathcal{H}$-description, polytope Q given in $\mathcal{V}$-description
Output: "Yes" if $P \subseteq Q$, "No" otherwise
Status (general): co$\mathcal{NP}$-complete
Status (fixed dim.): Polynomial time

Freund and Orlin [20] proved that this problem is co$\mathcal{NP}$-complete. Note that the reverse question whether $Q \subseteq P$ is trivial. The questions where either both P and Q are given in $\mathcal{H}$-description or both in $\mathcal{V}$-description can be solved by linear programming (Problem 24), see Eaves and Freund [17]. For fixed dimension, one can enumerate all vertices of P in polynomial time (see Problem 1) and compare the descriptions of P and Q (after removing redundant points).

Related problems: 3

5. Face Lattice of Geometric Polytopes

Input: Polytope P in $\mathcal{H}$-description
Output: Hasse-diagram of the face lattice of P
Status (general): Polynomial total time
Status (fixed dim.): Polynomial time

See comments on Problem 1. Many algorithms for the VERTEX ENUMERATION PROBLEM in fact compute the whole face lattice of the polytope. Swart [60], analyzing an algorithm of Chand and Kapur [10], proved that there exists a polynomial total time algorithm for this problem. For a faster algorithm see Seidel [56]. Fukuda, Liebling, and Margot [22] gave an algorithm which uses working space (without space for the output) bounded polynomially in the input size, but it has to solve many linear programs.

For fixed dimension, the size of the output is polynomial in the size of the input; hence, a polynomial total time algorithm becomes a polynomial algorithm in this case.

The problem of enumerating the k-skeleton of P seems to be open, even if k is fixed. Note that, for fixed k, the latter problem can be solved by linear programming (Problem 24) in polynomial time if the polytope is given in $\mathcal{V}$-description rather than in $\mathcal{H}$-description.

Related problems: 1, 2, 3, 13, 14

6. Degeneracy Testing

Input: Polytope P in $\mathcal{H}$-description
Output: "Yes" if P not simple (degenerate), "No" otherwise
Status (general): Strongly $\mathcal{NP}$-complete
Status (fixed dim.): Polynomial time

Independently proved to be $\mathcal{NP}$-complete in the papers of Chandrasekaran, Kabadi, and Murty [11] and Dyer [14]. Fukuda, Liebling, and Margot [22] proved that the problem is *strongly* $\mathcal{NP}$-complete. For fixed dimension, one can enumerate all vertices in polynomial time (see Problem 1) and check whether they are simple or not.

Bremner, Fukuda, and Marzetta [8] noted that if P is given in $\mathcal{V}$-description the problem is polynomial time solvable: enumerate the edges (1-skeleton, see Problem 5) and apply the Lower Bound Theorem.

Erickson [19] showed that in the worst case $\Omega(m^{\lceil d/2\rceil - 1} + m \log m)$ sideness queries are required to test whether a polytope is simple. For odd d this matches the upper bound. A *sideness query* is a question of the following kind: given $d + 1$ points $p_0, \ldots, p_d$ in $\mathbb{R}^d$, does p_0 lie "above", "below", or on the oriented hyperplane determined by $p_1, p_2, \ldots, p_d$.

Related problems: 1, 5

7. NUMBER OF VERTICES

Input: Polytope P in $\mathcal{H}$-description
Output: Number of vertices of P
Status (general): $\#\mathcal{P}$-complete
Status (fixed dim.): Polynomial time

Dyer [14] and Linial [40] independently proved that NUMBER OF VERTICES is $\#\mathcal{P}$-complete. It follows that the problem of computing the f-vector of P is $\#\mathcal{P}$-hard. Furthermore, Dyer [14] proved that the decision version ("Given a number k, does P have at least k vertices?") is strongly $\mathcal{NP}$-hard and remains $\mathcal{NP}$-hard when restricted to simple polytopes. It is unknown whether the decision problem is in $\mathcal{NP}$.

If the dimension is fixed, one can enumerate all vertices in polynomial time (see Problem 1).

Related problems: 1, 14

8. FEASIBLE BASIS EXTENSION

Input: Polytope P given as $\{x \in \mathbb{R}^s : Ax = b, x \geq 0\}$, a set $S \subseteq \{1, \ldots, s\}$
Output: "Yes" if there is a feasible basis with an index set containing S, "No" otherwise
Status (general): $\mathcal{NP}$-complete
Status (fixed dim.): Polynomial time

See Murty [49] (Garey and Johnson [24], Problem [MP4]). For fixed dimension, one can enumerate all bases in polynomial time.

The problem can be reformulated as follows. Let P be defined by a finite set $\mathcal{H}$ of affine halfspaces and let S be a subset of $\mathcal{H}$. Does $\bigcap\{H \in \mathcal{H} : H \notin S\}$ contain a vertex which is also a vertex of P?

9. Recognizing Integer Polyhedra

Input: Polytope P in $\mathcal{H}$-description
Output: "Yes" if P has only integer vertices, "No" otherwise
Status (general): Strongly co$\mathcal{NP}$-complete
Status (fixed dim.): Polynomial time

The hardness-proof is by Papadimitriou and Yannakakis [51]. For fixed dimension, one can enumerate all vertices (Problem 1) and check whether they are integral in polynomial time.

10. Diameter

Input: Polytope P in $\mathcal{H}$-description
Output: The diameter of P
Status (general): $\mathcal{NP}$-hard
Status (fixed dim.): Polynomial time

Frieze and Teng [21] gave the proof of $\mathcal{NP}$-hardness. For fixed dimension, one can enumerate all vertices (Problem 1), construct the graph and then compute the diameter in polynomial time.
The complexity status is unknown for simple polytopes. For simplicial polytopes the problem can be solved in polynomial time: Since simplicial polytopes have at most as many vertices as facets, one can enumerate their vertices (see Problem 1), and finally compute the graph (and hence the diameter) from the vertex-facet incidences in polynomial time.
If P is given in $\mathcal{V}$-description, one can compute the graph (1-skeleton, see Problem 5) and hence the diameter in polynomial time.

11. Minimum Triangulation

Input: Polytope P in $\mathcal{V}$-description, positive integer K
Output: "Yes" if P has a triangulation of size K or less, "No" otherwise
Status (general): $\mathcal{NP}$-complete
Status (fixed dim.): $\mathcal{NP}$-complete

A *triangulation* $\mathcal{T}$ of a d-polytope P is a collection of d-simplices, whose union is P, their vertices are vertices of P, and any two simplices intersect in a common face (which might be empty). In particular, $\mathcal{T}$ is a (pure) d-dimensional geometric simplicial complex (see Section 7). The size of $\mathcal{T}$ is the number of its d-simplices. Every (convex) polytope admits a triangulation.
Below, De Loera, and Richter-Gebert [4,5] proved that Minimum Triangulation is $\mathcal{NP}$-complete for (fixed) $d \geq 3$. Furthermore, it is $\mathcal{NP}$-hard to compute a triangulation of minimal size for (fixed) $d \geq 3$.

12. VOLUME

Input: Polytope P in $\mathcal{H}$-description
Output: The volume of P
Status (general): $\#\mathcal{P}$-hard, FPRAS
Status (fixed dim.): Polynomial time

Dyer and Frieze [15] showed that the general problem is $\#\mathcal{P}$-hard (and $\#\mathcal{P}$-easy as well). Dyer, Frieze, and Kannan [16] found a *fully polynomial randomized approximation scheme* (*FPRAS*) for the problem, i.e., a family $(A_\varepsilon)_{\varepsilon>0}$ of randomized algorithms, where, for each $\varepsilon > 0$, A_ε computes a number V_ε with the property that the probability of $(1-\varepsilon)\operatorname{vol}(P) \le V_\varepsilon \le (1+\varepsilon)\operatorname{vol}(P)$ is at least $\frac{3}{4}$, and the running times of the algorithms A_ε are bounded by a polynomial in the input size and $\frac{1}{\varepsilon}$.

For fixed dimension, one can first compute all vertices of P (see Problem 1) and its face lattice (see Problem 5) both in polynomial time. Then one can construct some triangulation (see Problem 11) of P (e.g., its barycentric subdivision) in polynomial time and compute the volume of P as the sum of the volumes of the (maximal) simplices in the triangulation.

The complexity status of the analogue problem with the polytope specified by a $\mathcal{V}$-description is the same.

3   Combinatorial Structure

In this section we collect problems that are concerned with computing certain combinatorial information from compact descriptions of the combinatorial structure of a polytope. Such compact encodings might be the vertex-facet incidences, or, for simple polytopes, the abstract graphs. An example of such a problem is to compute the dimension of a polytope from its vertex-facet incidences. Initialize a set S by the vertex set of an arbitrary facet. For each facet F compute the intersection of S with the vertex set of F. Replace S by a maximal one among the proper intersections and continue. The dimension is the number of "rounds" performed until S becomes empty.

All data is meant to be purely combinatorial. For all problems in this section it is unknown if the "integrity" of the input data can be checked, proved, or disproved in polynomial time. For instance, it is rather unlikely that one can efficiently prove or disprove that a lattice is the face lattice of some polytope (see Problems 29, 30).

Sometimes, it might be worthwhile to exchange the roles of vertices and facets by duality of polytopes. Our choices of view points have mainly been guided by personal taste.

Some orientations of the abstract graph $\mathcal{G}_P$ of a simple polytope P play important roles (although such orientations can also be considered for non-simple polytopes, they have not yet proven to be useful in the more general context). An orientation is called a *unique-sink orientation* (*US-orientation*) if it induces a unique sink on every subgraph of $\mathcal{G}_P$ corresponding to a nonempty face of P. A US-orientation is called an *abstract objective function orientation* (*AOF-orientation*) if it is acyclic. General US-orientations of graphs of cubes have recently received some attention (Szabó and Welzl [61]). AOF-orientations were used, e.g., by Kalai [35]. Since their linear extensions are precisely the shelling orders of the dual polytope, they have been considered much earlier.

13. Face Lattice of Combinatorial Polytopes

Input: Vertex-facet incidence matrix of a polytope P
Output: Hasse-diagram of the face lattice of P
Status (general): Polynomial total time
Status (fixed dim.): Polynomial time

Solvable in $\mathcal{O}(\min\{m, n\} \cdot \alpha \cdot \varphi)$ time, where m is the number of facets, n is the number of vertices, α is the number of vertex-facet incidences, and φ is the size of the face lattice [33]. Note that φ is exponential in d (for fixed d it is polynomial in m and n). Without (asymptotically) increasing the running time it is also possible to label each node in the Hasse diagram by the dimension and the vertex set of the corresponding face.

It follows from [33] that one can compute the Hasse-diagram of the *k-skeleton* (i.e., all faces of dimensions at most k) of P in $\mathcal{O}(n \cdot \alpha \cdot \varphi^{\leq k})$ time, where $\varphi^{\leq k}$ is the number of faces of dimensions at most k. Since the latter number is in $\mathcal{O}(n^{k+1})$, the k-skeleton can be computed in polynomial time (in the input size) for fixed k.

Related problems: 5, 14

14. *f*-Vector of Combinatorial Polytopes

Input: Vertex-facet incidence matrix of a polytope P
Output: f-vector of P
Status (general): Open
Status (fixed dim.): Polynomial time

By the remarks on Problem 13, it is clear that the first k entries of the f-vector can be computed in polynomial time for every fixed k.

If the polytope is simplicial and a shelling (or a partition) of its boundary complex is available (see Problems 17 and 18), then one can compute the entire f-vector in polynomial time [65, Chap. 8].

Related problems: 7, 13, 17, 18, 32

15. RECONSTRUCTION OF SIMPLE POLYTOPES

Input: The (abstract) graph $\mathcal{G}_P$ of a simple polytope P
Output: The family of the subsets of nodes of $\mathcal{G}_P$ corresponding to the vertex
    sets of the facets of P
Status (general): Open
Status (fixed dim.): Open

Blind and Mani [6] proved that the entire combinatorial structure of a simple
polytope is determined by its graph. This is false for general polytopes (of dimension at least four), which is the main reason why we restrict our attention
to simple polytopes for the remaining problems in this section. Kalai [35] gave
a short, elegant, and constructive proof of Blind and Mani's result. However,
the algorithm that can be derived from it has a worst-case running time that
is exponential in the number of vertices of the polytope.

In [32] it is shown that the problem can be formulated as a combinatorial
optimization problem for which the problem to find an AOF-orientation of $\mathcal{G}_P$
(see Problem 17) is strongly dual in the sense of combinatorial optimization.
In particular, the vertex sets of the facets of P have a *good characterization*
in terms of $\mathcal{G}_P$ (in the sense of Edmonds [18]). The problem is polynomial
time equivalent to computing the cycles in $\mathcal{G}_P$ that correspond to the 2-faces
of P.
The problem can be solved in polynomial time in dimension at most three
by computing a planar embedding of the graph, which can be done in linear
time (Hopcroft and Tarjan [30], Mehlhorn and Mutzel [45]).

Related problems: 16, 17, 18

16. FACET SYSTEM VERIFICATION FOR SIMPLE POLYTOPES

Input: The (abstract) graph $\mathcal{G}_P$ of a simple polytope P and a family $\mathcal{F}$ of
    subsets of nodes of $\mathcal{G}_P$
Output: "Yes" if $\mathcal{F}$ is the family of subsets of nodes of $\mathcal{G}_P$ that correspond
    to the vertex sets of the facets of P, "No" otherwise
Status (general): Open
Status (fixed dim.): Open

In [32] it is shown that both the "Yes"- as well as the "No"-answer can be
proved in polynomial time in the size of $\mathcal{G}_P$ (provided that the integrity
of the input data is guaranteed). Any polynomial time algorithm for the
construction of an AOF- or US-orientation (see Problems 17 and 18) of $\mathcal{G}_P$
would yield a polynomial time algorithm for this problem (see [32]).
Up to dimension three the problem can be solved in polynomial time (see the
comments to Problems 15 and 17).

Related problems: 15, 17, 18, 30

17. AOF-Orientation

Input: The (abstract) graph $\mathcal{G}_P$ of a simple polytope P
Output: An AOF-orientation of $\mathcal{G}_P$
Status (general): Open
Status (fixed dim.): Open

(Simple) polytopes admit AOF-orientations because every linear function in general position induces an AOF-orientation.

In [32] it is shown that one can formulate the problem as a combinatorial optimization problem, for which a strongly dual problem in the sense of combinatorial optimization exists (see the comments to Problem 15). Thus, the AOF-orientations of $\mathcal{G}_P$ have a good characterization (see Problem 16) in terms of $\mathcal{G}_P$, i.e., there are polynomial size proofs for both cases an orientation being an AOF-orientation or not (provided that the integrity of the input data is guaranteed). However, it is unknown if it is possible to check in polynomial time if a given orientation is an AOF-orientation.

In dimensions one and two the problem is trivial. For a three-dimensional polytope P the problem can be solved in polynomial time, e.g., by producing a plane drawing of $\mathcal{G}_P$ with convex faces (see Tutte [62]) and sorting the nodes with respect to a linear function (in general position).

A polynomial time algorithm would lead to a polynomial algorithm for Problem 16 (see [32]).

By duality of polytopes, the problem is equivalent to the problem of finding a shelling order of the facets of a simplicial polytope from the upper two layers of its face lattice. It is unknown whether it is possible to find in polynomial time a shelling order of the facets, even if the polytope is given by its entire face lattice. With this larger input, however, it is possible to check in polynomial time whether a given ordering of the facets is a shelling order.

Related problems: 16, 18, 34

18. US-Orientation

Input: The (abstract) graph $\mathcal{G}_P$ of a simple polytope P
Output: A US-orientation of $\mathcal{G}_P$
Status (general): Open
Status (fixed dim.): Open

Since AOF-orientations are US-orientations, it follows from the remarks on Problem 17 that (simple) polytopes admit US-orientations and that the problem can be solved in polynomial time up to dimension three. By slight adaptions of the arguments given in [32], one can prove that a polynomial time algorithm for this problem would yield a polynomial time algorithm for Problem 16 as well.

In contrast to Problem 17, no good characterization of US-orientations is known.

It is not hard to see that, by duality of polytopes, the problem is equivalent to the problem of finding from the upper two layers a partition of the face lattice of a simplicial polytope into intervals whose upper bounds are the facets (i.e., a partition in the sense of Stanley [58]). Similar to the situation with shelling orders, it is even unknown whether such a partition can be found in polynomial time if the polytope is specified by its entire face lattice. Again, with the entire face lattice as input it can be checked in polynomial time whether a family of subsets of the face lattice is a partition in that sense.

Related problems: 16, 17, 35

4   Isomorphism

Two polytopes $P_1 \subset \mathbb{R}^{d_1}$ and $P_2 \subset \mathbb{R}^{d_2}$ are *affinely equivalent* if there is a one-to-one affine map $T : \mathrm{aff}(P_1) \longrightarrow \mathrm{aff}(P_2)$ between the affine hulls of P_1 and P_2 with $T(P_1) = P_2$. Two polytopes are *combinatorially equivalent* (or *isomorphic*) if their face lattices are isomorphic. It is not hard to see that affine equivalence implies combinatorial equivalence.

As soon as one starts to investigate structural properties of polytopes by means of computer programs, algorithms for deciding whether two polytopes are isomorphic become relevant.

Some problems in this section are known to be hard in the sense that the *graph isomorphism problem* can polynomially be reduced to them. Although this problem is not known (and even not expected) to be $\mathcal{NP}$-complete, all attempts to find a polynomial time algorithm for it have failed so far. Actually, the same holds for a lot of problems that can be polynomially reduced to the graph isomorphism problem (see, e.g., Babai [3]).

19. AFFINE EQUIVALENCE OF $\mathcal{V}$-POLYTOPES

Input: Two polytopes P and Q given in $\mathcal{V}$-description
Output: "Yes" if P is affinely equivalent to Q, "No" otherwise
Status (general): Graph isomorphism hard
Status (fixed dim.): Polynomial time

The graph isomorphism problem can polynomially be reduced to the problem of checking the affine equivalence of two polytopes [34]. The problem remains graph isomorphism hard if $\mathcal{H}$-descriptions are additionally provided as input data and/or if one restricts the input to simple or simplicial polytopes.
For polytopes of bounded dimension the problem can be solved in polynomial time by mere enumeration of affine bases among the vertex sets.

Related problems: 20

20. COMBINATORIAL EQUIVALENCE OF $\mathcal{V}$-POLYTOPES

Input: Two polytopes P and Q given in $\mathcal{V}$-description
Output: "Yes" if P is combinatorially equivalent to Q, "No" otherwise
Status (general): co$\mathcal{NP}$-hard
Status (fixed dim.): Polynomial time

Swart [60] describes a reduction of the subset-sum problem to the negation of the problem.
For polytopes of bounded dimension the problem can be solved in polynomial time (see Problems 2 and 22).

Related problems: 2, 19, 22

21. POLYTOPE ISOMORPHISM

Input: The face lattices $\mathcal{L}_P$ and $\mathcal{L}_Q$ of two polytopes P and Q, respectively
Output: "Yes" if $\mathcal{L}_P$ is isomorphic to $\mathcal{L}_Q$, "No" otherwise
Status (general): Open
Status (fixed dim.): Polynomial time

The problem can be solved in polynomial time in constant dimension (see Problem 22). In general, the problem can easily be reduced to the graph isomorphism problem

Related problems: 22, 23

22. ISOMORPHISM OF VERTEX-FACET INCIDENCES

Input: Vertex-facet incidence matrices A_P and A_Q of polytopes P and Q, respectively
Output: "Yes" if A_P can be transformed into A_Q by row and column permutations, "No" otherwise
Status (general): Graph isomorphism complete
Status (fixed dim.): Polynomial time

The problem remains graph isomorphism complete even if $\mathcal{V}$- and $\mathcal{H}$-descriptions of P and Q are part of the input data [34].
In constant dimension the problem can be solved in polynomial time by a reduction [34] to the graph isomorphism problem for graphs of bounded degree, for which a polynomial time algorithm is known (Luks [41]).
Problem 21 can polynomially be reduced to this problem. For polytopes of bounded dimension both problems are polynomial time equivalent.

Related problems: 21, 20

23. SELFDUALITY OF POLYTOPES

Input: Face Lattice of a polytope P
Output: "Yes" if P is isomorphic to its dual, "No" otherwise
Status (general): Open
Status (fixed dim.): Polynomial time

This is a special case of problem 21. In particular, it is solvable in polynomial time in bounded dimensions.
It is easy to see that deciding whether a general $0/1$-matrix A (not necessarily a vertex-facet incidence matrix of a polytope) can be transformed into A^T by permuting its rows and columns is graph isomorphism complete.

Related problems: 21

5   Optimization

In this section, next to the original linear programming problem, we describe some of its relatives. In particular, combinatorial abstractions of the problem are important with respect to polytope theory (and, more general, discrete geometry). We pick out the aspect of combinatorial cube programming here (and leave aside abstractions like general combinatorial linear programming, LP-type problems, and oriented matroid programming), since it has received considerable attention lately.

24. LINEAR PROGRAMMING

Input: $\mathcal{H}$-description of a polyhedron $P \subset \mathbb{Q}^d$, $\boldsymbol{c} \in \mathbb{Q}^d$
Output: $\inf \left\{ \boldsymbol{c}^T \boldsymbol{x} \mid \boldsymbol{x} \in P \right\} \in \mathbb{Q} \cup \{-\infty, \infty\}$ and, if the infimum is finite, a point where the infimum is attained.
Status (general): Polynomial time; no *strongly* polynomial time algorithm known
Status (fixed dim.): Linear time in m (the number of inequalities)

The first polynomial time algorithm was a variant of the *ellipsoid algorithm* due to Khachiyan [38]. Later, also *interior point methods* solving the problem in polynomial time were discovered (Karmarkar [37]).
Megiddo found an algorithm solving the problem for a fixed number d of variables in $\mathcal{O}(m)$ arithmetic operations (Megiddo [44]).
No strongly polynomial time algorithm (performing a number of arithmetic operations that is bounded polynomially in d and the number of half-spaces rather than in the coding lengths of the input coordinates) is known. In particular, no polynomial time variant of the *simplex algorithm* is known. However, a randomized version of the simplex algorithm solves the problem in (expected) subexponential time (Kalai [36], Matoušek, Sharir, and Welzl [42]).

Related problems: 25, 26, 27

25. OPTIMAL VERTEX

Input: $\mathcal{H}$-description of a polyhedron $P \subset \mathbb{Q}^d$, $c \in \mathbb{Q}^d$
Output: $\inf \left\{ c^T v \mid v \text{ vertex of } P \right\} \in \mathbb{Q} \cup \{\infty\}$ and, if the infimum is finite, a vertex where the infimum is attained.
Status (general): Strongly $\mathcal{NP}$-hard
Status (fixed dim.): Polynomial time

Proved to be strongly $\mathcal{NP}$-hard by Fukuda, Liebling, and Margot [22]. By linear programming this problem can be solved in polynomial time if P is a polytope. In fixed dimension one can compute all vertices of P in polynomial time (see Problem 1).

Related problems: 1, 24, 26

26. VERTEX WITH SPECIFIED OBJECTIVE VALUE

Input: $\mathcal{H}$-description of a polyhedron $P \subset \mathbb{Q}^d$, $c \in \mathbb{Q}^d$, $C \in \mathbb{Q}$
Output: "Yes" if there is a vertex v of P with $c^T v = C$; "No" otherwise
Status (general): Strongly $\mathcal{NP}$-complete
Status (fixed dim.): Polynomial time

Proved to be $\mathcal{NP}$-complete by Chandrasekaran, Kabadi, and Murty [11] and strongly $\mathcal{NP}$-complete by Fukuda, Liebling, and Margot [22]. The problem remains strongly $\mathcal{NP}$-complete even if the input is restricted to polytopes [22].

Related problems: 24, 25

27. AOF CUBE PROGRAMMING

Input: An oracle for a function $\sigma : \{0,1\}^d \longrightarrow \{+,-\}^d$ defining an AOF-orientation of the graph of the d-cube
Output: The sink of the orientation
Status (general): Open
Status (fixed dim.): Constant time

The problem can be solved in a subexponential number of oracle calls by the *random facet* variant of the simplex algorithm due to Kalai [36]. For a derivation of the explicit bound $e^{2\sqrt{d}} - 1$ see Gärtner [25].
In fixed dimension the problem is trivial by mere enumeration.
The problem generalizes linear programming problems whose sets of feasible solutions are combinatorially equivalent to cubes.

Related problems: 24, 28

28. USO Cube Programming

Input: An oracle for a function $\sigma : \{0,1\}^d \longrightarrow \{+,-\}^d$ defining a US-orientation of the graph of the d-cube
Output: The sink of the orientation
Status (general): Open
Status (fixed dim.): Constant time

Szabó and Welzl [61] describe a randomized algorithm solving the problem in an expected number of $\mathcal{O}(\alpha^d)$ oracle calls with $\alpha = \sqrt{43/20} < 1.467$ and a deterministic algorithm that needs $\mathcal{O}(1.61^d)$ oracle calls. Plugging an optimal algorithm for the three-dimensional case (found by Günter Rote) into their algorithm, Szabó and Welzl even obtain an $\mathcal{O}(1.438^d)$ randomized algorithm. The problem not only generalizes Problem 27, but also certain linear complementary problems and smallest enclosing ball problems.
In fixed dimension the problem is trivial by mere enumeration.

Related problems: 27

6   Realizability

In this section problems are discussed which bridge the gap from combinatorial descriptions of polytopes to geometrical descriptions, i.e., it deals with questions of the following kind: given combinatorial data, does there exist a polytope which "realizes" this data? E.g., given a 0/1-matrix is this the matrix of vertex-facet incidences of a polytope? The problems of computing combinatorial from geometrical data is discussed in Section 2.

The problems listed in this section are among the first ones asked in (modern) polytope theory, going back to the work of Steinitz and Radermacher in the 1930's [59].

29. Steinitz Problem

Input: Lattice $\mathcal{L}$
Output: "Yes" if $\mathcal{L}$ is isomorphic to the face lattice of a polytope, "No" otherwise
Status (general): $\mathcal{NP}$-hard
Status (fixed dim.): $\mathcal{NP}$-hard

If $\mathcal{L}$ is isomorphic to the face lattice of a polytope, it is ranked, atomic, and coatomic. These properties can be tested in polynomial time in the size of $\mathcal{L}$. Furthermore, in this case, the dimension d of a candidate polytope has to be $\operatorname{rank} \mathcal{L} - 1$.
The problem is trivial for dimension $d \leq 2$. Steinitz's Theorem allows to solve $d = 3$ in polynomial time: construct the (abstract) graph G, test if the facets

can consistently be embedded in the plane (linear time [30,45]) and check for 3-connectedness (in linear time, see Hopcroft and Tarjan [29]).

Mnëv proved that the Steinitz Problem for d-polytopes with $d + 4$ vertices is $\mathcal{NP}$-hard [47]. Even more, Richter-Gebert [53] proved that for (fixed) $d \geq 4$ the problem is $\mathcal{NP}$-hard.

For fixed $d \geq 4$ it is neither known whether the problem is in $\mathcal{NP}$ nor whether it is in co$\mathcal{NP}$. It seems unlikely to be in $\mathcal{NP}$, since there are 4-polytopes which cannot be realized with rational coordinates of coding length which is bounded by a polynomial in $|\mathcal{L}|$ (see Richter-Gebert [53]).

Related problems: 30

30. SIMPLICIAL STEINITZ PROBLEM

Input: Lattice $\mathcal{L}$
Output: "Yes" if $\mathcal{L}$ is isomorphic to the face lattice of a simplicial polytope, "No" otherwise
Status (general): $\mathcal{NP}$-hard
Status (fixed dim.): Open

As for Problem 29, $\mathcal{L}$ is ranked, atomic, and coatomic if the answer is "Yes." In this case, the dimension d of any matched polytope is rank $\mathcal{L} - 1$.

As for general polytopes (Problem 29), this problem is polynomial time solvable in dimension $d \leq 3$.

The problem is $\mathcal{NP}$-hard, which follows from the above mentioned fact that the Steinitz problem for d-polytopes with $d + 4$ vertices is $\mathcal{NP}$-hard and a construction (Bokowski and Sturmfels [7]) which generalizes it to the *simplicial* case (but increases the dimension). It is, however, open whether the problem is $\mathcal{NP}$-hard for *fixed* dimension. For fixed $d \geq 4$, it is neither known whether the problem is in $\mathcal{NP}$ nor whether it is in co$\mathcal{NP}$.

The following question is interesting in connection with Problem 16 (see also the notes there): Given an (abstract) graph G, is G the graph of a simple polytope? If we do not restrict the question to simple polytopes the problem is also interesting.

Related problems: 16, 29

7   Beyond Polytopes

This section is concerned with problems on finite abstract simplicial complexes. Some of the problems listed are direct generalizations of problems on polytopes. Most of the basic notions relevant in our context can be looked up in [65]; for topological concepts like *homology* we refer to Munkres' book [48].

A *finite abstract simplicial complex* Δ is a non-empty set of subsets (the *simplices* or *faces*) of a finite set of *vertices* such that $F \in \Delta$ and $G \subset F$ imply

$G \in \Delta$. The *dimension* of a simplex $F \in \Delta$ is $|F| - 1$. The *dimension* $\dim(\Delta)$ of Δ is the largest dimension of any of the simplices in Δ. If all its maximal simplices with respect to inclusion (i.e., its *facets*) have the same cardinality, then Δ is *pure*. A pure d-dimensional finite abstract simplicial complex whose *dual graph* (defined on the facets, where two facets are adjacent if they share a common $(d-1)$-face) is connected is a *pseudo-manifold* if every $(d-1)$-dimensional simplex is contained in at most two facets. The boundary of a simplicial $(d+1)$-dimensional polytope induces a d-dimensional pseudo-manifold.

Throughout this section a finite abstract simplicial complex Δ is given by its list of facets or by the complete list of all simplices. In the first case, the input size can be measured by n and m, the numbers of vertices and facets.

31. EULER CHARACTERISTIC

Input: Finite abstract simplicial complex Δ given by a list of facets
Output: Euler characteristic $\chi(\Delta) \in \mathbb{Z}$
Status (general): Open
Status (fixed dim.): Polynomial time

It is unknown whether the decision version "$\chi(\Delta) = 0$?" of this problem is in $\mathcal{NP}$. The problem is easy if Δ is given by a list of all of its simplices. For fixed dimension, one can enumerate all simplices of Δ and compute the Euler characteristic in polynomial time.
Currently the fastest way to compute the Euler characteristic is to determine $\mathcal{V} = \{S : S \text{ is an intersection of facets of } \Delta\}$ and then compute $\chi(\Delta)$ in time $\mathcal{O}(|\mathcal{V}|^2)$ by a Möbius function approach, see Rota [54]. Usually $\mathcal{V}$ is much smaller than the whole face lattice of Δ. $\mathcal{V}$ can be listed lexicographically by an algorithm of Ganter [23], in time $\mathcal{O}(\min\{m, n\} \cdot \alpha \cdot |\mathcal{V}|)$, where α is the number of vertex-facets incidences.

Related problems: 32

32. f-VECTOR OF SIMPLICIAL COMPLEXES

Input: Finite abstract simplicial complex Δ given by a list of facets
Output: The f-vector of Δ
Status (general): #$\mathcal{P}$-hard
Status (fixed dim.): Polynomial time

If Δ is given by all of its simplices the problem is trivial. Clearly, for fixed k, the first k entries of the f-vector can be computed in polynomial time, since the number of k-simplices in Δ is polynomial in n. Hence the problem is polynomial time solvable for fixed dimension $\dim(\Delta)$.
It is unknown whether the decision problem "Given the list of facets of Δ and some $\varphi \in \mathbb{N}$; is φ the total number of faces of Δ?" is contained in $\mathcal{NP}$.

This problem is only known to be in $\mathcal{NP}$ for partitionable (see Problem 18) simplicial complexes (see Kleinschmidt and Onn [39]).

To the best of our knowledge, no proof of $\#\mathcal{P}$-hardness of the general problem has appeared in the literature. Therefore we include one here.

Consider an instance of SAT, i.e., a formula in conjunctive normal form (CNF-formula) $C_1 \wedge \cdots \wedge C_m$ with variables $x_1, \ldots, x_n$ (each C_i contains only disjunctions of literals). It is well known (Valiant [63]) that computing the number of satisfying truth assignments is $\#\mathcal{P}$-complete. Define $E = \{t_1, f_1, \ldots, t_n, f_n\}$.

Part I. First, let E be the vertex set of a simplicial complex Δ defined by the minimal non-faces (*circuits*) $C_1', \ldots, C_m', P_1, \ldots, P_n$, where $P_i = \{t_i, f_i\}$ for every i. Here for any clause C, $C' := \{f_j : x_j$ literal in $C\} \cup \{t_j : \overline{x}_j$ literal in $C\}$, e.g., for $C = x_1 \vee x_2 \vee \overline{x}_3$ we have $C' = \{f_1, f_2, t_3\}$. The idea is that t_i corresponds to the assignment of a true-value and f_i corresponds to the assignment of a false-value to variable x_i. The circuits exclude subsets of E which include both t_i and f_i for all variables x_i and exclude truth-assignments to variables which would not satisfy a clause C_j. It is, however, allowed that for some variable x_i neither t_i nor f_i is included in a face. But every $(n-1)$-face (n-subset of E) (if there exists any) corresponds to a truth-assignment to the variables (which uses exactly one value for each variable) and satisfies the formula. These subsets are counted by $f_{n-1}(\Delta)$. Hence computing f_{n-1} is $\#\mathcal{P}$-complete and computing the f-vector of Δ is $\#\mathcal{P}$-hard. Moreover this shows that computing the dimension of a simplicial complex given by the minimal non-faces is $\mathcal{NP}$-hard.

Part II. We now construct a simplicial complex $\overline{\Delta}$ (the *dual* complex) which is given by facets. Define $\overline{\Delta}$ by the facets $\overline{C_1'}, \ldots, \overline{C_m'}, \overline{P_1}, \ldots, \overline{P_n}$, where for $S \subseteq E$, $\overline{S} := E \setminus S$. We have that a set $S \subseteq E$ is a face of Δ if and only if $\overline{S}$ is *not* a face of $\overline{\Delta}$. Hence, $f_{n-1}(\Delta) + f_{n-1}(\overline{\Delta}) = \binom{2n}{n}$. It follows that one can efficiently compute $f_{n-1}(\Delta)$ from $f_{n-1}(\overline{\Delta})$.

Related problems: 14, 31

33. HOMOLOGY

Input: Finite abstract simplicial complex Δ given by a list of facets, $i \in \mathbb{N}$
Output: The i-th homology group of Δ, given by its rank and its torsion coefficients
Status (general): Open
Status (fixed dim.): Polynomial time

There exists a polynomial time algorithm if Δ is given by the list of all simplices, since the Smith normal form of an integer matrix can be computed efficiently (Iliopoulos [31]). For fixed i or $\dim(\Delta) - i$, the sizes of the boundary matrices are polynomial in the size of Δ and the Smith normal form can again be computed efficiently.

Related problems: 31, 32

34. SHELLABILITY

Input: Finite abstract pure simplicial complex Δ given by a list of facets
Output: "Yes" if Δ is shellable, "No" otherwise
Status (general): Open
Status (fixed dim.): Open

Given an ordering of the facets of Δ, it can be tested in polynomial time whether it is a shelling order. Hence, the problem in $\mathcal{NP}$.

The problem can be solved in polynomial time for one-dimensional complexes, i.e., for graphs: a graph is shellable if and only if it is connected. Even for $\dim(\Delta) = 2$, the status is open. In particular, it is unclear if the problem can be solved in polynomial time if Δ is given by a list of all simplices.

For two-dimensional pseudo-manifolds the problem can be solved in linear time (Danaraj and Klee [13]).

Related problems: 17, 35

35. PARTITIONABILITY

Input: Finite abstract simplicial complex Δ given by a list of facets
Output: "Yes" if Δ is partionable, "No" otherwise
Status (general): Open
Status (fixed dim.): Open

As in Problem 18, partitionability is meant in the sense of Stanley [58] (see also [65]). Noble [50] proved that the problem is in $\mathcal{NP}$.

PARTITIONABILITY can be solved in polynomial time for one-dimensional complexes, i.e., for graphs: a graph is partitionable if and only if at most one of its connected components is a tree. For two-dimensional complexes the complexity status is open. In particular, it is unclear if the problem can be solved in polynomial time if Δ is given by a list of all simplices.

Related problems: 18, 34

Table of Problems

References

1. D. Avis, D. Bremner, and R. Seidel. How good are convex hull algorithms? *Comput. Geom.*, 7(5-6):265–301, 1997.
2. D. Avis and K. Fukuda. A pivoting algorithm for convex hull and vertex enumeration of arrangements and polyhedra. *Discrete Comput. Geom.*, 8(3):295–313, 1992.
3. L. Babai. Automorphism groups, isomorphism, reconstruction. In R. L. Graham, M. Grötschel, and L. Lovász, editors, *Handbook of Combinatorics*, volume 2, pages 1447–1540. Elsevier (North-Holland), Amsterdam, 1995.
4. A. Below, J. A. De Loera, and J. Richter-Gebert. The complexity of finding small triangulations of convex 3-polytopes. Journal of Algorithms, to appear; arXiv:math.CO/0012177, 2000.
5. A. Below, J. A. De Loera, and J. Richter-Gebert. Finding minimal triangulations of convex 3-polytopes is $\mathcal{NP}$-hard. In *Proceedings of the 11th annual ACM-SIAM symposium on Discrete algorithms, San Francisco, CA, USA*, pages 65–66. SIAM, Philadelphia, 2000.
6. R. Blind and P. Mani-Levitska. On puzzles and polytope isomorphisms. *Aequationes Math.*, 34:287–297, 1987.
7. J. Bokowski and B. Sturmfels. *Computational Synthetic Geometry*. Number 1355 in Lecture Notes in Mathematics. Springer-Verlag, 1989.
8. D. Bremner, K. Fukuda, and A. Marzetta. Primal-dual methods for vertex and facet enumeration. *Discrete Comput. Geom.*, 20(3):333–357, 1998.
9. T. M. Chan. Output-sensitive results on convex hulls, extreme points, and related problems. *Discrete Comput. Geom.*, 16:369–387, 1996.
10. D. R. Chand and S. S. Kapur. An algorithm for convex polytopes. *J. Assoc. Comput. Mach.*, 17(78–86), 1970.
11. R. Chandrasekaran, S. N. Kabadi, and K. G. Murty. Some $\mathcal{NP}$-complete problems in linear programming. *Oper. Res. Lett.*, 1(3):101–104, 1982.
12. B. Chazelle. An optimal convex hull algorithm in any fixed dimension. *Discrete Comput. Geom.*, 10, 1993.
13. G. Danaraj and V. Klee. A representation of 2-dimensional pseudomanifolds and its use in the design of a linear-time shelling algorithm. *Ann. Discrete Math.*, 2:53–63, 1978.
14. M. E. Dyer. The complexity of vertex enumeration methods. *Math. Oper. Res*, 8:381–402, 1983.
15. M. E. Dyer and A. M. Frieze. On the complexity of computing the volume of a polyhedron. *SIAM J. Comput.*, 17(5):967–974, 1988.
16. M. E. Dyer, A. M. Frieze, and R. Kannan. A random polynomial-time algorithm for approximating the volume of convex bodies. *J. Assoc. Comput. Mach.*, 38(1):1–17, 1991.
17. B. C. Eaves and R. M. Freund. Optimal scaling of balls and polyhedra. *Math. Program.*, 23:138–147, 1982.
18. J. Edmonds. Paths, trees, and flowers. *Can. J. Math.*, 17:449–467, 1965.
19. J. Erickson. New lower bounds for convex hull problems in odd dimensions. *SIAM J. Comput.*, 28(4):1198–1214, 1999.
20. R. M. Freund and J. B. Orlin. On the complexity of four polyhedral set containment problems. *Math. Program.*, 33:139–145, 1985.

21. A. M. Frieze and S.-H. Teng. On the complexity of computing the diameter of a polytope. *Comput. Complexity*, 4:207–219, 1994.
22. K. Fukuda, T. M. Liebling, and F. Margot. Analysis of backtrack algorithms for listing all vertices and all faces of a convex polyhedron. *Comput. Geom.*, 8:1–12, 1997.
23. B. Ganter. Algorithmen zur Formalen Begriffsanalyse. In B. Ganter, R. Wille, and K. E. Wolff, editors, *Beiträge zur Begriffsanalyse*, pages 241–254. B.I. Wissenschaftsverlag, 1987.
24. M. R. Garey and D. S. Johnson. *Computers and Intractability: A Guide to the Theory of $\mathcal{NP}$-Completeness*. W. H. Freeman and Company, San Francisco, 1979.
25. B. Gärtner. The random facet simplex algorithm on combinatorial cubes. Manuscript, February 2001.
26. E. Gawrilow and M. Joswig. Homepage of polymake. http://www.math.tu-berlin.de/diskregeom/polymake/.
27. E. Gawrilow and M. Joswig. polymake: A framework for analyzing convex polytopes. In G. Kalai and G. M. Ziegler, editors, *Polytopes — Combinatorics and Computation*, pages 43–74. Birkhäuser, 2000.
28. E. Gawrilow and M. Joswig. polymake: an approach to modular software design in computational geometry. In *Proc. 17th ACM Ann. Symp. Comput. Geom.*, pages 222–231, 2001.
29. J. E. Hopcroft and R. E. Tarjan. Dividing a graph into triconnected components. *SIAM J. Comput.*, 2:135–158, 1973.
30. J. E. Hopcroft and R. E. Tarjan. Efficient planarity testing. *J. Assoc. Comput. Mach.*, 21:549–568, 1974.
31. C. S. Iliopoulos. Worst-case complexity bounds on algorithms for computing the canonical structure of finite Abelian groups and the Hermite and Smith normal forms of an integer matrix. *SIAM J. Comput.*, 18(4):658–669, 1989.
32. M. Joswig, V. Kaibel, and F. Körner. On the k-systems of a simple polytope. *Israel J. Math.*, 129:109–117, 2002.
33. V. Kaibel and M. E. Pfetsch. Computing the face lattice of a polytope from its vertex-facet incidences. *Comput. Geom.*, 23(3):281–290, 2002.
34. V. Kaibel and A. Schwartz. On the complexity of isomorphism problems related to polytopes. To appear in *Graphs Comb.*
35. G. Kalai. A simple way to tell a simple polytope from its graph. *J. Comb. Theory, Ser. A*, 49(2):381–383, 1988.
36. G. Kalai. A subexponential randomized simplex algorithm. In *Proc. 24th Ann. ACM Symp. Theory Comput.*, pages 475–482, Victoria, 1992. ACM Press.
37. N. Karmarkar. A new polynomial-time algorithm for linear programming. *Combinatorica*, 4:373–395, 1984.
38. L. Khachiyan. A polynomial algorithm in linear programming. *Sov. Math., Dokl.*, 20:191–194, 1979.
39. P. Kleinschmidt and S. Onn. Signable posets and partitionable simplicial complexes. *Discrete Comput. Geom.*, 15(4):443–466, 1996.
40. N. Linial. Hard enumeration problems in geometry and combinatorics. *SIAM J. Alg. Disc. Math.*, 7(2):331–335, 1986.
41. E. M. Luks. Isomorphism of graphs of bounded valence can be tested in polynomial time. *J. Comput. Syst. Sci.*, 25:42–65, 1982.
42. J. Matoušek, M. Sharir, and E. Welzl. A subexponential bound for linear programming. *Algorithmica*, 16(4-5):498–516, 1996.

43. P. McMullen. The maximum numbers of faces of a convex polytope. *Mathematika*, 17:179–184, 1970.
44. N. Megiddo. Linear programming in linear time when the dimension is fixed. *J. Assoc. Comput. Mach.*, 31:114–127, 1984.
45. K. Mehlhorn and P. Mutzel. On the embedding phase of the Hopcroft and Tarjan planarity testing algorithm. *Algorithmica*, 16(2):233–242, 1996.
46. H. Minkowski. Geometrie der Zahlen (Geometry of numbers). Teubner Verlag, Leipzig, 1886 and 1910; reprinted by Chelsea, New York, 1953, and by Johnson, New York, 1968. (German).
47. N. E. Mnëv. The universality theorems on the classification problem of configuration varieties and convex polytopes varieties. In O. Y. Viro, editor, *Topology and Geometry – Rohlin Seminar*, number 1346 in Lecture Notes in Mathematics, pages 527–543. Springer-Verlag, 1988.
48. J. R. Munkres. *Elements of Algebraic Topology*. Addison-Wesley, Menlo Park CA, 1984.
49. K. G. Murty. A fundamental problem in linear inequalities with applications to the travelling salesman problem. *Math. Program.*, 2:296–308, 1972.
50. S. D. Noble. Recognising a partitionable simplicial complex is in $\mathcal{NP}$. *Discrete Math.*, 152:303–305, 1996.
51. C. H. Papadimitriou and M. Yannakakis. On recognizing integer polyhedra. *Combinatorica*, 10(1):107–109, 1990.
52. J. S. Provan. Efficient enumeration of the vertices of polyhedra associated with network LP's. *Math. Program.*, 63:47–64, 1994.
53. J. Richter-Gebert. *Realization Spaces of Polytopes*. Number 1643 in Lecture Notes in Mathematics. Springer-Verlag, 1996.
54. G.-C. Rota. On the foundations of combinatorial theory – I. Theory of Möbius functions. *Z. Wahrscheinlichkeitstheorie*, 2:340–368, 1964.
55. A. Schrijver. *Theory of linear and integer programming*. Wiley, 1986.
56. R. Seidel. Constructing higher-dimensional convex hulls at logarithmic cost per face. In *Proc. 18th Ann. ACM Sympos. Theory Comput.*, pages 404–413, 1986.
57. R. Seidel. Convex hull computations. In J. Goodman and J. O'Rouke, editors, *Handbook of Discrete and Computational Geometry*, chapter 19. CRC Press, Boca Raton, 1997.
58. R. P. Stanley. Balanced Cohen-Macaulay complexes. *Trans. Am. Math. Soc.*, 249:139–157, 1979.
59. E. Steinitz and H. Rademacher. *Vorlesungen über die Theorie der Polyeder*. Springer Verlag, 1934. Reprint 1976.
60. G. F. Swart. Finding the convex hull facet by facet. *J. Algorithms*, 6:17–48, 1985.
61. T. Szabó and E. Welzl. Unique sink orientations of cubes. Technical report, ETH Zürich, 2001. To appear in: Proc. 42nd Ann. Sympos. Found. Computer Science, Las Vegas, Oct 14–17, 2001.
62. W. T. Tutte. How to draw a graph. *Proc. Lond. Math. Soc., III. Ser.*, 13:743–768, 1963.
63. L. G. Valiant. The complexity of enumeration and reliability problems. *SIAM J. Comput.*, 8:410–421, 1979.
64. H. Weyl. Elementare Theorie der konvexen Polyeder. *Comment. Math. Helv.*, 7:290–306, 1935.
65. G. M. Ziegler. *Lectures on Polytopes*. Springer-Verlag, 1995. Revised edition 1998.

Computing Triangulations
Using Oriented Matroids

Julian Pfeifle[1] and Jörg Rambau[2]

[1] Dept. of Mathematics, MA 6-2, TU Berlin, 10623 Berlin, Germany
[2] Zuse-Institute Berlin, Takustr. 7, 14195 Berlin, Germany

Abstract. Oriented matroids are combinatorial structures that encode the combinatorics of point configurations. The set of all triangulations of a point configuration depends only on its oriented matroid. We survey the most important ingredients necessary to exploit oriented matroids as a data structure for computing all triangulations of a point configuration, and report on experience with an implementation of these concepts in the software package TOPCOM. Next, we briefly overview the construction and an application of the secondary polytope of a point configuration, and calculate some examples illustrating how our tools were integrated into the POLYMAKE framework.

1 Introduction

This paper surveys a selection of efficient combinatorial methods to compute triangulations of point configurations. We present results obtained for the first time by a software implementation (TOPCOM [20]) of these ideas. It turns out that a subset of all triangulations of a point configuration has a structure useful in different areas of mathematics, and we highlight one particular instance of such a connection. Finally, we calculate some examples by integrating TOPCOM into the POLYMAKE [7] framework. See also [29] for more related information on the combinatorics of triangulations.

Let us begin by motivating the use of triangulations and providing a precise definition.

1.1 Why Triangulations?

Triangulations are widely used as a standard tool to decompose complicated objects into simple objects. A solution to a problem on a complicated object can sometimes be found by gluing solutions on the simple objects. Some examples are the following:

o Numerics: Finite Elements Method
o Algebraic Topology: computation of topological invariants
o Computer Graphics: Raytracing

Besides these applications, structures on whole sets of triangulations have interesting connections to seemingly distant disciplines, among them:

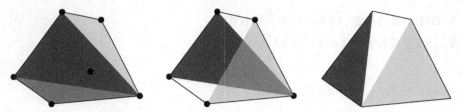

Figure 1. A correct triangulation, an unwanted intersection (IP not met), and an incomplete triangulation (UP not met)

- Algebraic Geometry: Connection to Toric Varieties
- Algebra: Polynomial System Solving
- Homotopy Theory: Structure of Loop Spaces

Therefore, the study of *spaces of triangulations* has become a subject in its own right in the field of discrete geometry [18].

1.2   What Exactly Are Triangulations?

For the rest of the paper, let $\mathcal{A}$ be a d-dimensional point configuration with n points. We assume that the points are labeled $1, 2, \ldots, n$, and denote the coordinate vector of the point i by a_i.

Definition 1.1. A subset T of $(d+1)$-subsets of $\mathcal{A}$ is a *triangulation of $\mathcal{A}$* if and only if

$$\bigcup_{\sigma \in T} \operatorname{conv} \sigma = \operatorname{conv} \mathcal{A} \tag{UP}$$

$$\operatorname{conv} \sigma \cap \operatorname{conv} \sigma' = \operatorname{conv}(\sigma \cap \sigma') \quad \forall \sigma, \sigma' \in T. \tag{IP}$$

Condition UP makes sure that the *union* of all (convex hulls of) simplices in T covers (the convex hull) of $\mathcal{A}$. Condition IP takes care of unwanted intersections. Note that we do not require all points to be used in a triangulation. Figure 1 provides a sketch of the situation.

2   The oriented matroid of a Point Configuration

In a naive approach, checking for non-empty interior intersection of two simplices or for a complete covering are linear programming problems. Since we need numerically exact results for our purposes, exact arithmetics is a must. However, linear programming with exact arithmetics is computationally expensive.

Fortunately, solving linear programs exactly can be avoided by rigorously separating exact arithmetic and combinatorial computations in simplicial complexes. It turns out that all arithmetic calculations can be reduced to computing $\binom{\#\mathcal{A}}{d+1}$ determinants of $((d + 1) \times (d + 1))$-matrices, possibly in a preprocessing step. All subsequent computations concerning triangulations of $\mathcal{A}$ that exclusively use information about the relative orientation of points can then be performed without using the coordinates ever again.

More specifically, we show in this section how the conditions of Definition 1.1 can be checked purely combinatorially, provided we have the *oriented matroid* of $\mathcal{A}$ at hand. The resulting combinatorial characterization has been formulated and extensively used, e.g., in [19]. More rigorous proofs can also be found there. Related characterizations and applications thereof can be found in [12,13]. More general information about oriented matroids can be found in [3].

The combinatorial concepts presented in this paper also hold for *triangulations of oriented matroids*, purely combinatorial objects that are not necessarily derived from any (realizable) point configuration [27]. For the computational methods it does not make any difference whether the oriented matroid under consideration is given implicitly by the coordinates of a "real" point configuration or directly by a list of combinatorial data of a general, maybe non-realizable oriented matroid.

2.1   Geometric Problem I: Proper Intersection of Simplices

We would like to use combinatorial data of $\mathcal{A}$ to check (IP). It turns out that there is a finite set of minimal obstructions, the *circuits* of $\mathcal{A}$, that describe the intersections of simplices in $\mathcal{A}$ completely. How does this work?

Assume that the convex hulls of the d-simplices σ and σ' intersect improperly, i.e., σ and σ' violate (IP). Then, by Radon's theorem, we find $Z^+ \subseteq \sigma$ and $Z^- \subseteq \sigma'$ such

(i) $Z^+ \cap Z^- = \emptyset$,
(ii) the relative interiors of Z^+ and Z^- intersect,
(iii) Z^+ and Z^- are inclusion minimal with these properties.

The pair (Z^+, Z^-) is called an *intersection circuit* of σ and σ'. In particular, a subset of $\mathcal{A}$ with properties (i)–(iii) is a *circuit* of $\mathcal{A}$.

Another interpretation of a circuit is that $\sum_{i \in Z^+} \lambda_i a_i = \sum_{i \in Z^-} \lambda_i a_i$ is an affine dependence with minimal support, for suitable $\lambda_i > 0$ with $\sum_{i \in Z^+} \lambda_i = \sum_{i \in Z^-} \lambda_i$, $i = 1, \ldots, n$. If we set $\Lambda := \sum_{i \in Z^+} \lambda_i$ then the (unique) intersection point in $\operatorname{conv} Z^+ \cap \operatorname{conv} Z^-$ is given by $\frac{1}{\Lambda} \sum_{i \in Z^+} \lambda_i a_i$.

This connection to affine dependences shows that we have exactly one circuit (modulo exchanging Z^+ and Z^-) for every affinely dependent set of points in $\mathcal{A}$. Since there are at most $\binom{n}{d+2}$ affinely dependent sets of points, there are at most that many circuits (modulo exchanging Z^+ and Z^-). Using

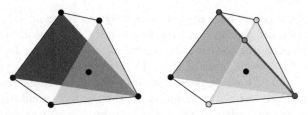

Figure 2. An unwanted intersection can be detected by a circuit

the set of all circuits of $\mathcal{A}$, we can easily check if (IP) holds. Namely, two simplices $\sigma, \sigma' \in T$ satisfy (IP) if and only if the following condition holds:

(IP') There exists no circuit (Z^+, Z^-) of $\mathcal{A}$ with $Z^+ \subseteq \sigma$ and $Z^- \subseteq \sigma'$.

Circuits are, in other words, obstructions to (IP). See Figure 2 for a sketch of two simplices intersecting improperly and a corresponding intersection circuit.

2.2   Geometric Problem II: Proper Covering by Simplices

In order to check (UP) purely combinatorially, we will assume (IP') for all simplices in T. A non-empty set of simplices satisfying (IP'), but not necessarily covering $\operatorname{conv} \mathcal{A}$, can be seen as a *partial triangulation* of $\mathcal{A}$. How can we detect an uncovered area in $\operatorname{conv} \mathcal{A}$ purely combinatorially? To this end, we look at facets of d-simplices in T. Such a facet is *an interior facet of T* if it is not a facet of $\mathcal{A}$. (A facet of $\mathcal{A}$ is a $(d-1)$-dimensional subset of $\mathcal{A}$ that is the intersection of $\mathcal{A}$ with a supporting hyperplane. A supporting hyperplane is an affine hyperplane that does not separate $\mathcal{A}$.)

The following is easy to see [19]: a partial triangulation T violates (UP) if and only if we find an interior facet of T that is not contained in any other simplex in T. Let us assume for the moment that we have a list of all facets of $\mathcal{A}$. Then we can simply go through the set of all facets of a partial triangulation T and count the number of simplices in T containing them. If there is an interior facet of T contained in only one simplex, then T violates (UP). This test is purely combinatorial.

Now, how do we get the list of all facets? It turns out that the set of *cocircuits* of $\mathcal{A}$ contains all the necessary information. Consider the set of all affine hyperplanes spanned by $(d-1)$-dimensional subsets of $\mathcal{A}$, oriented arbitrarily. Each such hyperplane defines a *signature* on the points in $\mathcal{A}$: the signature of a point is *zero* if it lies *on* the hyperplane; it is *positive* if it lies strictly on the *positive side* of the hyperplane; it is *negative* if it lies strictly on the *negative side* of the hyperplane. Such a signature on $\mathcal{A}$ is called a *cocircuit* of $\mathcal{A}$. We denote such a cocircuit by (C^+, C^-), where C^+ contains all points with positive signature and C^- those with negative signature.

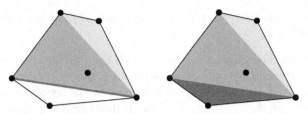

Figure 3. An interior facet covered by only one simplex detects a (UP)-violation; another simplex containing it is required

Figure 4. A facet F in T, i.e., a facet of some simplex in T, is interior if the cocircuit of $\mathcal{A}$ spanned by F has both positive and negative elements

A subset of $\mathcal{A}$ is a facet of $\mathcal{A}$ if and only if it is the zero set of a cocircuit having no positive elements, or if it is the zero set of a cocircuit having no negative elements. There are at most $\binom{n}{d}$ different hyperplanes spanned by subsets of a d-dimensional point configuration. Therefore, there are at most $\binom{n}{d}$ cocircuits of $\mathcal{A}$ modulo reversing signs. These give us all facets of $\mathcal{A}$ and thus all interior facets of T.

Summarizing this section, we can use the set of all cocircuits of $\mathcal{A}$ to check (UP). Namely, a partial triangulation T of $\mathcal{A}$ satisfies (UP) if and only if the following condition holds:

(UP') There is no interior facet of T lying in only one simplex of T.

In other words: interior facets (determined by the cocircuits) incident to exactly one simplex are obstructions for (UP).

2.3   Triangulations Depend Only on the Oriented Matroid

The considerations in the previous sections justify the name 'combinatorial characterization of triangulations' in the following theorem (see, e.g., [19]):

Theorem 2.1 (Combinatorial Characterization of Triangulations).
A subset T of $(d+1)$-subsets of $\mathcal{A}$ is a triangulation *of $\mathcal{A}$ if and only if*

(IP')   *For every pair of simplices $\sigma, \sigma' \in T$ there is no circuit (Z^+, Z^-) of $\mathcal{A}$ with $Z^+ \subseteq \sigma$ and $Z^- \subseteq \sigma'$.*

(UP') *For every interior facet F of T there are at least two simplices in T containing F.*

The following fact allows a unified view on the previous two sections: The circuits of $\mathcal{A}$ determine, purely combinatorially, the cocircuits of $\mathcal{A}$, and vice versa. The *oriented matroid* of $\mathcal{A}$ is now defined by the set of circuits or by the set of cocircuits of $\mathcal{A}$, depending on what is more convenient in a particular situation.

Therefore, the set of triangulations of $\mathcal{A}$ depends only on the oriented matroid of $\mathcal{A}$.

2.4 An Interface from Geometry to Combinatorics: The Chirotope

How do we compute the circuits and the cocircuits from the coordinates of the points in $\mathcal{A}$? There are actually several ways to do this. The most commonly used approach is the computation of a third equivalent combinatorial structure of $\mathcal{A}$: the chirotope.

The *chirotope* of $\mathcal{A}$ is the following alternating function on the set of $(d+1)$-subsets of $\mathcal{A}$:

$$\chi : \left\{ \begin{array}{c} \binom{\mathcal{A}}{d+1} \to \{+,-,0\} \\ (i_1, i_2, \ldots, i_{d+1}) \mapsto \mathrm{sign}\big(\det(a_{i_1}, a_{i_2}, \ldots, a_{i_{d+1}})\big) \end{array} \right. \tag{1}$$

That means, in particular, that the chirotope assigns to each ordered basis of $\mathcal{A}$ its orientation. The orientation is usually normalized such that the affine standard basis of $\mathbb{R}^d$ has orientation $+$. The chirotope value on $(d+1)$-subsets of $\mathcal{A}$ that are not independent is zero. One of many ways to accelerate the computation of the chirotope is the *elimination tree method* of TOPCOM, indicated in Figure 5. Column normal forms for all k-subsets of columns are maintained as nodes in computation tree for all $k \le d+1$; they can be reused in order to save elimination steps. The tree is traversed in depth-first-search; determinants are produced in the leaves in lexicographic order w.r.t. the indices of the point coordinates.

While in a naïve way of of computing determinants we need $O((d+1)^3)$ arithmetic operations per determinant (if we use an elimination algorithm), the elimination tree method just needs $O\big(\frac{n}{n-d}(d+1)^2\big)$ arithmetic operations per determinant [22].

There is another way of exploiting structure on the set of all determinants of submatrices of $\mathcal{A}$: the *exterior algebra method*. See [12] for a description; it is also used in the context of generating oriented matroids [4,6]. The basic idea is that there are certain relations between determinants coming from the distributive law in the exterior algebra. These relations correspond to the *Grassmann-Plücker-relations*.

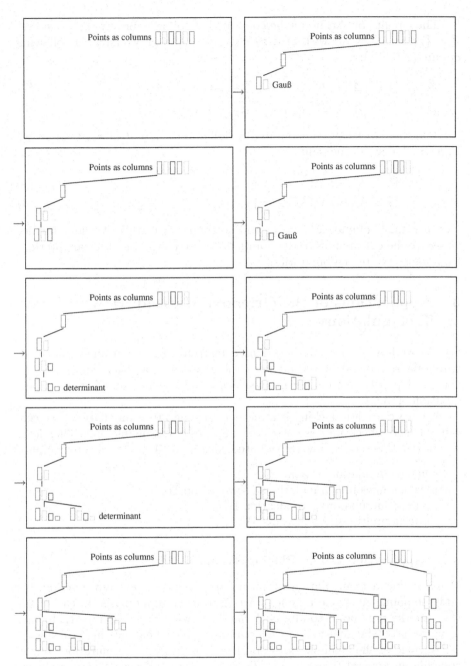

Figure 5. Building the computation tree for the chirotope: the matrices in the nodes are all transformed in column normal form before new columns are added (indicated are regions of possible non-zeroes in the columns); the determinants can be computed easily at the leaves

The circuits of $\mathcal{A}$ can now be computed as follows: for any $(d+2)$-subset $\underline{Z} = \{z_1, z_2, \ldots, z_{d+2}\}$ with $z_1 < z_2 < \cdots < z_{d+2}$ of $\mathcal{A}$, we have the following circuit (Z^+, Z^-):

$$Z^+ = \left\{ z_k \in \underline{Z} \mid (-1)^k \chi(z_1, \ldots, z_{k-1}, z_{k+1}, \ldots, z_{d+2}) = + \right\} \tag{2}$$

$$Z^- = \left\{ z_k \in \underline{Z} \mid (-1)^k \chi(z_1, \ldots, z_{k-1}, z_{k+1}, \ldots, z_{d+2}) = - \right\} \tag{3}$$

The cocircuits can be computed as follows: for any d-subset C of $\mathcal{A}$, we have the following cocircuit:

$$C^+ = \left\{ i \in \mathcal{A} \setminus C \mid \chi(C, i) = + \right\} \tag{4}$$

$$C^- = \left\{ i \in \mathcal{A} \setminus C \mid \chi(C, i) = - \right\} \tag{5}$$

Note that computing circuits and cocircuits is possible without further access to the coordinates of the points in $\mathcal{A}$: the chirotope forms an interface from geometry to combinatorics.

3   Applications of the Oriented Matroid: How to Find Triangulations

So far, we have seen that the oriented matroid of $\mathcal{A}$ determines the set of triangulations of $\mathcal{A}$. Moreover, once the chirotope has been computed, we can check purely combinatorially whether or not a set of simplices is a triangulation of $\mathcal{A}$.

What we are still lacking is a method to generate triangulations; we certainly cannot check all possible sets of simplices in reasonable time. Therefore, we show in this section how the oriented matroid of $\mathcal{A}$ can be used to compute

- o some triangulation (Application I),
- o local changes in triangulations (Application II),
- o many triangulations (Application III),
- o all triangulations (Application IV).

3.1   Application I: The Placing Triangulation

Constructing a triangulation by placing [11] is an incremental method: we add the points in $\mathcal{A}$ one by one, starting from an affine basis. In each step, we cone the new point to all *visible facets* of what we have already. In this way, we get a triangulation for each permutation of the points in $\mathcal{A}$. In special cases, some of the triangulations obtained from distinct permutations might coincide, in general they do not. We will see that all necessary ingredients can be computed via the oriented matroid.

For a more accurate definition of the placing triangulation, assume that we already have a triangulation T' of the subconfiguration $\mathcal{A}' := \mathcal{A} \setminus \{a\}$ for some $a \in \mathcal{A}$. There are two cases: $a \in \operatorname{conv} \mathcal{A}'$ or $a \notin \operatorname{conv} \mathcal{A}'$.

In the first case we do nothing: T' is also a triangulation of $\mathcal{A}$. Indeed: (IP') holds because no new simplex has been added; for the same reason (UP') holds because all facets of $\mathcal{A}'$ are still facets of $\mathcal{A}$.

In the second case we define visible facets of T': a facet F, supported by the hyperplane H_F, of a d-simplex of T' is called a *facet of T' visible from a* if

○ a is not on H_F.
○ $\mathcal{A}$ and a are not on the same side of H_F.

In this case, we add all cones of visible facets with apex a to T':

$$T := T' \cup \{F \cup \{a\} \mid F \text{ is a facet of } T' \text{ visible from } a\}$$

Why does this give us a triangulation of $\mathcal{A}$? First, the new simplices intersect properly with each other, since they form a cone over properly intersecting simplices. Second, the interior intersection between a new and an old simplex must be empty because none of the new simplices has an interior intersection with conv $\mathcal{A}'$: it is separated from conv $\mathcal{A}'$ by the supporting hyperplane of the corresponding facet of $\mathcal{A}$. Therefore, (IP) holds.

(UP) holds as well: The only facets in T' that become interior facets in conv $\mathcal{A}$ are those visible from a, by construction.

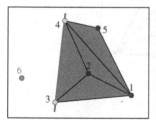 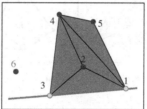 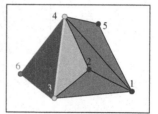

Figure 6. Cocircuits of $\mathcal{A}$ detect visible facets: the cocircuit spanned by $\{3,4\}$ induces different signs on the new point 6 and the old points. Thus, $\{3,4\}$ is a visible facet. The cocircuit spanned by $\{1,3\}$ induces the same signs on the new and the old points: $\{1,3\}$ is not visible. Note that the addition of the cone over $\{3,4\}$ removes an obstruction for (UP) because $\{3,4\}$ is an interior facet of the new configuration including 6 and was included in only one simplex so far

Now, we want to solely use the oriented matroid data in the step from T' to T. The only thing we need is the set of visible facets, a subset of the boundary facets of T' (which are those facets of T' that are contained in exactly one simplex in T'). Visibility can now be checked for every boundary facet F in T' by looking at the cocircuit C_F spanned by F; namely, F is visible from a if the following holds:

• The new point a must not be in C_F^0.
• If one of the points in $\mathcal{A}'$ is in C_F^+ then a must be in C_F^-.

An example for these conditions is indicated in Figure 6.

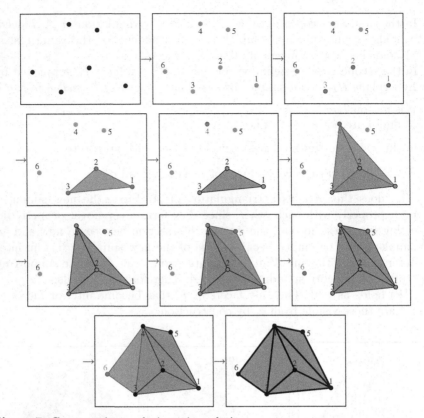

Figure 7. Constructing a placing triangulation

Note that it suffices to find a non-zero signature of only one point a' in $\mathcal{A}'$; then all points in $\mathcal{A}'$ with non-zero signature automatically have the same signature as a'.

By induction, we can construct a triangulation of the whole point configuration by this placing operation. Furthermore, to do this we only need the cocircuits of the oriented matroid. Figure 7 shows the corresponding steps for our running example configuration.

3.2 Application II: Flips

Now that we know how to produce one triangulation we will show that flipping—the standard method to produce new triangulations from old ones—can also be done by using the oriented matroid.

A flip in a two-dimensional triangulation is, e.g., an edge flip: replacing the diagonal in a convex quadrilateral. But in our context also the removal of an interior point lying in the convex hull of exactly three edges is a flip.

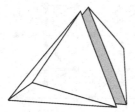 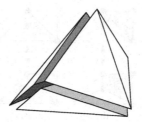

Figure 8. The two possible triangulations of five points in general position in dimension three; Z^- contains, w.l.o.g., the three vertices of the interior cutting face in the left picture; Z^+ contains the two vertices of the interior cutting edge in the right picture

In general a flip is the following: Consider a circuit $Z := (Z^+, Z^-)$ in $\mathcal{A}$. Let $\underline{Z} := Z^+ \cup Z^-$. Then it is easy to see that there are exactly two triangulations $T^+(Z)$ and $T^-(Z)$ of $\underline{Z}$. The triangulations are:

$$T^+(Z) := \{\underline{Z} \setminus \{z_-\} | z_- \in Z^-\} \tag{6}$$

$$T^-(Z) := \{\underline{Z} \setminus \{z_+\} | z_+ \in Z^+\} \tag{7}$$

Figure 8 sketches T^+ and T^- in dimension three for a circuit on five points in general position. Note that T^+ and T^- differ in their number of simplices.

Assume that a triangulation of $\mathcal{A}$ contains one of $T^+(Z)$ and $T^-(Z)$ as a subcomplex, say $T^+(Z)$. If Z spans a full-dimensional subconfiguration, then flipping simply means replacing $T^+(Z)$ by $T^-(Z)$ in T.

If the circuit spans a lower-dimensional subconfiguration then things are more complicated: still, the circuit itself has exactly two triangulations $T^+(Z)$ and $T^-(Z)$. These are lower-dimensional triangulations. If one of them, say $T^+(Z)$, is a subcomplex in T we cannot simply replace $T^+(Z)$ by $T^-(Z)$ and still get a triangulation. If, however, we encounter that every maximal simplex in $T^+(Z)$ has the same link $L(Z)$ in T then we can replace $T^+(Z) * L(Z)$ by $T^-(Z) * L(Z)$ in T. (Here, "$*$" denotes the simplicial join.)

So, finding a flip in terms of oriented matroids amounts to the following:

1. Pick a circuit Z.
2. Check whether $T^+(Z)$ or $T^-(Z)$ is a full-dimensional subcomplex of T.
3. If $T^+(Z)$ is in T, then replacing $T^+(Z)$ by $T^-(Z)$ is a valid flip.
4. If not, check the links of all maximal simplices of $T^+(Z)$ or $T^-(Z)$.
5. If the links, w.l.o.g. the ones in $T^+(Z)$, are identical, say equal to $L(Z)$, then replacing $T^+(Z) * L(Z)$ by $T^-(Z) * L(Z)$ is a valid flip.

Figure 9 lists all flip types in dimension two.

Thus, the only information about the point configuration we need for flipping is its set of circuits. Given the notion of flipping, one can define the *flip graph* of $\mathcal{A}$:

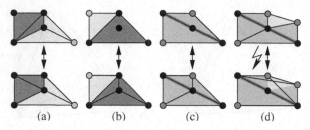

Figure 9. All possible flip types in dimension two; the rightmost picture shows a situation where the links of all maximal simplices are not equal, whence the corresponding flip is impossible

○ The nodes of the flip graph are all triangulations.
○ Two nodes are connected by an edge if the corresponding triangulations can be flipped into each other by a single flip.

3.3   Application III: Computing a Component of the Flip-graph

So far, we have the following:

○ We can compute one triangulation.
○ We construct new triangulations from old ones by flipping.

Using standard enumeration techniques like breadth-first-search or depth-first-search on the flip graph we can now enumerate all triangulations in the component of the placing triangulation.

An interesting feature of the flip graph is that its equivariant version—i.e., the one consisting of combinatorial symmetry classes only—is compatible with a breadth-first-search. Thus, the combinatorial symmetries of the point configuration can be exploited in this framework.

Let us—in a short excursion—describe the ideas of exploiting symmetries. We assume that we already have all those permutations of $\mathcal{A}$ that maintain the combinatorial structure of $\mathcal{A}$. This means we know all permutations π with the following property: whenever (Z^+, Z^-) is a circuit in $\mathcal{A}$, then $(\pi(Z^+), \pi(Z^-))$ is also a circuit in $\mathcal{A}$.

In an ordinary breadth-first-search (BFS) in the flip graph we would like to maintain only two sets of nodes: the set of currently *open* (unprocessed) nodes C and the set of new open nodes N. A step in the BFS corresponds to the following: for all triangulations in C we flip all their unmarked flips, add the resulting triangulation to N, and mark all the "backward"-flips; finally, we set $C := N$ and $N := \emptyset$. In other words, we forget about closed nodes, nodes all of whose edges have been processed already.

In the equivariant BFS, we do not want to store the complete orbits of triangulation. We rather want to store just those representatives that we meet first in the enumeration process. For every new triangulation we check whether one of the stored representatives is in the orbit of the new triangulation.

However, then it is not at all clear that we can never re-enter a symmetry class whose representative we have forgotten already. The trick is that we additionally mark flips that are equivalent to "backward flips" via the same combinatorial symmetry that transforms the new triangulation into one of the known representatives. More accurately, we do the following: whenever we find via the flip f a triangulation T with $\pi(T) = T'$ for some stored representative $T' \in C$ and some combinatorial symmetry π, we mark $\pi(f^{-1})$ in T'; one can prove that this way we can forget about all nodes whose edges have been processed completely; we will never enter their symmetry class again. See the appendix for an example.

Moreover, whenever a flip in a triangulation node is marked, we also mark all equivalent flips corresponding to the automorphism group of the triangulation: they would hit the same symmetry classes as the originally marked flip.

3.4   Application IV: Computing All Triangulations

For a long time it was an open problem whether or not the flip graph of any point configuration is connected. Recently, in [25] an example of a six-dimensional point configuration with a triangulation without flips was presented. The configuration has 324 points; the triangulations has over $25,000$ simplices [21]. Therefore, in general, we cannot compute all triangulations by flipping. Very recently, Santos has presented two brandnew constructions of point configurations with disconnected flip graph with 50 and 26 points, resp., in dimension five that can even be tweaked into convex position [26].

Here is another method of enumerating triangulations using the combinatorial characterization. We build up triangulations by adding one simplex at a time. Then we try to add new simplices maintaining (IP') for the complex built so far. As soon as (UP') is satisfied we have reached a triangulation.

Trying all possibilities by backtracking in this method yields an enumeration tree of all partial triangulations with triangulations sitting in some of the leaves. It can be seen as an enumeration of independent sets in the graph of forbidden intersections between simplices [30,10]. Since there are far more partial triangulations than triangulations, this method is not as fast as the flipping method, whenever the number of simplices in a triangulation is large (say at least ten).

4   Implementing the Ideas: TOPCOM

All of the above (and some more) has been implemented in the package TOPCOM (*Triangulations of Point Configurations and Oriented Matroids*).

TOPCOM is software under the GNU public license [20] and contains collection of clients for computations in point configurations, oriented matroids, and triangulations. Given a point configuration, TOPCOM can compute, for example,

- o its chirotope,
- o its circuits,
- o its cocircuits,
- o its facets,
- o a placing triangulation,
- o flips in a triangulation,
- o the flip graph component of a triangulation,
- o all triangulations.

This is done fairly quickly by using data structures custom-made for exploiting the combinatorial algorithms presented in this exposition.

Some numbers computed with TOPCOM that were unknown before can be found in Table 1. The number of triangulations of lattice points in dimension two can by now be computed much more efficiently by a special two-dimensional method [1]; for $(3 \times n)$- and $(4 \times n)$-grids a recursion formula for the number of triangulations was found by Ziegler.

Checking the result of Santos requires sophisticated use of a *lazy* chirotope: preprocessing is prohibitive for 324 points in dimension six.

The latest addition in TOPCOM allows to check *regularity* or *coherence* of a triangulation with the help of an LP solver with exact arithmetics, like, e.g., cdd [5]. This concept is—in contrast to all the other ones presented in this exposition—not combinatorial: whether or not a triangulation is regular depends on the specific coordinates of the points.

Why this concept is important is the subject of the remaining sections of this paper.

Table 1. Some figures computed by TOPCOM for the first time (1.8GHz Pentium IV, 1GByte RAM, SuseLinux 7.3); cyclic polytopes have connected flip graphs [19]; so have two-dimensional point sets; fine triangulations use all the vertices

What	Configuration	Description	Result	CPU/s
# triangulations	$C(12,5)$	cyclic polytope	5,049,932	2,104
# triangulations	$C(13,7)$	cyclic polytope	6,429,428	4,055
# fine triangulations	(4×5)-lattice	two-dim. lattice, 20 points	2,822,648	501
# flip graph component	$\Delta_3 \times \Delta_3$	product of tetrahedra	4,533,408	499
# flip graph component	C^4	four-dimensional cube	92,487,256	15,624
check (IP') & (UP')	Santos triang.	six-dim. construction	okay	70
# flips	Santos triangulation	six-dim. construction	0	30

5   Exploring Further Structures

We have seen in Section 3.2 that flipping is a natural way to locally modify a given triangulation. In this section, we will build on this concept and present one of the most striking and beautiful constructions of the theory of polyhedral subdivisions, which shows that a certain subclass of triangulations of a point configuration carries quite strong structural properties. Namely, we will outline the construction of the *secondary polytope* of a point configuration, briefly sketch one situation where it can be useful, and indicate how to calculate some interesting examples by integrating TOPCOM and the POLYMAKE [7] framework.

5.1   The Convex Hull of Triangulations: Secondary Polytopes

Let us define a *regular* or *coherent* triangulation of a point configuration $\mathcal{A}$ in $\mathbb{R}^d$ as one that arises by projecting the "lower" facets (with respect to some fixed direction) of a $(d+1)$-dimensional polytope $\widetilde{P}$ to $\mathbb{R}^d$, in such a way that the "lower" vertices of $\widetilde{P}$ project exactly to the points in $\mathcal{A}$. Another way of stating this condition is to ask for a convex lifting function from $\mathbb{R}^d$ to $\mathbb{R}^{d+1}$ that is linear on the simplices of the triangulation. More generally, we will also consider *regular subdivisions* of $\mathcal{A}$, that is to say collections of affinely independent subsets (*cells*) of $\mathcal{A}$ of cardinality at least $d+1$ that satisfy (UP), (IP), and (the obvious adaptation of) the regularity property. See Figure 10.

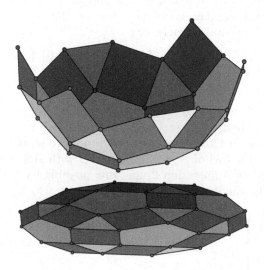

Figure 10. Convex lifting function and regular subdivision [23]

Regular triangulations correspond in a one-to-one fashion to the vertices of a convex polytope $\Sigma(\mathcal{A})$ that only depends on the point configuration, the so-called *secondary polytope of* $\mathcal{A}$. Moreover, this correspondence is not just bijective, but structural: Two regular triangulations T and T' are connected by a flip if and only if the vertices v_T and $v_{T'}$ lie on an edge of the convex hull of $\Sigma(\mathcal{A})$. It turns out that this correspondence extends to the whole face lattice of the secondary polytope, such that to each face F of $\Sigma(\mathcal{A})$ there corresponds some regular subdivision $\sigma(F)$ of $\mathcal{A}$. Furthermore, if $F \subset G$ are two faces of $\Sigma(\mathcal{A})$, then $\sigma(F)$ is a *refinement* of $\sigma(G)$, which means that any cell of $\sigma(G)$ is the union of cells of $\sigma(F)$.

As an example, let us construct the secondary polytope of the point configuration $\mathcal{A}$ formed by the vertices of a prism P over a triangle. The homogeneous coordinates of P are given by the columns of the following matrix.

$$A = \begin{pmatrix} 0 & 1 & 0 & 0 & 1 & 0 \\ 0 & 0 & 1 & 0 & 0 & 1 \\ 0 & 0 & 0 & 1 & 1 & 1 \\ 1 & 1 & 1 & 1 & 1 & 1 \end{pmatrix}$$

Any triangulation of P must contain one of the tetrahedra formed by the base $\{1,2,3\}$ and one vertex i in the set $\{4,5,6\}$, where the point labels correspond to the column indices of A. This leaves two choices for the apex of the tetrahedron with base $\{4,5,6\}$, and each one determines the last tetrahedron of the triangulation uniquely. We see that there are six distinct triangulations of P in total, namely,

$$\{\{1,2,3,4\},\{2,3,4,5\},\{3,4,5,6\}\}, \quad \{\{1,2,3,4\},\{2,3,4,6\},\{2,4,5,6\}\},$$
$$\{\{1,2,3,5\},\{1,3,4,5\},\{3,4,5,6\}\}, \quad \{\{1,2,3,5\},\{1,3,5,6\},\{1,4,5,6\}\},$$
$$\{\{1,2,3,6\},\{1,2,4,6\},\{2,4,5,6\}\}, \quad \{\{1,2,3,6\},\{1,2,5,6\},\{1,4,5,6\}\}.$$

It turns out that all these triangulations are regular, and therefore we know that they all correspond to vertices of $\Sigma(\mathcal{A})$.

One way to construct the secondary polytope is to start by calculating a basis for the (right) kernel of A, i.e., a matrix B with $AB = 0$. Since A has full rank, its kernel has dimension 2, and one possible basis is given by the columns of the following matrix B.

$$B = \begin{pmatrix} 1 & 1 \\ -1 & 0 \\ 0 & -1 \\ -1 & -1 \\ 1 & 0 \\ 0 & 1 \end{pmatrix}$$

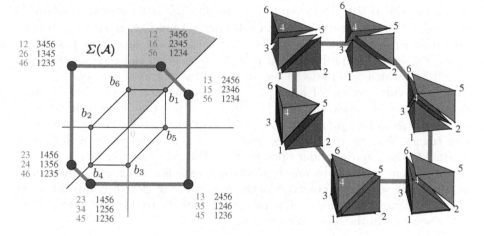

Figure 11. The hexagon as the secondary polytope of the prism P. *Left:* One maximal cone of the secondary fan is highlighted. Pairs of digits inside such a cone σ index vertices b_i in whose positive span σ lies, and the complementary 4-tuples label the simplex of the triangulation of $\mathcal{A}$ that σ corresponds to. *Right:* Triangulations corresponding to vertices of $\Sigma(P)$. Edges of $\Sigma(P)$ representing flips between triangulations.

By interpreting the *rows* of B as six points $b_1, b_2, \ldots, b_6$ in $\mathbb{R}^2$, we arrive at the *Gale transform* $\mathcal{A}^*$ of $\mathcal{A}$. In general, if $\mathcal{A}$ consists of n points in d-space (and $\mathcal{A}$ does not lie in any lower-dimensional subspace), then $\mathcal{A}^*$ is made up of n points in $(n - d - 1)$-space. Now consider the set $\mathcal{C}(\mathcal{A})$ of all full-dimensional positive cones spanned by the points in $\mathcal{A}^*$ with apex in 0, together with the set R of all their facets. The *chamber complex* $\widetilde{\mathcal{C}}(\mathcal{A})$ of $\mathcal{C}(\mathcal{A})$ is the union of all full-dimensional polyhedral cones whose facets are facets of cones in $\mathcal{C}(\mathcal{A})$, but whose relative interior is not crossed by any member of R. In our two-dimensional example, the set R consists of the six rays $\mathbb{R}_{\geq 0}\langle b_i \rangle$, $1 \leq i \leq 6$, so $\widetilde{\mathcal{C}}(\mathcal{A})$ is given by the following list of cones. See Figure 11 (left).

$$\widetilde{\mathcal{C}}(\mathcal{A}) = \left\{ \mathbb{R}_{\geq 0}\langle b_1, b_6 \rangle, \ \mathbb{R}_{\geq 0}\langle b_6, b_2 \rangle, \ \mathbb{R}_{\geq 0}\langle b_2, b_4 \rangle, \right.$$
$$\left. \mathbb{R}_{\geq 0}\langle b_4, b_3 \rangle, \ \mathbb{R}_{\geq 0}\langle b_3, b_5 \rangle, \ \mathbb{R}_{\geq 0}\langle b_5, b_1 \rangle \right\}$$

We now consider each cone $\sigma \in \widetilde{\mathcal{C}}(\mathcal{A})$ in turn, and write down the generators of all cones in $\mathcal{C}(\mathcal{A})$ that contain σ. For instance, $\sigma = \mathbb{R}_{\geq 0}\langle b_1, b_6 \rangle$ lies in the cones $\mathbb{R}_{\geq 0}\langle b_5, b_6 \rangle$, $\mathbb{R}_{\geq 0}\langle b_1, b_6 \rangle$, and $\mathbb{R}_{\geq 0}\langle b_1, b_2 \rangle$ of $\mathcal{C}(\mathcal{A})$, and the *complements* $\{1, 2, 3, 4\}$, $\{2, 3, 4, 5\}$, and $\{3, 4, 5, 6\}$ of these index sets correspond precisely to a triangulation of P! Since there are six maximal cones in $\widetilde{\mathcal{C}}(\mathcal{A})$, we expect each one of them to correspond to one of the six regular triangulations of P.

In fact this is true, and even more: The set $\widetilde{\mathcal{C}}(\mathcal{A})$ is a *complete polyhedral fan*, which means that the cones in $\widetilde{\mathcal{C}}(\mathcal{A})$ intersect precisely in common faces, and together span all of $\mathbb{R}^{n-d-1}$. This fan is called the *secondary fan* of $\mathcal{A}$. It has the additional property that it is the *normal fan* of a certain polytope in $\mathbb{R}^{n-d-1}$, which says that the vectors contained in a fixed cone of $\widetilde{\mathcal{C}}(\mathcal{A})$ are just the normal vectors of hyperplanes supporting exactly one face of this polytope. It now comes as no surprise that this polytope is the one defined to be the *secondary polytope* $\Sigma(\mathcal{A})$ of $\mathcal{A}$. Of course, this construction only determines $\Sigma(\mathcal{A})$ up to *normal equivalence*, i.e., any polytope with the same normal fan is also a secondary polytope of $\mathcal{A}$. In any case, passing from one maximal cone of $\widetilde{\mathcal{C}}(\mathcal{A})$ to an adjacent one corresponds to going from one vertex of the secondary polytope to the next, and therefore to a flip between these triangulations. This is illustrated in Figure 11 (right). We summarize our discussion in the following theorem.

Theorem 5.1. (Gel'fand, Kapranov, and Zelevinsky [8])

1. *The dimension of the secondary polytope $\Sigma(\mathcal{A})$ of a configuration of n points in $\mathbb{R}^d$ is $n - d - 1$.*
2. *The faces of $\Sigma(\mathcal{A})$ correspond to the regular subdivisions of $\mathcal{A}$.*
3. *If $F \subset G$ are faces of $\Sigma(\mathcal{A})$, then the subdivision of $\mathcal{A}$ corresponding to F refines the subdivision corresponding to G. In particular, the vertices of $\Sigma(\mathcal{A})$ encode the regular triangulations of $\mathcal{A}$.*

5.2 Hypergeometric Differential Equations and Secondary Polytopes

In this section, we briefly present the connection between secondary polytopes and (initial ideals of) certain systems of partial differential equations.

To the matrix A from the preceding section we can associate the following ideal in the (commutative) polynomial ring of differential operators $k[\partial] = k[\partial_1, \partial_2, \ldots, \partial_n]$ with $n = 6$:

$$I_A = \langle \partial^u - \partial^v : Au = Av, \ u, v \in \mathbb{N}^6 \rangle$$
$$= \langle \partial_1 \partial_5 - \partial_2 \partial_4, \ \partial_1 \partial_6 - \partial_3 \partial_4, \ \partial_3 \partial_5 - \partial_2 \partial_6 \rangle,$$

which corresponds to the system of differential equations

$$\frac{\partial^2}{\partial x_1 \partial x_5} f(x_1, x_2, \ldots, x_6) = \frac{\partial^2}{\partial x_2 \partial x_4} f(x_1, x_2, \ldots, x_6),$$

$$\frac{\partial^2}{\partial x_1 \partial x_6} f(x_1, x_2, \ldots, x_6) = \frac{\partial^2}{\partial x_3 \partial x_4} f(x_1, x_2, \ldots, x_6), \tag{9}$$

$$\frac{\partial^2}{\partial x_3 \partial x_5} f(x_1, x_2, \ldots, x_6) = \frac{\partial^2}{\partial x_2 \partial x_6} f(x_1, x_2, \ldots, x_6)$$

for a (formal) power series f in six variables. Notice how the differential operators that generate I_A correspond to elements of the kernel of A. The general theory developed in [24] tells us that a first step in constructing a series solution of (9) is to calculate the initial ideals $in_\omega(I_A)$ for all term orders $\prec_\omega$ on $k[\partial]$ induced by weight vectors $\omega \in \mathbb{Z}^n$. The positive hull of the weight vectors that select a given initial ideal of I_A is a polyhedral cone in $\mathbb{R}^n$, and it is readily seen that the set of all such cones forms a polyhedral fan, the *Gröbner fan* of I_A. It is then also clear that the weight vectors in the *maximal* cones of the Gröbner fan select *monomial* initial ideals, while those in lower-dimensional cones lead to initial ideals whose generators have more than one term.

Just as for the secondary fan, there exists an equivalence class of polytopes whose normal fan coincides with the Gröbner fan. Any representative from this class is called a *state polytope* [28, Chapter 2] of A. By the preceding paragraph, the vertices of the state polytope exactly correspond to the monomial initial ideals of I_A.

In general [28, Prop. 8.15], the Gröbner fan refines the secondary fan; an equivalent way of putting this is to say that the secondary polytope is a *Minkowski summand* of the state polytope. However, for a certain subclass of point configurations it is known that the Gröbner fan coincides with the secondary fan, and that therefore also the state polytope and the secondary polytope are the same (up to normal equivalence). These are the *unimodular* point configurations: those configurations *all* of whose triangulations are unimodular, i.e., entirely made up of simplices of unit volume (appropriately normalized for the dimension of the ambient space).

Therefore, for differential ideals coming from unimodular point configurations A, we can calculate the Gröbner fan via geometrical techniques. We only need to enumerate all triangulations T of A, and for each of them construct the following ideal, called the *Stanley-Reisner ideal* of T:

$$\left\langle \prod_{j \in J} \partial_j : J \text{ does not index a face of } T \right\rangle = \bigcap_{\sigma \in T} \langle \partial_j : j \notin \sigma \rangle \subset k[\partial].$$

By unraveling definitions, this is exactly the initial ideal of I_A selected by any weight vector in the cone of the secondary fan which is dual to the vertex v_T of $\Sigma(A)$. See Figure 12. Note that this initial ideal is square-free by construction.

If the point configuration is not unimodular, i.e., if it admits some triangulation with at least one simplex of non-unit volume, then as we saw above the Gröbner fan is a proper refinement of the secondary fan. The Stanley-Reisner ideal of such a triangulation T is then only the *radical* of the initial ideals selected by weight vectors in those Gröbner cones that refine the cone corresponding to T. It therefore does not properly reflect the algebraic structure of I_A anymore. However, calculating all regular triangulations of A at least gives a lower bound for the number of monomial initial ideals of I_A.

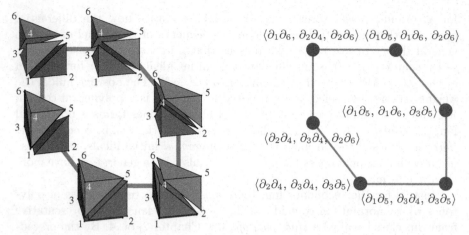

Figure 12. How to construct geometrically the six initial ideals of the unimodular ideal $I_{\mathcal{A}} = \langle \partial_1\partial_5 - \partial_2\partial_4, \partial_1\partial_6 - \partial_3\partial_4, \partial_3\partial_5 - \partial_2\partial_6 \rangle$: The generators of each initial ideal are precisely the minimal non-faces of the corresponding regular triangulation of $\mathcal{A}$.

We now present a way of actually computing the secondary polytope of a point configuration.

5.3 The GKZ Vectors

The original construction of the secondary polytope—presented by Gelfand and Zelevinsky in 1989—remained somewhat mysterious; as we will see, it gives rise to a straightforward recipe for calculating secondary polytopes, but it is not at all so straightforward to understand what is happening geometrically. In 1992, Billera and Sturmfels [2] finally presented secondary polytopes as the *fiber polytopes* of the projection of the $(n-1)$-dimensional simplex to a configuration $\mathcal{A}$ of n points. We will not develop this theory here, but instead refer the interested reader to [2], where also the formulation in terms of Gale transforms was first given, and especially to Chapter 9 of [31].

The GKZ construction proceeds as follows. We associate an n-dimensional vector v_T to any given triangulation T of $\mathcal{A}$, in such a way that the i-th coordinate of v_T is the sum of the volumes of all simplices in T incident to the point i.

$$(v_T)_i = \sum_{\sigma:\, \sigma \in T,\, i \in \sigma} \operatorname{vol} \operatorname{conv} \sigma$$

This gives us one n-dimensional vector for each triangulation of $\mathcal{A}$. The *secondary polytope* $\Sigma(\mathcal{A}) \subset \mathbb{R}^n$ of $\mathcal{A}$ is then defined as the convex hull of all such vectors obtained by considering every possible triangulation of $\mathcal{A}$.

$$\Sigma(\mathcal{A}) = \operatorname{conv}\{v_T : T \text{ triangulation of } \mathcal{A}\}$$

It turns out that the secondary polytope defined in this way is not full-dimensional, but resides in an $(n - d - 1)$-dimensional subspace. However, the fact that this polytope coincides, up to scaling and normal equivalence, with the secondary polytope as defined earlier definitely comes as a surprise!

5.4   How to Find the Face Lattice of the Secondary Polytope

To actually calculate the face lattice of the secondary polytope of a point configuration in $\mathbb{R}^d$, we combine the methods presented in Sections 3.1, 3.3, and 5.3. First, we calculate a placing triangulation of $\mathcal{A}$, which is known to be regular. Now we could proceed as in Section 3.4 by completing partial triangulations, but in fact the faster method described in Section 3.3 of generating the connected component of the flip graph that contains the placing triangulation is sufficient for our purposes. Since flips correspond to edges of the secondary polytope, and the 1-skeleton of any convex polytope of dimension at least 2 is connected, we know that this component contains at least all regular triangulations of $\mathcal{A}$—possibly along with some non-regular ones.

Next, we embed the nodes of the flip graph in $\mathbb{R}^n$ via their GKZ coordinates, project the resulting point configuration to $\mathbb{R}^{n-d-1}$, and calculate (the vertex-facet incidence matrix of) the convex hull of the result.

We have achieved an embedding not only of the vertices corresponding to regular triangulations, but of the entire flip graph. This allows to investigate, for example, the *flip distance* of a non-regular triangulation to the nearest regular one. Of course, our "fast" procedure following Section 3.3 misses all connected components of the flip graph that do not contain any regular triangulation.

6   Implementing the Ideas: Software Integration with polymake

We have integrated the tools available in TOPCOM and POLYMAKE by writing clients that interchange between the respective data representations and implement the procedure described in the previous section using the standard POLYMAKE rule base.

To calculate the secondary polytope of a configuration $\mathcal{A}$ of n points in $\mathbb{R}^d$, the user executes the command **secondary point-conf** , where **point-conf** is the name of a POLYMAKE file containing the homogeneous coordinates of $\mathcal{A}$. The client **secondary** converts this data to TOPCOM format, requests a list of all triangulations of $\mathcal{A}$, and for each one calculates the coordinates of its GKZ embedding in $\mathbb{R}^n$. Next, the client asks for these points to be projected to $\mathbb{R}^{n-d-1}$, and then to calculate the convex hull of these projected points. These requests are all answered by the POLYMAKE server, which in turn calls the appropriate clients for each task as specified in the rule base. Finally, the **secondary** client outputs the flip graph of $\mathcal{A}$ both with its embedding and

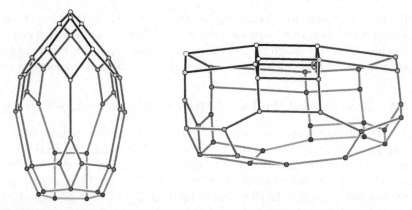

Figure 13. *Left:* Secondary polytope of the cyclic 4-polytope with 8 vertices. All 40 triangulations are regular. *Right:* Secondary of a different neighborly 4-polytope with 8 vertices [9, Ch. 7], which has one nonregular triangulation (circled; in the interior of the convex hull) among 41

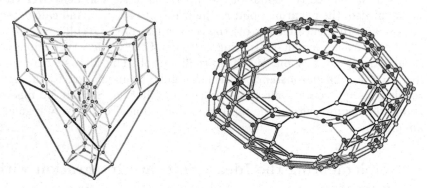

Figure 14. *Left:* Schlegel diagram of the secondary polytope of the 3-cube [16]. All 74 triangulations are regular. *Right:* Secondary of the cyclic 8-polytope with 12 vertices, realized on the Carathéodory curve [15]. There are 42 nonregular triangulations among 244 in all

as an abstract (`.gml-`)graph, and marks points corresponding to non-regular triangulations. If dimension permits, this embedded flip graph can then be visualized, e.g. with *JavaView* [17].

In Figures 13 and 14, we present the results of four such calls to `secondary`. All running times, excluding the computation of the convex hull, remained well under one minute on a Sun Blade. The bottleneck is calculating the convex hull: The longest such computation with `cdd` [5] for the secondary of the cyclic 8-polytope with 12 vertices realized on the Carathéodory curve took 2 minutes.

With a view towards future developments, we remark that the computation of the entire vertex-facet incidence matrix of the secondary polytope seems wasteful if all one is interested in is the information which edges of the embedded flip graph actually lie on the convex hull of the secondary polytope.

Moreover, while TOPCOM is fine-tuned to exploit symmetries of a point configuration, at this moment there is no convex hull code available that could do likewise. However, implementing an algorithm that simultaneously inserts all points of an orbit under a given symmetry group may well prove to be a non-trivial task.

In the future, perhaps it will be possible to exploit the fact that TOPCOM not only provides a list of points corresponding to triangulations, but in fact a connected component of the flip-graph that includes the entire 1-skeleton of the polytope.

There have been other approaches to computing the edge graph of the secondary polyope by reverse search, e.g. in [14]. Unfortunately, we do not have access to any code based on these algorithms. If memory limitations dominate then reverse search is certainly the method of choice.

7 Conclusion

Oriented matroids are a suitable interface between calculations in coordinates and computations in combinatorial geometry. In this exposition, we have presented the computation of triangulations of a point configuration using oriented matroids. After the computation of the chirotope, all other required operations can be performed in a purely combinatorial way. The package TOPCOM implements this concept. Regular triangulations provide a beautiful connection to algebraic structures. Their handling, however, requires the integration of additional software; we have shown examples that POLYMAKE is a suitable tool for this.

Acknowledgement: We would like to thank Michael Joswig for many stimulating discussions and his unfailing support.

References

1. Oswin Aichholzer, *The path of a triangulation*, Proceedings of the 15th Annual ACM Symposium on Computational Geometry (Miami Beach, Florida, USA), 1999, pp. 14–23.
2. Louis J. Billera and Bernd Sturmfels, *Fiber polytopes*, Annals of Mathematics **135** (1992), 527–549.
3. Anders Björner, Michel Las Vergnas, Bernd Sturmfels, Neil White, and Günter M. Ziegler, *Oriented matroids*, Encyclopedia of Mathematics, vol. 46, Cambridge University Press, Cambridge, 1993.

4. Jürgen Bokowski and A. Guedes de Oliveira, *On the generation of oriented matroids*, Discrete & Computational Geometry **24** (2000), 197–208, Branko Grünbaum Birthday Issue.
5. Komei Fukuda, *cdd—an implementation of the double description method with LP solver*, http://www.ifor.math.ethz.ch/~fukuda/cdd_home/cdd.html.
6. Komei Fukuda and Lukas Finschi, *Generation of oriented matroids—a graph theoretical approach*, Discrete & Computational Geometry **27** (2002), 117–136.
7. Ewgenij Gawrilow and Michael Joswig, *Polymake—a versatile tool for the algorithmic treatment of polytopes and polyhedra*, 2001.
8. Izrail M. Gelfand, Mikhail M. Kapranov, and Andrei V. Zelevinsky, *Discriminants, resultants, and multidimensional determinants*, Mathematics: Theory & Applications, Birkhäuser, Boston, 1994.
9. Branko Grünbaum, *Convex Polytopes*, Interscience, London, 1967.
10. E. Lawler, J. Lenstra, and A. H. G. Rinnooy Kan, *Generating all maximal independent sets: NP-hardness and polynomial-time algorithms*, SIAM Journal on Computing **9** (1980), 558–565.
11. Carl Lee, *Triangulations of polytopes*, CRC Handbook of Discrete and Computational Geometry (Jacob E. Goodman and Joseph O'Rourke, eds.), CRC Press LLC, Boca Raton, 1997, pp. 271–190.
12. Jesús A. de Loera, *Triangulations of polytopes and computational algebra*, Ph.D. thesis, Cornell University, 1995.
13. Jesús A. de Loera, Serkan Hoşten, Francisco Santos, and Bernd Sturmfels, *The polytope of all triangulations of a point configuration*, Documenta Mathematika **1** (1996), 103–119.
14. Tomonari Masada, Hiroshi Imai, and Keiko Imai, *Enumeration of regular triangulations*, Proceedings of the 12th Annual Symposium on Computational Geometry (Philadelphia, PA, USA), ACM Press, 1996, pp. 224–233.
15. Julian Pfeifle, *Secondary polytope of $c_4(8)$ on the carathéodory curve*, Electronic Geometry Models (2000), http://www.eg-models.de/2000.09.032.
16. Julian Pfeifle, *Secondary polytope of the 3-cube*, Electronic Geometry Models (2000), http://www.eg-models.de/2000.09.031.
17. Konrad Polthier et al., *Javaview*, http://www.javaview.de, 2001.
18. Jörg Rambau, *Projections of polytopes and polyhedral subdivisions*, Berichte aus der Mathematik, Shaker, Aachen, 1996, Dissertation, TU Berlin.
19. Jörg Rambau, *Triangulations of cyclic polytopes and higher Bruhat orders*, Mathematika **44** (1997), 162–194.
20. Jörg Rambau, *TOPCOM—Triangulations of Point Configurations and Oriented Matroids*, Software under the Gnu Public Licence, http://www.zib.de/rambau/TOPCOM.html, 1999.
21. Jörg Rambau, *Point configuration and a triangulation without flips as constructed by Santos*, Electronic Geometry Models (2000), http://www.eg-models.de/2000.08.005.
22. Jörg Rambau, *Topcom: Triangulations of point configurations and oriented matroids*, ZIB-Report 02-17, 2002, Proceedings of the International Congress of Mathematical Software, to appear.
23. Ulrich Reitebuch, *Rhombicosidodecahedron*, Electronic Geometry Models (2000), http://www.eg-models.de/2000.09.013.
24. Mutsumi Saito, Bernd Sturmfels, and Nobuki Takayama, *Gröbner Deformations of Hypergeometric Differential Equations*, Algorithms and Computation in Mathematics, vol. 6, Springer, 2000.

25. Francisco Santos, *A point configuration whose space of triangulations is disconnected*, Journal of the American Mathematical Society **13** (2000), 611–637.
26. Francisco Santos, *Non-connected toric Hilbert schemes*, Preprint CO/0204044 v1, arXiv:math, April 2002.
27. Francisco Santos, *Triangulations of oriented matroids*, Memoirs of the American Mathematical Society **156** (2002).
28. Bernd Sturmfels, *Gröbner Bases and Convex Polytopes*, University Lecture Series, vol. 8, AMS, 1996.
29. Fumihiko Takeuchi, *Combinatorics of triangulations*, Ph.D. thesis, Graduate School of Information Science, University of Tokyo, 2001.
30. Fumihiko Takeuchi and Hiroshi Imai, *Enumerating triangulations for products of two simplices and for arbitrary configurations of points*, Computing and Combinatorics, 1997, pp. 470–481.
31. Günter M. Ziegler, *Lectures on polytopes*, Graduate Texts in Mathematics, vol. 152, Springer, New York, 1995, *Updates, Corrections, and more* available at http://www.math.tu-berlin.de/~ziegler.

A   Equivariant BFS: An Example Run for the Six-gon

Here, we show how the equivariant BFS behaves on the graph of triangulations of the six-gon. We invite the interested reader to look for the symmetries that induce the marking operations on flips. Only the marking in picture 12 is critical and prevents the algorithm from returning to an old symmetry class. The other marking operations hit already marked edges.

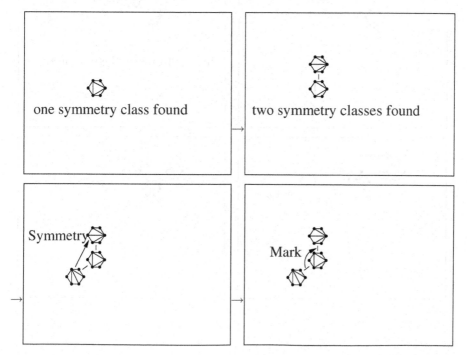

one symmetry class found

two symmetry classes found

Symmetry

Mark

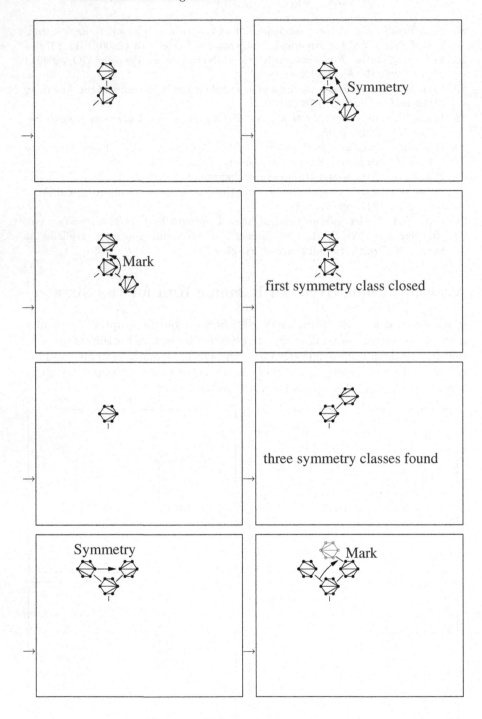

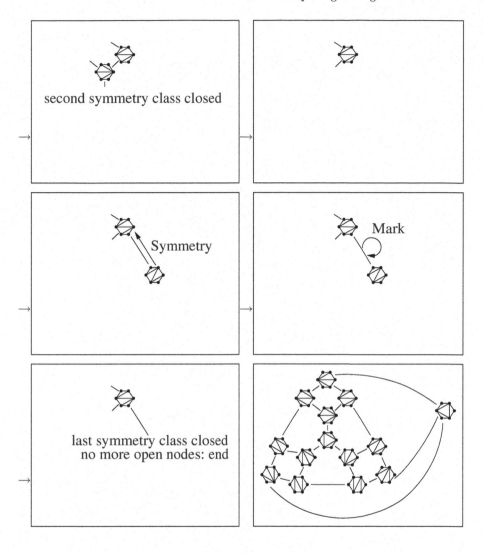

Discrete Geometry for Algebraic Elimination

Ioannis Z. Emiris

INRIA Sophia-Antipolis, France, and
Dept. of Informatics & Telecommunications, University of Athens, Greece
emiris@di.uoa.gr, http://www.di.uoa.gr/~iemiris

Abstract. Multivariate resultants provide efficient methods for eliminating variables in algebraic systems. The theory of toric (or sparse) elimination generalizes the results of the classical theory to polynomials described by their supports, thus exploiting their sparseness. This is based on a discrete geometric model of the polynomials and requires a wide range of geometric notions as well as algorithms. This survey introduces toric resultants and their matrices, and shows how they reduce system solving and variable elimination to a problem in linear algebra. We also report on some practical experience.

1 Introduction

The problem of computing all common zeros of a system of polynomials is of fundamental importance in a wide variety of scientific and engineering applications. This article surveys efficient methods based on the toric (or sparse) resultant for computing all *isolated* solutions of an arbitrary system of n polynomials in n unknowns. In particular, we construct matrix formulae which yield nontrivial multiples of the resultant thus reducing root-finding to the eigendecomposition of a square matrix or to multivariate polynomial factorization. Other elimination problems treated by resultants include the determination of predicates and the implicitization of parametric varieties.

Our methods can strongly exploit a type of sparseness of the polynomials. This is an advantage as compared to other algebraic methods, such as Gröbner bases and characteristic sets. All approaches have complexity exponential in n, but Gröbner bases suffer in the worst case by a quadratic exponent, even if we focus on zero-dimensional ideals. Moreover, they are discontinuous with respect to perturbations in the input coefficients, unlike resultant matrix methods in general and the more recent methods that compute directly a normal form. Of course, Gröbner bases provide an arithmetic over algebraic varieties and have been well developed, including public domain stand-alone implementations or as part of standard computer algebra systems. There is also a number of numerical approaches for solving algebraic systems, such as topological degree and exclusion methods, homotopy continuation, and Newton-like iterative methods. Further information on the aforementioned techniques is found in [5,6,16,20,29,24] and the references thereof.

Our methods find their natural application in problems arising in a variety of fields, including problems expressed in terms of geometric and kinematic constraints in robotics, vision and geometric modeling where several problems are sparse in the present sense. Another motivation comes from systems that must be repeatedly solved for different coefficients, in which case the resultant matrix can be computed exactly once. This occurs, for instance, in parallel robot calibration, where 10,000 instances may have to be solved.

The next section describes briefly the main steps in the theory of toric elimination, which aspires to generalize the results and algorithms of its mature counterpart, classical elimination. Section 3 presents the construction of toric resultant matrices of Sylvester-type. Section 4 reduces solution of arbitrary algebraic systems to numerical linear algebra, thus yielding methods which avoid any issues of convergence. The last section reports on a few experiments with our incremental algorithm and attempts a comparison to Gröbner bases software, on the systems of the cyclic-n family. All algorithms discussed have been implemented either in Maple and/or in C, and are publicly available through the author's webpage. Previous work and open questions are mentioned in the corresponding sections.

2   Toric Elimination Theory

Toric elimination generalizes several results of classical elimination theory on multivariate polynomial systems of arbitrary degree by considering their structure. This leads to stronger algebraic and combinatorial results in general [5,22,28,29]. Assume that the number of variables is n; roots in $(\overline{K}^{*})^{n}$ are called *toric*, where $\overline{K}$ is the algebraic closure of the coefficient field. We use x^{e} to denote the monomial $x_{1}^{e_1} \cdots x_{n}^{e_n}$, where $e = (e_1, \ldots, e_n) \in \mathbb{Z}^{n}$. Let the input *Laurent* polynomials be

$$f_1, \ldots, f_n \in K[x_1^{\pm 1}, \ldots, x_n^{\pm 1}]. \tag{1}$$

Let the *support* $\mathcal{A}_i = \{a_{i1}, \ldots, a_{im_i}\} \subset \mathbb{Z}^{n}$ denote the set of exponent vectors corresponding to monomials in f_i with nonzero coefficients: $f_i = \sum_{a_{ij} \in \mathcal{A}_i} c_{ij} x^{a_{ij}}$, for $c_{ij} \neq 0$. The *Newton polytope* $Q_i \subset \mathbb{R}^{n}$ of f_i is the convex hull of support $\mathcal{A}_i$. For arbitrary sets A and $B \subset \mathbb{R}^{n}$, their *Minkowski sum* is $A + B = \{a + b \mid a \in A, b \in B\}$.

Definition 2.1. Given convex polytopes $A_1, \ldots, A_n, A_k' \subset \mathbb{R}^{n}$, the *mixed volume* $\mathrm{MV}(A_1, \ldots, A_n)$ is the unique real-valued non-negative function, invariant under permutations, such that, $\mathrm{MV}(A_1, \ldots, \mu A_k + \rho A_k', \ldots, A_n) = \mu \, \mathrm{MV}(A_1, \ldots, A_k, \ldots, A_n) + \rho \mathrm{MV}(A_1, \ldots, A_k', \ldots, A_n)$, for $\mu, \rho \in \mathbb{R}_{\geq 0}$. Moreover, we set $\mathrm{MV}(A_1, \ldots, A_n) = n! \, \mathrm{Vol}(A_1)$, when $A_1 = \cdots = A_n$, where $\mathrm{Vol}(\cdot)$ denotes euclidean volume in $\mathbb{R}^{n}$.

If the polytopes have integer vertices, their mixed volume takes integer values. An equivalent definition [19,27] is the following.

Definition 2.2. For $\lambda_1, \ldots, \lambda_n \in \mathbb{R}_{\geq 0}$ and convex polytopes $Q_1, \ldots, Q_n \subset \mathbb{R}^n$, the *mixed volume* $\mathrm{MV}(Q_1, \ldots, Q_n)$ is precisely the coefficient of $\lambda_1 \lambda_2 \cdots \lambda_n$ in $\mathrm{Vol}(\lambda_1 Q_1 + \cdots + \lambda_n Q_n)$ expanded as a polynomial in $\lambda_1, \ldots, \lambda_n$.

One may verify that mixed volume scales in the same way as the generic number of common roots of a polynomial system. We are now ready to state a slight generalization of Bernstein's theorem, also known as BKK bound [5,22].

Theorem 2.3. *Given system (1), the cardinality of common isolated zeros in $(\overline{K}^*)^n$, counting multiplicities, is bounded by $\mathrm{MV}(Q_1, \ldots, Q_n)$, regardless of the dimension of the variety. Equality holds when a certain subset of the coefficients corresponding to the vertices of the Q_i's are generic.*

Newton polytopes provide a "sparse" counterpart of total degree. The same holds for mixed volume vis-à-vis Bézout's bound (equal to the product of all total degrees), where the former is usually significantly smaller for systems encountered in engineering applications. It is possible to generalize the notion of mixed volume to that of *stable mixed volume,* thus extending the bound to affine roots [18,23].

The mixed volume computation is tantamount to enumerating all *mixed cells* in a mixed (tight coherent polyhedral) subdivision of $Q_1 + \cdots + Q_n$. This also identifies the integer points comprising a monomial basis of the quotient ring of the ideal defined by the input polynomials. Mixed, or stable mixed, cells also correspond to start systems (of binomial equations, hence with an immediate solution) for a *toric homotopy* to the original system's roots [21,24,31].

Definition 2.4. The *resultant* of a polynomial system of $n + 1$ polynomials with indeterminate coefficients in n variables is a polynomial in these indeterminates, whose vanishing provides a necessary and sufficient condition for the existence of common roots of the system.

Alternatively, the resultant can be expressed by Poisson's formula, namely $C \prod_\alpha f_0(\alpha)$, where f_0 is one of the polynomials, evaluated at all common roots α of the other n equations, and C is a function of the coefficients of these n polynomials. It is then easy to see that the resultant is homogeneous in the coefficients of each polynomial. The history of resultants (and elimination theory) includes such luminaries as Euler, Bézout, Cayley, Macaulay. Different resultants exist depending on the space of the roots we wish to characterize, namely projective, affine, toric or residual [1].

The simplest case, where the classical projective and toric resultants coincide, is that of a linear system of $n + 1$ equations in n variables. The determinant of the coefficient matrix is the system's resultant and, under the assumption on the non-vanishing of certain minors, it becomes zero exactly when there is a (unique) common root.

The question of whether two polynomials $f_1(x), f_2(x) \in K[x]$ have a common root leads to a condition that has to be satisfied by the coefficients of both polynomials; again classical and toric resultants coincide. The system's *Sylvester matrix* is of dimension $\deg f_1 + \deg f_2$ and its determinant is the system's resultant, provided the leading coefficients are nonzero. This matrix rows contain the coefficient vectors of polynomials $x^k f_j$, for $k = 0, \ldots, \deg f_i - 1$ and $\{i, j\} = \{1, 2\}$. *Bézout* developed a method for computing the resultant as a determinant of a matrix of dimension equal to $\max\{\deg f_1, \deg f_2\}$.

Example 2.5. Consider the following bivariate system from [6].

$$f_1 = x_1^2 + x_1 x_2 + 2 x_1 \qquad + x_2 - 1 = 0,$$
$$f_2 = x_1^2 \qquad + 3 x_1 - x_2^2 + 2 x_2 - 1 = 0.$$

The Sylvester resultant of f_1 and f_2, treated as polynomials in x_2, is:

$$\det \begin{bmatrix} x_1 + 1 & x_1^2 + 2 x_1 - 1 & 0 \\ 0 & x_1 + 1 & x_1^2 + 2 x_1 - 1 \\ -1 & 2 & x_1^2 + 3 x_1 - 1 \end{bmatrix} = -x_1^3 - 2 x_1^2 + 3 x_1.$$

Notice that the rows of the matrix correspond to polynomials $x_2 f_1, f_1, f_2$, whereas its columns to monomials $x_2^2, x_2, 1$. The roots of the resultant are $\{-3, 0, 1\}$; they are the x_1-coordinates of the common zeros of f_1 and f_2, namely $(-3, 1), (0, 1), (1, -1)$. Furthermore, right kernel vectors contain the values of the column monomials at the x_2-coordinates of the roots. When, e.g., $x_1 = 1$, then

$$\begin{bmatrix} 2 & 2 & 0 \\ 0 & 2 & 2 \\ -1 & 2 & 3 \end{bmatrix} \begin{bmatrix} 1 \\ -1 \\ 1 \end{bmatrix} = \begin{bmatrix} 0 \\ 0 \\ 0 \end{bmatrix}.$$

Below is the Bézout matrix, which yields the resultant as

$$\det \begin{bmatrix} x_1 + 1 & x_1^2 + 2 x_1 - 1 \\ -x_1^2 - 4 x_1 - 1 & -(x_1 + 1)(x_1^2 + 3x_1 - 1) \end{bmatrix}.$$

Its first row contains f_1 and its columns are indexed by $x_2, 1$. The construction of the second row goes beyond the scope of this review.

Toric resultants express the existence of toric roots. Formally,

$$f_0, \ldots, f_n \in K[x_1^{\pm 1}, \ldots, x_n^{\pm 1}] \tag{2}$$

f_i corresponding to generic point $c_i = (c_{i1}, \ldots, c_{im_i})$ in the space of polynomials with support $\mathcal{A}_i$. This space is identified with projective space $\mathbb{P}_K^{m_i - 1}$. Then system (2) can be thought of as point $c = (c_0, \ldots, c_n)$. Let Z denote the Zariski closure, in the product of projective spaces, of the set of all c such that the system has a solution in $(\overline{K}^*)^n$. Note that Z is an irreducible variety.

Definition 2.6. The *toric (or sparse) resultant* $R = R(\mathcal{A}_0, \ldots, \mathcal{A}_n)$ of system (2) is a polynomial in $\mathbb{Z}[c]$. If $\mathrm{codim}(Z) = 1$ then R is the defining irreducible polynomial of the hypersurface Z. If $\mathrm{codim}(Z) > 1$ then $R = 1$.

Set $\mathrm{MV}_{-i} = \mathrm{MV}(Q_0, \ldots, Q_{i-1}, Q_{i+1}, \ldots, Q_n)$, then R is homogeneous in the coefficients of f_i with $\deg_{f_i} R = \mathrm{MV}_{-i}$.

3 Matrix Formulae

Different means of expressing each resultant are possible, split into Sylvester, Bézout and hybrid-type formulae [5,6,15,29]. Ideally, we wish to express it as a matrix determinant, a quotient of two determinants, or a divisor of a determinant where the quotient is a nontrivial extraneous factor. This section discusses matrix formulae for the toric resultant known as *toric resultant matrices*. We restrict ourselves to Sylvester-type matrices; such matrices for the toric resultant are also known as *Newton matrices* because they depend on the input Newton polytopes. For other types of resultant matrices see [1,5,6,10,15,29] and the references thereof. Sylvester-type matrices generalize the coefficient matrix of a linear system and Macaulay's matrix. The latter extends Sylvester's construction to arbitrary systems of homogeneous polynomials, and its determinant is a nontrivial multiple of the projective resultant. Its construction is presented below; for an illustration, see example 4.1.

The transpose of a Sylvester-type matrix corresponds to the linear transformation:

$$(g_0, \ldots, g_n) \quad \mapsto \quad \sum_{i=0}^{n} g_i f_i, \tag{3}$$

where the space of polynomials g_i depends on the kind of matrix used. Each row expresses the product of a monomial with an input polynomial; its entries are coefficients of that product, each corresponding to the monomial indexing the corresponding column. The degree of $\det M$ in the coefficients of f_i equals the number of rows with coefficients of f_i. This must be greater or equal to $\deg_{f_i} R$. It is possible to pick any one polynomial so that there is an optimal number of rows containing its coefficients; this number is obviously $\deg_{f_i} R$. This is true both in the case of Macaulay's matrix and in the case of the Newton matrix constructions below.

There are two main approaches to construct a well-defined, square, generically nonsingular matrix M, such that $R \mid \det M$. The first approach [2,4,28], relies on a *mixed subdivision* of the Minkowski sum of the Newton polytopes $Q = Q_0 + \cdots + Q_n$. Each cell σ in this subdivision is written uniquely as the Minkowski sum $\sigma = F_0 + \cdots + F_n$, where each F_i is a face of Q_i, so that $\sum_i \dim F_i = \dim \sigma$. This implies that at least one face is a vertex.

The algorithm uses a subset of $(Q+\delta) \cap \mathbb{Z}^n$ to index the rows and columns of M, where $\delta \in K^n$ is an arbitrarily small and *sufficiently generic* vector.

This vector must perturb all integer points indexing some row (or column) of the matrix in the strict interior of a maximal cell. It can be chosen randomly and the validity of our choice can be confirmed by the matrix construction algorithm. The probability of error for a vector with uniformly distributed entries is bounded in [2].

Now consider an integer point p in a maximal cell σ, and suppose that one of its Minkowski summands is vertex $a_{ij} \in Q_i$. Then the row indexed by p contains the coefficients of polynomial f_i multiplied by monomial $x^{p-a_{ij}}$. For those cells whose Minkowski sum has more than one vertices, the algorithm uses some rule to avoid index i for the matrix to have the minimum number of rows containing f_i, namely $\deg_{f_i} R$.

Example 3.1. Consider a system of 3 polynomials in 2 unknowns:

$$f_1 = c_{11} + c_{12}xy + c_{13}x^2y + c_{14}x,$$
$$f_2 = c_{21}y + c_{22}x^2y^2 + c_{23}x^2y + c_{24}x,$$
$$f_3 = c_{31} + c_{32}y + c_{33}xy + c_{34}x.$$

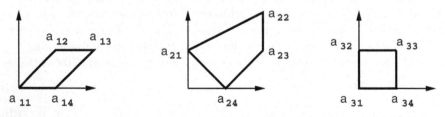

Figure 1. The Newton polytopes and the exponent vectors a_{ij} corresponding to the nonzero terms in Example 3.1.

The Newton polytopes are shown in Figure 1. The mixed volumes are $MV(Q_1, Q_2) = 4$, $MV(Q_2, Q_3) = 4$, $MV(Q_3, Q_1) = 3$, so the toric resultant's total degree is 11. Compare this with the Bézout numbers of these subsystems: $8, 6, 12$; hence the projective resultant's total degree is 26.

Assume that the lifting functions are $l_1(x, y) = Lx + L^2y, l_2(x, y) = -L^2x - y, l_3(x, y) = x - Ly$, where $L \gg 1$. These functions are sufficiently generic since they define a mixed subdivision where every cell is uniquely defined as the Minkowski sum of faces $F_i \subset Q_i$. The algorithm defines lifted Newton polytopes by mapping each point (x, y) in Newton polytope Q_i to a point $(x, y, l_i(x, y))$.

The lower envelope of the Minkowski sum of the lifted Q_i's is then projected to the plane, yielding generically a mixed subdivision of Q. Figure 2 shows $Q + \delta$ and the integer points it contains; notice that every point belongs to a unique maximal cell. Every maximal cell σ is labeled by the indices of the Q_i vertex or vertices appearing in the unique Minkowski sum

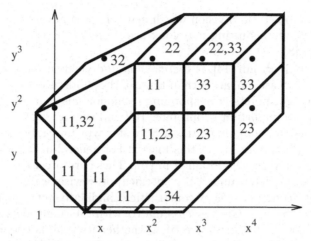

Figure 2. A mixed subdivision of Q perturbed by $(-3/8, -1/8)$, in Example 3.1.

$\sigma = F_0 + \cdots + F_n$, with ij denoting vertex $a_{ij} \in Q_i$. For instance, point $(1,0)$ belongs to a maximal cell $\sigma = a_{11} + F + F'$, where F, F' are the edges $(a_{24}, a_{23}) \subset Q_2$ and (a_{31}, a_{34}) respectively. The corresponding row in the matrix will be filled in with the coefficient vector of $x^{(1,0)} f_1$.

The Newton matrix M appears below with rows and columns indexed by the integer points in the perturbed Minkowski sum. M contains, by construction, the minimum number of f_1 rows, namely 4. The total number of rows is $4 + 4 + 7 = 15$, i.e., the determinant degree is higher than optimal by 1 and 3, respectively, in the coefficients of f_2 and f_3.

	1,0	2,0	0,1	1,1	2,1	3,1	0,2	1,2	2,2	3,2	4,2	1,3	2,3	3,3	4,3
1,0	c_{11}	c_{14}	0	0	c_{12}	c_{13}	0	0	0	0	0	0	0	0	0
2,0	c_{31}	c_{34}	0	c_{32}	c_{33}	0	0	0	0	0	0	0	0	0	0
0,1	0	0	c_{11}	c_{14}	0	0	0	c_{12}	c_{13}	0	0	0	0	0	0
1,1	0	0	0	c_{11}	c_{14}	0	0	0	c_{12}	c_{13}	0	0	0	0	0
2,1	c_{24}	0	c_{21}	0	c_{23}	0	0	0	c_{22}	0	0	0	0	0	0
3,1	0	c_{24}	0	c_{21}	0	c_{23}	0	0	0	c_{22}	0	0	0	0	0
0,2	0	0	c_{31}	c_{34}	0	0	c_{32}	c_{33}	0	0	0	0	0	0	0
1,2	0	0	0	c_{31}	c_{34}	0	0	c_{32}	c_{33}	0	0	0	0	0	0
2,2	0	0	0	0	0	0	0	0	c_{11}	c_{14}	0	0	0	c_{12}	c_{13}
3,2	0	0	0	0	c_{31}	c_{34}	0	0	c_{32}	c_{33}	0	0	0	0	0
4,2	0	0	0	0	0	c_{24}	0	0	c_{21}	c_{23}	0	0	0	0	c_{22}
1,3	0	0	0	0	0	0	0	c_{31}	c_{34}	0	0	c_{32}	c_{33}	0	0
2,3	0	0	0	c_{24}	0	0	c_{21}	0	c_{23}	0	0	0	c_{22}	0	0
3,3	0	0	0	0	0	0	0	0	c_{31}	c_{34}	0	0	c_{32}	c_{33}	0
4,3	0	0	0	0	0	0	0	0	0	c_{31}	c_{34}	0	0	c_{32}	c_{33}

The greedy version [4] produces a matrix with dimension 14 which can be obtained by deleting the row and the column corresponding to point $(1,3)$. The incremental algorithm below yields a matrix of dimension 13.

The subdivision-based approach can be coupled with a recent important development, namely the existence of a minor in the Newton matrix that divides the determinant so as to yield the exact toric resultant [7]. This generalizes Macaulay's famous quotient formula that yields the exact projective resultant [25]. An explicit algorithm for the toric quotient formula requires an appropriate mixed subdivision which may rely on a new lifting algorithm. A glimpse of what this lifting may look like is offered by the hybrid matrix constructed in [8].

The second algorithm [14], is *incremental* and yields usually smaller matrices and, in any case, no larger than those of the subdivision algorithm. The selection of integer points corresponding to monomials multiplying the row polynomials is an important subproblem [13] and uses a vector $v \in (\mathbb{Q}^*)^n$. In those cases where a minimum matrix of Sylvester type provably exists [30], the incremental algorithm produces this matrix. For general multi-homogeneous systems, the best vector is obtained in [10]. These are precisely the systems for which v can be deterministically specified; otherwise, a random v can be used. The algorithm proceeds by constructing candidate rectangular matrices in the input coefficients. Given such a matrix with the coefficients specialized to generic values, the algorithm verifies whether its rank is complete. If so, any square nonsingular submatrix can be returned as M; otherwise, new rows (and columns) are added to the candidate.

The choice of vector v is important. So, in practice, different choices can be tried out and the smallest matrix used for subsequent computations. Other techniques to reduce matrix size (and mixed volumes) include the introduction of new variables to express subexpressions which are common to several input polynomials. For an illustration, see [12].

The asymptotic bit complexity of these algorithms is singly exponential in n, proportional to the total degree of the toric resultant, and polynomial in the number of Q_i vertices, provided all $MV_i > 0$. The relation between mixed volume and number of integer points in the Minkowski sum lies behind these bounds [11].

Newton matrices, including the candidates constructed by the incremental algorithm, are characterized by a structure that generalizes the Toeplitz structure and has been called *quasi-Toeplitz* [17] (cf. [3]). By exploiting this structure, determinant evaluation has quasi-quadratic arithmetic complexity and quasi-linear space complexity in the matrix dimension. The efficient implementation of this structure is open today and is important for the competitiveness of the entire approach.

4   Algebraic Solving by Linear Algebra

To solve well-constrained system (1) by the resultant method we define an overconstrained system and apply the resultant matrix construction. For a more comprehensive discussion the reader may refer to [5,16]. Matrix M need only be computed once for all systems with the same supports. So this step can be carried out offline, while the matrix operations to approximate all isolated roots for each coefficient specialization are online. Numerical issues for the latter are discussed in [12,16].

Resultant matrices reduce system solving to certain standard operations in computer algebra. In particular, univariate or multivariate determinants can be computed by evaluation and interpolation techniques. However, the determinant development in the monomial basis may be avoided because there are algorithms for univariate polynomial solving as well as multivariate polynomial factorization which require only the values of these polynomials at specific points; cf. e.g. [3,26] and the references thereof. All of these evaluations would exploit the quasi-Toeplitz structure of Sylvester-type matrices [3,17].

We present two ways of defining an overconstrained system. The first method *adds an extra polynomial* f_0 to the given system thus yielding a well-studied object, the u-resultant.

$$f_0 = u_0 + u_1 x_1 + \cdots + u_n x_n \in (K[u_0, \ldots, u_n])[x_1^{\pm 1}, \ldots, x_n^{\pm 1}].$$

Coefficients $u_1, \ldots, u_n$ may be randomly specialized or left as indeterminates; in the latter case, solving reduces to factorizing the u-polynomial. It is known that the u-resultant factorizes into linear factors $u_0 + u_1 \alpha_1 + \cdots + u_n \alpha_n$ where $(\alpha_1, \ldots, \alpha_n)$ is an isolated root of the original system. This is an instance of Poisson's formula. u_0 is usually an indeterminate that we shall denote by x_0 below for uniformity of notation. M will describe the multiplication map for f_0 in the coordinate ring of the ideal defined by (1).

Example 4.1. The system of the previous example is augmented by generic linear form $f_0 = u_1 x_1 + u_2 x_2 + u_0$. The Macaulay matrix is

$$M = \begin{bmatrix} 1 & 1 & 2 & 0 & 1 & -1 & 0 & 0 & 0 & 0 \\ 0 & 1 & 0 & 1 & 2 & 0 & 0 & 1 & -1 & 0 \\ 0 & 0 & 1 & 0 & 1 & 2 & 0 & 0 & 1 & -1 \\ 1 & 0 & 3 & -1 & 2 & -1 & 0 & 0 & 0 & 0 \\ 0 & 1 & 0 & 0 & 3 & 0 & -1 & 2 & -1 & 0 \\ 0 & 0 & 1 & 0 & 0 & 3 & 0 & -1 & 2 & -1 \\ 0 & u_1 & 0 & u_2 & u_0 & 0 & 0 & 0 & 0 & 0 \\ 0 & 0 & u_1 & 0 & u_2 & u_0 & 0 & 0 & 0 & 0 \\ 0 & 0 & 0 & 0 & u_1 & 0 & 0 & u_2 & u_0 & 0 \\ 0 & 0 & 0 & 0 & 0 & u_1 & 0 & 0 & u_2 & u_0 \end{bmatrix}$$

The matrix columns are indexed by column vector v such that

$$v^T = [x_1^3, x_1^2 x_2, x_1^2, x_1 x_2^2, x_1 x_2, x_1, x_2^3, x_2^2, x_2, 1].$$

Vector Mv expresses the polynomials indexing the rows of M, which are multiples of f_0, f_1, f_2 by various monomials, namely $\{x_1, x_2, 1\} f_1$, $\{x_1, x_2, 1\} f_2$, and $\{x_1 x_2, x_1, x_2, 1\} f_0$. These monomial sets form the supports of polynomials g_i in (3). The number of f_0 rows is optimal, namely 4, equal to the Bézout bound. The matrix determinant

$$\det M = (u_1 - u_2 + u_0)(-3u_1 + u_2 + u_0)(u_2 + u_0)(u_1 - u_2)$$

gives the u-resultant, which corresponds to the affine solutions $(1, -1), (-3, 1)$, $(0, 1)$, and one solution $(0 : 1 : -1)$ at infinity.

One may delete row 5 and column 7, thus obtaining a smaller matrix. Its determinant is a nontrivial multiple of the toric resultant and, in this case, equals the above determinant within a constant. The number of rows per polynomial are all optimal except for f_1 which appears 3 times whereas the corresponding BKK bound is 2. This means that the determinant will equal the toric resultant multiplied by an extraneous factor linear in the coefficients of f_1.

An alternative way to obtain an overconstrained system is by *hiding* one of the original variables in the coefficient field and consider the system as follows (we modify previous notation to unify the subsequent discussion):

$$f_0, \ldots, f_n \in (K[x_0]) \, [x_1^{\pm 1}, \ldots, x_n^{\pm 1}].$$

M is a matrix polynomial in x_0, and may not be linear.

An important issue concerns the degeneracy of the input coefficients. This may result in the trivial vanishing of the toric resultant or of $\det M$ when there is an infinite number of common roots (in the torus or at toric infinity) or simply due to the matrix constructed. An infinitesimal perturbation has recently been proposed [9] which respects the structure of Newton polytopes and is computed at no extra asymptotic cost.

The perturbed determinant is a polynomial in the perturbation variable, whose leading coefficient is nonzero whereas the least significant coefficient is $\det M$. Irrespective of which coefficients vanish, there is always a trailing nonzero coefficient which vanishes when x_0 takes its values at the system's isolated roots, even in the presence of positive-dimensional components. This univariate polynomial is known as a *projection operator* because it projects the roots to the x_0-coordinate. Univariate polynomial solving thus yields these coordinates. Again, the u-resultant allows us to recover all coordinates via multivariate factoring.

4.1   Eigenproblems

Most of the computation is numeric, yet the method has global convergence and avoids issues related to the starting point of iterative methods. Our implementations use double precision floating point numbers and the LA-PACK library because it implements state-of-the-art algorithms, offering the choice of a tradeoff between speed and accuracy, and provides efficient ways for computing estimates on the condition numbers and error bounds. Since the matrix construction involves exact computation, any error in computing the roots comes from this phase.

A basic property of resultant matrices is that right vector multiplication expresses evaluation of the row polynomials. Specifically, multiplying by a column vector containing the values of column monomials q at some $\alpha \in (\overline{K}^*)^n$ produces the values of the row polynomials $\alpha^p f_{i_p}(\alpha)$. Letting $\mathcal{E}$ be the monomial set indexing the matrix rows and columns,

$$M(x_0) \begin{bmatrix} \vdots \\ \alpha^q \\ \vdots \end{bmatrix} = \begin{bmatrix} \vdots \\ \alpha^p f_{i_p}(\alpha) \\ \vdots \end{bmatrix}, \quad q, p \in \mathcal{E}, i_p \in \{0, \ldots, n\}.$$

Computationally it is preferable to have to deal with as small a matrix as possible. To this end we partition M into four blocks M_{ij} so that the upper left submatrix M_{11} is square, independent of x_0, and of maximal dimension so that it remains well-conditioned. Our code concentrates all constant columns to the left and within these columns permutes all zero rows to the bottom.

Once M_{11} is specified, let $A(x_0) = M_{22}(x_0) - M_{21}(x_0)M_{11}^{-1}M_{12}(x_0)$ To avoid computing M_{11}^{-1}, we use its LU (or QR) decomposition to solve $M_{11}X = M_{12}$ and compute $A = M_{22} - M_{21}X$. Let $\mathcal{B} \subset \mathcal{E}$ index A. If $(\alpha_0, \alpha) \in \overline{K}^{n+1}$ is a common root with $\alpha \in \overline{K}^n$, then $\det A(\alpha_0) = 0$ and, for any vector $v' = [\cdots \alpha^q \cdots]$, where q ranges over $\mathcal{B}$, $A(\alpha_0)v' = 0$. Moreover,

$$\begin{bmatrix} M_{11} & M_{12}(\alpha_0) \\ 0 & A(\alpha_0) \end{bmatrix} \begin{bmatrix} v \\ v' \end{bmatrix} = \begin{bmatrix} 0 \\ 0 \end{bmatrix} \Rightarrow M_{11}v + M_{12}(\alpha_0)v' = 0,$$

determines v once v' has been computed. Vector $[v, v']$ contains the values of every monomial in $\mathcal{E}$ at α.

It can be shown that $\mathcal{E}$ affinely spans $\mathbb{Z}^n$ and an affinely independent subset can be computed in polynomial time [11]. Given v, v' and these points, we can compute the coordinates of α. If all independent points are in $\mathcal{B}$ then v' suffices. To find the vector entries that will allow us to recover the root coordinates it is typically sufficient to search in $\mathcal{B}$ for pairs of entries corresponding to q_1, q_2 such that $q_1 - q_2 = (0, \ldots, 0, 1, 0, \ldots, 0)$. This lets us compute the i-th coordinate, if the unit appears at the i-th position. In general, the problem of choosing the best vector entries for computing the root coordinates is open, and different choices may lead to different accuracy.

To reduce the problem to an eigendecomposition, let r be the dimension of $A(x_0)$, and $d \geq 1$ the highest degree of x_0 in any entry. We wish to find all values of x_0 at which

$$A(x_0) = x_0^d A_d + x_0^{d-1} A_{d-1} + \cdots + x_0 A_1 + A_0$$

becomes singular. These are the eigenvalues of the *matrix polynomial*. Furthermore, for every eigenvalue λ, there is a basis of the kernel of $A(\lambda)$ defined by the *right eigenvectors* of the matrix polynomial associated to λ. If A_d is nonsingular then the eigenvalues and right eigenvectors of $A(x_0)$ are the eigenvalues and right eigenvectors of *monic* matrix polynomial $A_d^{-1}A(x_0)$. This is always the case when adding an extra linear polynomial, since $d = 1$

and $A_1 = I$ is the $r \times r$ identity matrix; then $A(x_0) = -A_1(-A_1^{-1}A_0 - x_0 I)$. Generally, the *companion matrix* of a monic matrix polynomial is a square matrix C of dimension rd:

$$C = \begin{bmatrix} 0 & I & \cdots & 0 \\ \vdots & & \ddots & \\ 0 & 0 & \cdots & I \\ -A_d^{-1}A_0 & -A_d^{-1}A_1 & \cdots & -A_d^{-1}A_{d-1} \end{bmatrix}.$$

The eigenvalues of C are precisely the eigenvalues λ of $A_d^{-1}A(x_0)$, whereas its right eigenvector $w = [v_1, \ldots, v_d]$ contains a right eigenvector v_1 of $A_d^{-1}A(x_0)$ and $v_i = \lambda^{i-1}v_1$, for $i = 2, \ldots, d$.

We now address the question of a singular A_d. The following *rank balancing* transformation in general improves the conditioning of A_d. If matrix polynomial $A(x_0)$ is not identically singular for all x_0, then there exists a transformation $x_0 \mapsto (t_1 y + t_2)/(t_3 y + t_4)$ for some $t_i \in \mathbb{Z}$, that produces a new matrix polynomial of the same degree and with nonsingular leading coefficient. The new matrix polynomial has coefficients of the same rank, for sufficiently generic t_i. The asymptotic as well as practical complexity of this stage is dominated by the eigendecomposition.

If a matrix polynomial with invertible leading matrix is found, then the eigendecomposition of the corresponding companion matrix is undertaken. If A_d is ill-conditioned for all linear rank balancing transformations, then we build the matrix pencil and apply a *generalized eigendecomposition* to solve $C_1 x + C_0$. This returns pairs (α, β) such that matrix $C_1 \alpha + C_0 \beta$ is singular with an associated right eigenvector. For $\beta \neq 0$ the eigenvalue is α/β, while for $\beta = 0$ we may or may not wish to discard the eigenvalue. The case $\alpha = \beta = 0$ occurs if and only if the pencil is identically zero within machine precision.

5   Experiments

This section reports on a few experiments with the incremental algorithm, which is implemented in C, and attempts a comparison to Gröbner bases computations. We examine the cyclic-n systems, a standard benchmark in computer algebra. Each system has a total of n equations in variables $x_1, \ldots, x_n$, defined by $f_k = \sum_{i=1}^{n} \prod_{j=i}^{i+k-1} x_j$, where $x_{n+t} = x_t$, for $k = 1, \ldots, n-1$, and $f_n = x_1 x_2 \cdots x_n - 1$.

A Gröbner basis has been computed over the integers with respect to a degree reverse-lexicographic monomial ordering (DRL), using Faugère's publicly available software Gb. There are other efficient programs for Gröbner bases computation, including Asir, Faugère's new algorithm FGb, Macaulay 2, Magma, the PoSSo engine, and Singular; for all of these, see [20] and the references thereof. The hardware used was a Pentium-III CPU with 512 MB at 935 MHz. The toric resultant computations were conducted on a SUN Ultra 5 architecture running at 400 MHz, with 128 MB of memory.

Table 1 shows, in this order, n, the dimension and the degree of the system's variety V, the time to compute the Gröbner basis by Gb, the total degree of the toric resultant R in the input coefficients, the dimension of matrix M obtained by our code, the time to construct this matrix for a fixed direction vector v, and, lastly, the time for the linear algebra operations and for solving the resulting eigenproblem.

We were not able to solve the cyclic-7 instance with Gb, due to some problem in the way space usage is optimized when storing the exponents. The timing reported in the last line of Table 1 is estimated from that in [20], where it is reported that cyclic-7 is solved in 5hrs17min on a Pentium-Pro architecture with 128 MB at 200 MHz. The same article reports a solution by FGb in 39.7sec. On the other hand, it is indicated in [20] that cyclic-7 had not been solved by most, if not all, of the aforementioned Gröbner bases programs.

Table 1. Comparative running times on the cyclic-n family.

n	$\dim V$	$\deg V$	Gb	$\deg R$	$\dim M$	construct	lin.alg.
4	1	4	0 sec	20	25	0 sec	0 sec
5	0	70	0 sec	82	147	3 sec	2 sec
6	0	156	1 sec	290	851	12 min	6 min
7	0	924	67 min	1356	–	–	–

The resultant matrix approach is limited by a dense representation of candidate matrices, on which successive rank tests are executed. For the vectors v tested, the matrix dimension goes beyond 3000. The candidate matrices are rectangular with more than 5000 rows, which makes their explicit storage and manipulation impossible on our hardware. Here, exploiting structure would be crucial in reducing timings and, more importantly, storage by almost an order of magnitude. Current work, e.g. [13], considers the enumeration of monomials for the matrix construction phase. Another alternative is to use more compact resultant matrices, namely of Bézout or hybrid types.

Acknowledgment: I would like to thank Philippe Trébuchet for conducting the experiments with Gb, and acknowledge partial support by the IST Programme of the EU as a Shared-cost RTD (FET Open) Project under Contract No IST-2000-26473 (ECG - Effective Computational Geometry for Curves and Surfaces).

References

1. L. Busé, M. Elkadi, and B. Mourrain. Residual resultant of complete intersection. *J. Pure & Applied Algebra*, 164:35–57, 2001.
2. J.F. Canny and I.Z. Emiris. A subdivision-based algorithm for the sparse resultant. *J. ACM*, 47(3):417–451, May 2000.
3. J.F. Canny, E. Kaltofen, and Y. Lakshman. Solving systems of non-linear polynomial equations faster. In *Proc. Annual ACM Intern. Symp. on Symbolic and Algebraic Computation*, pages 121–128, 1989. ACM Press.
4. J. Canny and P. Pedersen. An algorithm for the Newton resultant. Technical Report 1394, Comp. Science Dept., Cornell University, 1993.
5. D. Cox, J. Little, and D. O'Shea. *Using Algebraic Geometry*. Number 185 in Graduate Texts in Mathematics. Springer-Verlag, New York, 1998.
6. A. Díaz, I.Z. Emiris, E. Kaltofen, and V.Y. Pan. Algebraic algorithms. In M.J. Atallah, editor, *Handbook of Algorithms and Theory of Computation*, chapter 16. CRC Press, Boca Raton, Florida, 1999.
7. C. D'Andrea. Macaulay-style formulas for the sparse resultant. *Trans. of the AMS*, 354:2595–2629, 2002.
8. C. D'Andrea and I.Z. Emiris. Hybrid resultant matrices of bivariate polynomials. In *Proc. Annual ACM Intern. Symp. on Symbolic and Algebraic Computation*, pages 24–31, London, Ontario, 2001. ACM Press.
9. C. D'Andrea and I.Z. Emiris. Computing sparse projection operators. In *Symbolic Computation: Solving Equations in Algebra, Geometry, and Engineering*, vol. 286, AMS Contemporary Mathematics, pp. 121–139, Providence, Rhode Island, AMS, 2001.
10. A. Dickenstein and I.Z. Emiris. Multihomogeneous resultant matrices. In *Proc. Annual ACM Intern. Symp. on Symbolic and Algebraic Computation*, pp. 46–54, Lille, France, 2002. ACM Press, 2002.
11. I.Z. Emiris. On the complexity of sparse elimination. *J. Complexity*, 12:134–166, 1996.
12. I.Z. Emiris. A general solver based on sparse resultants: Numerical issues and kinematic applications. Technical Report 3110, Projet SAFIR, INRIA Sophia-Antipolis, France, 1997.
13. I.Z. Emiris. Enumerating a subset of the integer points inside a Minkowski sum. *Comp. Geom.: Theory & Appl., Spec. Issue*, 20(1-3):143–166, 2002.
14. I.Z. Emiris and J.F. Canny. Efficient incremental algorithms for the sparse resultant and the mixed volume. *J. Symbolic Computation*, 20(2):117–149, 1995.
15. I.Z. Emiris and B. Mourrain. Computer algebra methods for studying and computing molecular conformations. *Algorithmica, Special Issue on Algorithms for Computational Biology*, 25:372–402, 1999.
16. I.Z. Emiris and B. Mourrain. Matrices in elimination theory. *J. Symbolic Computation, Special Issue on Elimination*, 28:3–44, 1999.
17. I.Z. Emiris and V.Y. Pan. Symbolic and numeric methods for exploiting structure in constructing resultant matrices. *J. Symbolic Computation*, 33:393–413, 2002.
18. I.Z. Emiris and J. Verschelde. How to count efficiently all affine roots of a polynomial system. *Discrete Applied Math., Special Issue on Comput. Geom.*, 93(1):21–32, 1999.

19. G. Ewald. *Combinatorial Convexity and Algebraic Geometry*. Springer, New York, 1996.
20. J.-C. Faugère. A new efficient algorithm for computing Gröbner Bases (F4). J. Pure & Applied Algebra, 139(1-3): 61–88, 1999.
21. T. Gao, T.Y. Li, and X. Wang. Finding isolated zeros of polynomial systems in C^n with stable mixed volumes. *J. Symbolic Computation*, 28:187–211, 1999.
22. I.M. Gelfand, M.M. Kapranov, and A.V. Zelevinsky. *Discriminants and Resultants*. Birkhäuser, Boston, 1994.
23. B. Huber and B. Sturmfels. Bernstein's theorem in affine space. *Discr. and Computational Geometry*, 17(2):137–142, March 1997.
24. T.Y. Li. Numerical solution of multivariate polynomial systems by homotopy continuation methods. *Acta Numerica*, pages 399–436, 1997.
25. F.S. Macaulay. Some formulae in elimination. *Proc. London Math. Soc.*, 1(33):3–27, 1902.
26. V.Y. Pan. Solving a polynomial equation: Some history and recent progress. *SIAM Rev.*, 39(2):187–220, 1997.
27. R. Schneider. *Convex Bodies: The Brunn-Minkowski Theory*. Cambridge University Press, Cambridge, 1993.
28. B. Sturmfels. On the Newton polytope of the resultant. *J. of Algebr. Combinatorics*, 3:207–236, 1994.
29. B. Sturmfels. Introduction to Resultants. In *Applic. Computational Algebr. Geom.*, D.A. Cox and B. Sturmfels, eds., vol. 53, Proc. Symp. Applied Math., AMS, pages 25–39, 1998.
30. B. Sturmfels and A. Zelevinsky. Multigraded resultants of Sylvester type. *J. of Algebra*, 163(1):115–127, 1994.
31. J. Verschelde and K. Gatermann. Symmetric Newton polytopes for solving sparse polynomial systems. *Adv. Appl. Math.*, 16(1):95–127, 1995.

Sparse Resultant Perturbations[*]

Carlos D' Andrea[1] and Ioannis Z. Emiris[2]

[1] Dept. of Mathematics, Univ. of California at Berkeley, USA,
   cdandrea@math.berkeley.edu
[2] Dept. of Informatics & Telecommunications, Univ. of Athens, Greece,
   emiris@di.uoa.gr, http://www.di.uoa.gr/~iemiris,
   and INRIA Sophia-Antipolis, France

Abstract. We consider linear infinitesimal perturbations on sparse resultants. This yields a family of projection operators, hence a general method for handling algebraic systems in the presence of "excess" components or other degenerate inputs. The complexity is simply exponential in the dimension and polynomial in the sparse resultant degree. Our perturbation generalizes Canny's Generalized Characteristic Polynomial (GCP) for the homogeneous case, while it provides a new and faster algorithm for computing Rojas' toric perturbation. We illustrate our approach through its Maple implementation applied to specific examples. This work generalizes the linear perturbation schemes proposed in computational geometry and is also applied to the problem of rational implicitization with base points.

1 Introduction

Resultants (or eliminants) offer an efficient approach (in terms of asymptotic complexity) for solving polynomial systems. Given a system of polynomial equations, they eliminate several variables simultaneously and reduce system solving over the complex numbers to univariate polynomial factorization or a matrix eigenproblem.

In [7], we extended the applicability of the sparse (or toric) resultant to non-generic instances. The sparse resultant subsumes the classical (or projective) resultant; for general information see [5,8,20]. Here, we overview this approach, and discuss two geometric applications: first, we show how our sparse resultant perturbations generalize the known linear perturbations in *computational geometry* and, second, how they are directly applicable to the problem of parametric *implicitization with base points*.

In computational geometry, predicates may sometimes make certain assumptions about the input. The treatment of cases violating these assumptions is tedious and intricate, thus seldom included in the theoretical discussion, yet it remains a nontrivial matter for implementors. For instance,

---

[*] Both authors partially supported by the IST Programme of the EU as a Shared-cost RTD (FET Open) Project IST-2000-26473 (ECG - Effective Computational Geometry for Curves and Surfaces), and by Action A00E02 of the ECOS-SeTCIP French-Argentina bilateral collaboration.

in constructing convex hulls in d dimensions, the *Orientation* predicate supposes that no $d + 1$ points lie on the same $(d - 1)$-dimensional hyperplane. A general method for eliminating the need of explicitly dealing with some of these special cases relies on linear infinitesimal perturbations [9,10,18].

The problem of computing the implicit, or algebraic, representation of a curve, surface, or hypersurface, given its rational parametric representation, lies at the heart of several algorithms in computer-aided geometric design and geometric modeling. The problem is harder in the presence of base points, i.e. when the parametric polynomials vanish for any value of the variables in the implicit equation. Several algorithms exist for implicitization, but fewer can handle base points in the case of surfaces or hypersurfaces, so the field is still of active research; see, for instance, [2,12,14,16] and the references thereof.

Let $\mathcal{A} = (\mathcal{A}_0, \ldots, \mathcal{A}_n)$, with $\mathcal{A}_i \subset \mathbb{Z}^n$ be finite sets. We assume that their union spans the affine lattice $\mathbb{Z}^n$; this technical assumption can be eventually removed. Consider a system of generic Laurent polynomials (i.e. with integer exponents) in n variables $x = (x_1, \ldots, x_n)$:

$$f_i(x) = \sum_{a \in \mathcal{A}_i} c_{ia}\, x^a, \quad c_{ia} \neq 0, \quad i = 0, 1, \ldots, n, \tag{1}$$

The *Newton polytope* of a polynomial f_i is the convex hull of the exponent vectors in its support $\mathcal{A}_i$; it provides the sparse counterpart to total degree. The *mixed volume* of n Newton polytopes (or the corresponding polynomials) is denoted by $\mathrm{MV}(\cdot)$. The cornerstone of sparse elimination theory is the theorem stating that the number of isolated roots in $(\mathbb{C}^*)^n$ of $f_1, \ldots, f_n$, is bounded by the mixed volume of their supports. Mixed volume generalizes Bézout's classical bound given by the product of total degrees.

The *sparse resultant* $\mathrm{Res}_{\mathcal{A}}(f_0, \ldots, f_n)$ is a polynomial in $\mathbb{Z}[c_{ia}]$, homogeneous in the coefficients of each f_i, with degree equal to $\mathrm{MV}(f_0, \ldots, f_{i-1}, f_{i+1}, \ldots, f_n)$. The sparse resultant vanishes after a specialization of the coefficients iff the specialized system (1) has a solution in the *toric variety* containing $(\mathbb{C}^*)^n$ and associated to the given Newton polytopes, where $\mathbb{C}^* := \mathbb{C} \setminus \{0\}$. Typically, it may be computed as an irreducible factor of a matrix determinant. For the sake of simplicity, we consider coefficients in a field whose algebraic closure is $\mathbb{C}$, but this field can be arbitrary.

Let us consider system (1) over n variables and m parameters; the latter may be the coefficients or any smaller parameter set on which the coefficients depend. The sparse resultant then defines a *projection operator* from a space of dimension $n + m$ to one of dimension m, and for generic coefficients it does not vanish. It is typically computed as the determinant, or a nontrivial divisor of the determinant of a *resultant matrix*. Such matrices include Sylvester's and Macaulay's formulae; our construction generalizes these matrices. However, there are three cases where this construction fails:

○ The resultant may be identically zero at the specialized coefficients due to lack of genericity, usually because the coefficients depend on few parameters.

○ The sparse resultant may also vanish because the well-constrained system has infinite solutions in the toric variety, forming "excess components".

○ The sparse resultant may be nonzero, but the resultant matrix determinant may be identically zero due to a vanishing extraneous factor.

The dense homogeneous case was settled in [3] by means of a homogeneous GCP. In the sparse context, the only existing approach [17] either requires substantial additional computation or has to perturb all given coefficients.

We settle the sparse case in its generality by an efficient generalization of the GCP. The main contribution of this paper is a simple perturbation scheme, based on a lifting of the polytopes. Only a subset of the input coefficients need be perturbed. It produces sparse projection operators defined by a determinant such that, after specialization, these operators are not identically zero but vanish on roots in the proper components of the variety, including all isolated roots. Thus, we avoid all shortcomings mentioned above. Furthermore,

○ the input supports $\mathcal{A}_i$ are not incremented (i.e., each perturbation polynomial has support in the respective $\mathcal{A}_i$), and

○ the overall complexity remains simply exponential in n and scales as a function of the sparsity parameters.

The main algorithm can be summarized in a single sentence: add linear terms in a perturbation variable, with random coefficients, to all monomials on the diagonal of any sparse resultant matrix.

The next section overviews the most relevant previous work. A very general context of linear perturbations is defined in section 3 to subsume all existing constructions on resultants. In section 4 we detail an efficient algorithm, trade randomization for efficiency, bound the asymptotic complexity, and derive bounds on the number of roots from these operators. In section 5 we examine our approach through its Maple implementation and some applications. Input perturbations have been applied in computational geometry in what concerns linear objects; section 6.1 shows that our scheme generalizes this approach, since the resultant may express geometric predicates among non-linear objects as well. Section 6.2 shows a direct application of our method to the problem of implicitizing rational parametric (hyper)surfaces in the presence of singularities or base points.

2   Related Work

The direct precursor of our work [3] settled the dense (classical) case with complexity $d^{O(n)}$, where d bounds the total degree of the polynomials. In terms of the matrix dimension N, the running time is in $O(N^4)$. By means of the GCP, Canny computes a projection operator which is not identically zero, but vanishes on all proper components of the system $f_i = 0$ in the dense case, i.e. when $\mathcal{A}_i = \{x \in \mathbb{Z}_{\geq 0}{}^n : x_1 + \cdots + x_n \leq d_i\}$. Macaulay's formula for the classical resultant allows computation of this projection operator as a quotient of two matrices both having a linear perturbation in the diagonal, just like the characteristic polynomial of square matrices.

We concentrate on the sparse resultant and its matrix, which generalizes the coefficient matrix of a linear system, Sylvester's matrix and Macaulay's matrix. The matrix is constructed by means of a mixed subdivision of the Minkowski (or vector) sum of the Newton polytopes and their associated lifting to $\mathbb{R}^{n+1}$ [4]. This has been the first algorithm for the sparse resultant to achieve complexity polynomial in its degree and simply exponential in n. But this approach relies on certain genericity assumptions, mainly that the resultant does not vanish identically; this hypothesis is removed in this paper. An additional source of improvement comes from D'Andrea's recent formula of the sparse resultant as the quotient of two determinants, the numerator being defined from a sparse resultant matrix and the denominator being one of its submatrices [6]. This generalizes Macaulay's classical formula.

Degeneracy for the sparse resultant has been studied by Rojas [17], who introduced a toric GCP. A randomized algorithm may simply perturb all coefficients. This is a general technique but contradicts the goal of efficiency which is to minimize the use of the perturbation variable. Our scheme perturbs only the subset of the coefficients appearing on the matrix diagonal. In the worst case, the perturbed matrix determinants defined by Rojas and in this paper have the same degree in the perturbation variable, namely equal to the matrix dimension. Alternatively, there is a deterministic method based on *fillings* of the Newton polytopes. One advantage of fillings, illustrated in section 5, is that the maximum degree of the perturbed determinant in the perturbation variable is usually smaller than with our construction. Unfortunately, finding such fillings requires time exponential in the system dimension, namely in $O(n^{2.616}k^{2n+2})$ arithmetic steps over $\mathbb{Q}$, where k is the total number of support elements [17, Main Thm 3].

The algebraic foundations of the present work, along with establishing the validity of the perturbation, have appeared in [7]. One can find there those proofs of theorems that were omitted here. That article also examined the question of optimality of the perturbations (by decreasing the number of perturbed coefficients) and the hardness of this problem. It would be interesting to re-examine the possibility of reducing the number of perturbed coefficients in the case of implicitization. Several problem-specific perturbation methods exist. One which is connected to resultants and has been applied to the implicitization problem was proposed in [16] in order to minimize the number of perturbed terms.

In [7], certain algorithms for computing the projection operators by means of matrix operations are suggested, most notably based on interpolation, randomization, and the quasi-Toeplitz structure of sparse resultant matrices [11]. Recent advances in determinant computation should largely accelerate these algorithms [15].

3   Sparse Projection Operators

This section discusses a framework for efficient perturbations, and the combinatorial properties of sparse systems used to define such perturbations. The *raison d'être* of linear perturbations is that they minimize the degree in the perturbation variable, thus reducing complexity. Suppose we have a family $p := (p_0(x) \ldots, p_n(x))$ of Laurent polynomials such that $\mathrm{supp}(p_i) \subset \mathcal{A}_i$, and

$$\mathrm{Res}_{\mathcal{A}}(p_0, \ldots, p_n) \neq 0.$$

Note that the $\mathcal{A}_i$ are fixed at input, so the inclusion of $\mathrm{supp}(p_i)$ is usually proper.

Definition 3.1. The *Toric Generalized Characteristic Polynomial* associated with the sequence $(p_0, \ldots, p_n)$ (*p*-GCP) is

$$C_p(\epsilon) := \mathrm{Res}_{\mathcal{A}}\left(f_0 - \epsilon p_0, \ldots, f_n - \epsilon p_n\right).$$

ϵ may be regarded as a positive infinitesimal indeterminate.

Proposition 3.2. $C_p(\epsilon)$ *is a polynomial in* ϵ *of degree*

$$r := \deg(\mathrm{Res}_{\mathcal{A}}(f_0, \ldots, f_n))$$

and not identically zero. Its leading coefficient is $\mathrm{Res}_{\mathcal{A}}(p_0, \ldots, p_n)$.

Proof. Because of the homogeneities of the sparse resultant, we have $C_p(\epsilon) = \mathrm{Res}_{\mathcal{A}}\left(\epsilon(\frac{1}{\epsilon}f_0 - p_0), \ldots, \epsilon(\frac{1}{\epsilon}f_n - p_n)\right) = \epsilon^r \mathrm{Res}_{\mathcal{A}}\left(\frac{1}{\epsilon}f_0 - p_0, \ldots, \frac{1}{\epsilon}f_n - p_n\right)$. In order to see that the latter resultant is nonzero, remark that it tends to $\mathrm{Res}_{\mathcal{A}}(p_0, \ldots, p_n)$, which is clearly nonzero, for ϵ sufficiently large.

Example 3.3 (Dense case). When all polynomials are dense with respect to their total degree, then $\mathcal{A}_i = \{x \in \mathbb{Z}_{\geq 0}{}^n : x_1 + \ldots + x_n \leq d_n\}$. We set $p_0 = 1$ and $p_i := x_i^{d_i}$, for $i = 1, \ldots, n$. Then $\mathrm{Res}_{\mathcal{A}}(1, x_1^{d_1}, \ldots, x_n^{d_n}) = 1$, and $C_p(\epsilon)$ is the Generalized Characteristic Polynomial constructed in [3].

The *u*-resultant of system $f_1, \ldots, f_n$ is the resultant of the overconstrained system where $f_0 := u_0 + u_1 x_1 + \cdots + u_n x_n$ and $\mathcal{A}_0 = \{0, e_1, \ldots, e_n\}$. We can relax the original requirement that $\mathrm{Res}_{\mathcal{A}}(p_0, \ldots, p_n) \neq 0$:

Proposition 3.4 (u-resultant). *Consider the u-resultant construction with f_0 linear. Require that $\mathrm{Res}_{\mathcal{A}}(f_0, p_1, \ldots, p_n) \neq 0$ and let $p_0 := 0$ in definition 3.1. Then $C_p(\epsilon)$ is a polynomial in ϵ of degree $\deg(\mathrm{Res}_{\mathcal{A}}(f_0, \ldots, f_n)) - \mathrm{MV}(f_1, \ldots, f_n)$ and not identically zero. Its leading coefficient is $\mathrm{Res}_{\mathcal{A}}(f_0, p_1, \ldots, p_n)$.*

Proof. The proof of proposition 3.2 can be adapted to this case, where the leading coefficient becomes $\mathrm{Res}_{\mathcal{A}}(f_0, p_1, \ldots, p_n)$ when ϵ tends to infinity. This is nonzero for generic choices of u_i, because it vanishes only if the u_i are such that there are solutions to the system $f_0 = p_1 = \cdots = p_n = 0$.

Example 3.5 (Rojas' toric GCP). Choose $p_1, \ldots, p_n$ with support in the family $(\mathcal{A}_1, \ldots, \mathcal{A}_n)$, set $p_0 := 0$ and $f_0 := \sum_{a \in \mathcal{A}_0} u_a \, x^a$, where $\{u_a\}$ are parameters. Now $C_p(\epsilon) \not\equiv 0$ if $\mathrm{Res}_{\mathcal{A}}(f_0, p_1, \ldots, p_n) \neq 0$ (see also [17, Main Thm. 2.4(1)]). When the overconstrained system has finite solutions in the toric compactification given by their supports, then $C_p(\epsilon)$ is the polynomial called Toric Generalized Characteristic Polynomial in [17]. Checking whether the system has a finite number of solutions may be accomplished by the combinatorial method of fillings or by using random polynomials with the same supports.

The following theorem states the main property of polynomials $C_p(\epsilon)$, and allows us to regard them as a generalization of the classical GCP. We start with $n + 1$ polynomials $f_i(y_1, \ldots, y_m, x_1, \ldots, x_n)$ with support (viewed as polynomials in $\mathbb{C}\left[x_1^{\pm 1}, \ldots, x_n^{\pm 1}\right]$) contained in $\mathcal{A}_i$.

Let Z be the algebraic set $V(f_0, \ldots, f_n) \subset \mathbb{C}^m \times \mathcal{T}$, where $\mathcal{T}$ is the toric variety associated to the respective Newton polytopes. Recall that $\mathcal{T}$ lies in $\mathbb{P}^N$ for some $N \in \mathbb{Z}_{>0}$, so w.l.o.g. we consider $Z \subset \mathbb{C}^m \times \mathbb{P}^N$.

Theorem 3.6. *Let W be a proper component of Z, so that the dimension of W is $m - 1$. Let $C_p(y_1, \ldots, y_m)(\epsilon)$ be the p-GCP of definition 3.1. Arranging $C_p(\epsilon)$ in powers of ϵ, let $C_{p,k}(y_1, \ldots, y_m)$ be its coefficient of lowest degree in ϵ, namely k. If $\pi_y : \mathbb{C}^m \times \mathcal{T} \to \mathbb{C}^m$ denotes projection on the y_i coordinates, then $C_{p,k}(\pi_y(s)) = 0$, for all $s \in W$.*

The proof, detailed in [7, Thm 4.1], is based on the resultant matrix properties demonstrated in [4].

The coefficient $C_{p,k}$ of minimum degree in ϵ in the p-GCP $C_p(\epsilon)$ of definition 3.1 is a suitable *projection operator*, i.e., a nonzero polynomial in the parameters y (typically the input coefficients), which vanishes when there is a solution of the specialized system (1) in some proper component. $C_{p,k}$ is not identically zero even after the specialization, thus fulfilling the first two conditions set in the Introduction. The $C_{p,k}$ is an analogue of the projection operator obtained in [17]. The difference is that our projection operator may be computed via a more flexible algebraic framework, thus allowing faster computation.

We recall now the mixed subdivision-based method for computing the sparse resultant from a matrix determinant. This discussion will allow us to construct a family $(p_0, \ldots, p_n)$ of sparse polynomials, with nonzero resultant.

Let $m_i := \#(\mathcal{A}_i)$, and set $m := \sum_{i=0}^n m_i$. For any $\omega := (\omega_0, \ldots, \omega_n) \in \mathbb{R}^m$, $\omega_i \in \mathbb{R}^{m_i}$, $\mathrm{init}_\omega(\mathrm{Res}_\mathcal{A})$, the *initial term* of $\mathrm{Res}_\mathcal{A}(f_0, \ldots, f_n)$ is a well-defined polynomial in the coefficients of $f_0, \ldots, f_n$ which is a monomial if ω is generic [20]. Let us pick a vector $\omega := (\omega_0, \ldots, \omega_n)$, which corresponds to functions $\omega_i : \mathcal{A}_i \to \mathbb{Q}$, for $i = 0, \ldots, n$. We require that $\omega_i(a) > 0$ for all $a \in \mathcal{A}_i$, $i = 0, \ldots, n$, and suppose that ω_i is the restriction of an affine function (in most cases we can take it to be linear). This hypothesis can be removed [20].

Consider the *lifted polytopes* $Q_{i,\omega} := \mathrm{conv}\{(a, \omega_i(a)) : a \in \mathcal{A}_i\} \subset \mathbb{R}^{n+1}$, and their Minkowski sum $Q_\omega := Q_{0,\omega} + \cdots + Q_{n,\omega}$. Taking the *upper* envelope of Q_ω, we get a coherent polyhedral (or mixed) *subdivision* Δ_ω of the Minkowski sum $Q := Q_0 + \ldots + Q_n$, by means of the natural projection $\mathbb{R}^{n+1} \to \mathbb{R}^n$ of Q_ω. Moreover, this projection, constrained to the upper envelope of Q_ω, becomes a bijection. Suppose that ω is sufficiently generic ([4,20]), then we have a tight mixed coherent decomposition, i.e. each facet in the upper envelope is of the form $F_\omega = F_{0,\omega} + \ldots + F_{n,\omega}$ where F_i is a face of Q_i and $n = \dim(F_0) + \cdots + \dim(F_n)$. This implies that at least one summand is zero-dimensional, i.e., a Newton polytope vertex. Furthermore, the sum expressing F_ω maximizes the aggregate lifting of any sum of n faces equal to F_ω; so it is sometimes called an optimal sum. The cell F, (the projection of F_ω) is said to be *mixed of type i* or *i-mixed* if $\dim F_i = 0$, $\dim F_j > 0$, $j \neq i$.

We use the matrix construction based on Δ_ω by fixing a small vector $\delta \in \mathbb{Q}^n$, and set $\mathcal{E} := \mathbb{Z}^n \cap (\delta + Q)$. The *row content* $\mathrm{RC}(F) = (i, a)$ for any cell of Δ_ω will be a pair (i, a) which satisfies: if $F_\omega = F_{0,\omega} + \ldots + F_{n,\omega}$ is the unique facet on the upper envelope of Q_ω projecting to F expressed by this optimal sum, let i be the *largest* index such that $\dim(F_i) = 0$, $F_i = \{a\}$. We also define the row content of all points $p \in F : p + \delta \in \mathcal{E}$ to be $\mathrm{RC}(p) = \mathrm{RC}(F)$.

Definition 3.7. Consider all monomials corresponding to the *matrix diagonal*: $\mathcal{B}_i := \{a \in \mathcal{A}_i : \mathrm{RC}(p) = (i, a), p \in \mathcal{E}\}$, $i = 0, \ldots, n$. Then define the perturbation polynomials using a parameter $\lambda \in \mathbb{C}^*$ as $p_{\lambda, i} := \sum_{a \in \mathcal{B}_i} \lambda^{\omega_i(a)} x^a$, $i = 0, \ldots, n$.

Proposition 3.8. [7] *For all but finite* $\lambda \in \mathbb{C}^*$, *we have that* $\mathrm{Res}_\mathcal{A}(p_\lambda) \neq 0$, *where* $p_\lambda := (p_{\lambda,0}, \ldots, p_{\lambda,n})$

To deterministically guarantee the validity of a given λ, it suffices to test the sparse resultant matrix and its determinant for p_λ.

4 Matrix Construction

Methods for computing the GCP as a (nonzero) determinant of a sparse resultant matrix are proposed in this section. We also derive precise bounds on the number of roots for specialized coefficients, and estimate the complexity of our algorithms.

Construct the square matrix $\mathcal{M}_{\omega,\delta} := (m_{p,p'})_{p,p'\in\mathcal{E}}$ with $m_{p,p'} = c_{ia'}$ for $\mathrm{RC}(p) = (i,a)$ and $p - a + a' = p'$. The matrix determinant is generically a nonzero multiple of $\mathrm{Res}_{\mathcal{A}}$. Its degree in the coefficients of f_i equals the number of rows containing this polynomial's coefficients. The particular definition of $\mathrm{RC}(\cdot)$ implies that the determinant degree in the coefficients of f_0 is $\deg_{f_0} \mathrm{Res}_{\mathcal{A}} = \mathrm{MV}(f_1,\ldots,f_n)$.

Theorem 4.1. [7] *Let* $p_{\lambda,i}$ *as in definition 3.7. Then, for every specialization of* $f_0,\ldots,f_n$ *and almost all* $\lambda \in \mathbb{C}^*$,

$$\det \mathcal{M}_{\omega,\delta} (f_0 - \epsilon p_{\lambda,0}, f_1 - \epsilon p_{\lambda,1}, \ldots, f_n - \epsilon p_{\lambda,n})$$

is a nonzero polynomial in ϵ *of degree* $\#\mathcal{E}$. *This determinant equals* $C_{p_\lambda}(\epsilon)$ *multiplied by a polynomial in* ϵ, *not depending on the coefficients of* f_0 *and* $p_{\lambda,0}$, *where* $C_{p_\lambda}(\epsilon)$ *is the* p_λ-*GCP. In the case of the* u-*resultant of* $(f_1,\ldots,f_n)$, f_0 *does not have to be perturbed.*

The obvious construction of p_λ consists in picking a random value for $\lambda \in \mathbb{C}^*$. Thus only a few random bits are needed.

Proposition 4.2 (Derandomization). *It is possible to remove randomization completely by replacing* $\epsilon p_{\lambda,i}$ *in the perturbation by* $p_{\epsilon,i}$.

By replacing the value λ by the perturbation parameter ϵ, this construction raises, in general, the perturbation degree in the resultant matrix and the subsequent computation.

Corollary 4.3. *If some very small probability of error is acceptable, then polynomials* $p_{\lambda,i}$ *do not have to depend on* λ *and their coefficients may be chosen completely at random. The probability that this choice be invalid equals the probability that the aggregate coefficient vector lies on the hypersurface defined by the vanishing of the determinant of the perturbation matrix.*

Let ϕ be the trailing coefficient of $\det \mathcal{M}_{\omega,\delta}(\epsilon)$. Then ϕ is a multiple of $C_{p,k}$. By [6], there exists a submatrix of the sparse resultant matrix such that the quotient of their determinants equals the sparse resultant. In any case, ϕ is a polynomial in the u_i indeterminates, in the u-resultant construction, and its factorization gives all the isolated roots of the given well-constrained system. Otherwise, if we have "hidden" some variable x_i in the space of coefficients, ϕ is a polynomial in x_i. and the zero set of $\phi(x_i)$ contains the coordinates of x_i at all isolated roots of the input system. Recall that if the variety of a system of n polynomials in n variables has positive dimension, then both the u−resultant and the resultant using hidden variables vanish identically. By theorem 3.6, it follows:

Corollary 4.4 (Root counting). *The degree of ϕ in the hidden variable or, in the case of the u-resultant, in any u_i indeterminate, bounds the number of isolated roots of the* specialized *system, which in some cases is tighter than the generic bound available by mixed volume or Bézout's theorem.*

If the degree of ϕ in the coefficients of some polynomial f_j is strictly smaller than the respective degree of $\text{Res}_{\mathcal{A}}$, then the variety of $f_0, \ldots, f_{j-1}$, $f_{j+1}, \ldots, f_n$ has positive dimension.

Theorem 4.5 (Complexity). [7] *All operations described in the above algorithm have asymptotic complexity dominated by the complexity of the matrix construction and root approximation algorithms, namely polynomial in* $\deg(\text{Res}_{\mathcal{A}})$ *and exponential with a linear exponent in* n.

5 Experiments

The perturbation has been implemented in Maple and can be found at the webpage of the second author. Below we report on specific applications.

Example 5.1. Consider the system in [20, Exam. 2.1]:

$$f_0 = \alpha_1 + \alpha_2 x_1^2 x_2^2 + \alpha_3 x_1 x_2^3, f_1 = \beta_1 + \beta_2 x_1^2 + \beta_3 x_1 x_2^2, f_2 = \gamma_1 x_1^3 + \gamma_2 x_1 x_2.$$

Setting $\delta := (0, -1/3)$, and affine lifting functions $\omega_1(a,b) := -a/2 - b/4 + 1$, $\omega_2(a,b) := 3a - 5b + 7$, $\omega_3(a,b) := 0$ yields

$$p_{\lambda,0} = \lambda + x_1 x_2^3, \ p_{\lambda,1} = \lambda^{13} x_1^2, \ p_{\lambda,2} = x_1^3 + x_1 x_2.$$

An explicit computation shows that $\text{Res}_{\mathcal{A}}(p_\lambda) = \pm\lambda^{93}$. The second simplification rule in section 4 allows to take $p_{\lambda,1} = x_1^2$, hence $\text{Res}_{\mathcal{A}}(p_\lambda) = \pm\lambda^2$. It is clear that $\lambda = 1$ is a good choice, and produces a good perturbation. We could also apply proposition 4.2 to avoid using λ altogether and, instead, perturb by $\epsilon^2 + \epsilon x_1 x_2^3$, ϵx_1^2, $\epsilon x_1^3 + \epsilon x_1 x_2$.

Consider now specialization

$$\alpha_1 = \beta_3 = -2, \ \alpha_2 = \alpha_3 = \gamma_1 = -\gamma_2 = 1, \ \beta_1 = u, \ \beta_2 = 2 - u.$$

In $\det \mathcal{M}(\epsilon)$, since the constant coefficient is zero, we conclude that the resultant of the input system vanishes for all u, hence the variety has positive dimension; indeed, it contains line $(1,1,u) \simeq \mathbb{C}$. The linear coefficient $\phi(u)$ yields the 6 projections of the isolated roots of the system on the u-axis. $\phi(u)$ can be computed by interpolating on $T = 8$ terms, instead of $\dim \mathcal{M} = 21$ terms.

Example 5.2. Consider the system in [4]:

$$\begin{aligned} f_0 &= c_{11} + c_{12}x_1 x_2 + c_{13}x_1^2 x_2 + c_{14}x_1, \\ f_1 &= c_{21}x_2 + c_{22}x_1^2 x_2^2 + c_{23}x_1^2 x_2 + c_{24}x_1, \\ f_2 &= c_{31} + c_{32}x_2 + c_{33}x_1 x_2 + c_{34}x_1. \end{aligned}$$

Setting $\delta := (-3/8, -1/8)$, $L \geq l^2 > l \gg 0$ and using $\omega_0 = (L - l)x_1 + (L - l^2)x_2$, $\omega_1 = (L + l^2)x_1 + (L + 1)x_2$, $\omega_2 = (L - 1)x_1 + (L + l)x_2$, we have

$$
\begin{aligned}
p_{\lambda,0} &= 1, \\
p_{\lambda,1} &= \lambda^{4L+2l^2+2}x_1^2 x_2^2 + \lambda^{3L+2l^2+1}x_1^2 x_2, \\
p_{\lambda,2} &= \lambda^{L+l}x_2 + \lambda^{2L+l-1}x_1 x_2 + \lambda^{L-1}x_1;
\end{aligned}
$$

and

$$
\pm\mathrm{Res}_{\mathcal{A}}(p_\lambda) = \lambda^{16L+3l^2+2} - 2\,\lambda^{17L+4l^2+l+2} + \lambda^{18L+5l^2+2l}.
$$

Observe that $\lambda = 1$ is a zero of this polynomial, so another value must be chosen. Any random value is valid with very high probability.

In the special case

$$
f_0 = 1 + x_1 x_2 + x_1^2 x_2 + x_1, \quad f_2 = 1 + x_2 + x_1 x_2 + x_1,
$$

the sparse resultant vanishes identically since the variety $V(f_0, f_1)$ has positive dimension (it is formed by the union of the isolated point $(1, -1)$ and the line $\{-1\} \times \mathbb{C}$). With $L = 10$ and $l = 3$, we have that the coefficient of ϵ^2 in the perturbed determinant is

$$
\phi = -(c_{22}c_{23})(c_{24} - c_{21} + c_{22} - c_{23})(c_{24} + c_{21} - c_{22} + c_{23}),
$$

and we can recover in the last two factors the value of f_1 at the isolated zero $(1, -1)$ and the point $(-1, -1)$ in the positive-dimensional component.

Example 5.3. This is the example of [17, Sect. 3.2]. To the system

$$
1 + 2x - 2x^2y - 5xy + x^2 + 3x^3y, \quad 2 + 6x - 6x^2y - 11xy + 4x^2 + 5x^3y,
$$

we add $f_0 := u_1x + u_2y + u_0$, which does not have to be perturbed, by theorem 4.1. We use function `spares` to construct a 16×16 matrix `Mue` in parameters u_0, u_1, u_2, ϵ. The number of rows per polynomial are, respectively, $4, 6, 6$, whereas the mixed volumes of the 2×2 subsystems are all equal to 4.

```
Mue := spares ([f0,f1,f2], [x,y]):
Due := det(Mue):              # in u0,u1,u2,e
degree (Due,e);               # 12
ldg := ldegree(Due,e);        # 1
phi := primpart(coeff(Due,e,ldg)): factor(phi);
```

For certain ω and δ, we have used $p_{\lambda,1} := -3x^2 + x^3y$, $p_{\lambda,2} := 2 + 5x^2$, not depending on any λ but instead with random coefficients. The perturbed determinant has maximum and minimum degree in ϵ, respectively, 12 and 1. Then $\phi(u_i)$ gives two factors corresponding to isolated solutions $(1/7, 7/4)$ and $(1, 1)$: $(49\,u_2 + 4\,u_1 + 28\,u_0)\,(u_2 + u_1 + u_0)$. Another two factors give points on the line $\{-1\} \times \mathbb{C}$ of solutions, but the specific points are very sensitive to the choice of ω and δ. One such choice yields: $(-u_0 + u_1)\,(27\,u_2 + 40\,u_1 - 40\,u_0)$. Rojas obtains a 17×17 matrix whose perturbed determinant has degree only 8 and the lowest term is again linear in ϵ, though different from the one above.

Example 5.4. In the robot motion problem implementation of Canny's *road-map algorithm* in [13], numerous "degenerate" systems are encountered. Let us examine a 3×3 system, where we hide x_3 to obtain dense polynomials of degrees $3, 2, 1$:

$$
\begin{aligned}
f_0 = {}& 54x_1{}^3 - 21.6x_1{}^2 x_2 - 69.12x_1 x_2{}^2 + 41.472x_2{}^3 + (50.625 + 75.45x_3)\, x_1{}^2 \\
& + (-92.25 + 32.88x_3)\, x_1 x_2 + (-74.592x_3 + 41.4)\, x_2{}^2 + \\
& + (131.25 + 19.04x_3{}^2 - 168x_3)x_1 + \left(-405 + 25.728x_3{}^2 + 126.4x_3\right) x_2 + \\
& + (-108.8\, x_3{}^2 + 3.75\, x_3 + 234.375), \\
f_1 = {}& -37.725\, x_1{}^2 - 16.44\, x_1 x_2 + 37.296\, x_2{}^2 + (-38.08x_3 + 84)\, x_1 + \\
& + (-63.2 - 51.456x_3)\, x_2 + (2.304x_3{}^2 + 217.6x_3 - 301.875), \\
f_2 = {}& 15\, x_1 - 12\, x_2 + 16\, x_3.
\end{aligned}
$$

Our Maple function `spares` applies an optimal perturbation to an identically singular 14×14 matrix in x_3. Now $\det \mathcal{M}_{\omega, \delta}(\epsilon)$ is of degree 14 and the trailing coefficient of degree 2, which provides a bound on the number of affine roots, tighter than Bézout's and the mixed volume, both equal to 6. We obtain

$$
\phi(x_3) = \left(x_3 - \frac{1434}{625} \right) \left(x_3 - \frac{12815703325}{21336} \right),
$$

the first solution corresponding to the unique isolated solution but the second one is superfluous, hence the variety has dimension zero and degree 1.

6   Geometric Applications

6.1   Predicates

Linear infinitesimal perturbations of the input have allowed computational geometers to avoid implementing the singular, or degenerate, configurations. The definition of what constitutes degenerate input depends on the problem and the predicates employed. The present work is a direct generalization of those approaches and in particular of the linear perturbation of [9,10,18].

As formalized in [18], the main requirement in designing a valid perturbation scheme is to define a set of input objects in generic position, whose defining coefficients shall be multiplied by ϵ and then added to the objects on which the predicate is evaluated. Let us take the standard example of $d + 1$ lines in d-dimensional space. *Transversality* decides whether they all meet at a point, which happens precisely when the determinant of a $(d+1)$-dimensional matrix $\Delta = [c_{ij}]$ equals zero, where the c_{ij}, $i, j \in \{1, \ldots, n+1\}$ are the coefficients of the line equations. If $\det \Delta \neq 0$, its sign indicates on which side of the $(d + 1)$st line lies the intersection of the other d lines. Applying the linear perturbation scheme of [9,10] we have

$$
c_{ij}(\epsilon) = c_{ij} + \epsilon\, i^j, \text{ or } c_{ij}(\epsilon) = c_{ij} + \epsilon\, (i^j \bmod q), \tag{2}
$$

for some prime q exceeding the number of input objects. Choosing between these two schemes depends on the predicates that must be handled, the second scheme being less generally applicable.

Evaluation of the perturbed predicate amounts to determining the sign of the trailing coefficient in the ϵ-polynomial representing the determinant of $\Delta(\epsilon)$. Then, $\Delta(\epsilon) = \Delta + \epsilon V$, where the (i, j) entry of V is i_i^j or i_i^j mod q. A merit of these schemes is that the proposed generic set facilitates computation of the perturbed predicate, by making use of Vandermonde matrices like V, and, more importantly, by reducing the evaluation of the perturbed determinant to computing a characteristic polynomial: $\det \Delta(\epsilon) = \det(\Delta + \epsilon V) = \det(-V)\det(-V^{-1}\Delta - \epsilon I)$, where I is the identity matrix.

Defining a polynomial system with nonzero sparse resultant satisfies precisely the aforementioned requirement for designing a valid scheme. But why is the resultant matrix expressing a geometric predicate? This shall depend on the applications that computational geometry wishes to tackle, in particular as it aspires to handle non-linear objects of increased algebraic complexity. Observe that existing predicates are instances of resultants, including Transversality, Orientation, and InCircle/InSphere. They are all typically expressed by determinant signs, and these are determinants of resultant matrices. For these simple examples, the sparse and classical projective resultants coincide. For instance, the Voronoi diagram of line segments gives rise to predicates expressed by quadratic polynomials (capturing distance). A resultant represents the precise predicate [1], which tests whether a segment intersects the Voronoi circle tangent to two segments and passing through a point.

When it comes to implementing an algorithm, all predicates must deal with identically perturbed input. This implies that we should anticipate all possible sparse resultant matrices for a given set of objects. In the worst case, we may perturb all monomials using random coefficients. The main limitation is that the projection operator only gives a necessary condition for solvability of the overconstrained system because we compute the projection operator multiplied by an extraneous factor. If we are able to guarantee that this factor has constant sign, then this approach can be useful. In certain cases, it should be possible to factor it out, especially by [6].

6.2   Implicitization with Base Points

The problem of switching from a rational parametric representation to an implicit, or algebraic, representation of a curve, surface, or hypersurface lies at the heart of several algorithms in computer-aided design and geometric modeling. This section describes a direct application of our perturbation method to the implicitization problem.

Given are rational parametric expressions

$$x_i = p_i(t)/q(t) \in K(t) = K(t_1, \ldots, t_n), \quad i = 0, \ldots, n,$$

in terms of the polynomials $p_0, \ldots, p_n, q$ over some field K of characteristic zero. The implicitization problem consists in computing the smallest algebraic surface in terms of $x = (x_0, \ldots, x_n)$ containing the closure of the image of the parametric map $t \mapsto x$. The most common case is for curve and surface implicitization, namely when $n = 1$ and $n = 2$ respectively. Several algorithms exist for this problem, cf. [14,19] and the references thereof.

Implicitization is equivalent to eliminating all parameters t from the polynomials $f_i(t) = p_i(t) - x_i q(t)$, regarded as polynomials in t. The resultant is well-defined for this system, and shall be a polynomial in x, equal to the implicit expression, provided that it does not vanish and the parametrization is generically one-to-one. In the latter case, the resultant is a power of the implicit equation. More subtle is the case where the resultant is identically zero, which happens precisely when there exist values of t, known as *base points*, for which the f_i vanish for all x. Base points forming a component of codimension 1 can be easily removed by canceling common factors in the numerator and denominator of the rational expressions for the x_j's. But higher codimension presents a harder problem.

Different methods have been proposed to handle base points, and the field is still of active research, e.g. [2,12,14,16] and the references thereof. These methods include resultant perturbations, since the trailing ϵ-coefficient yields the implicit equation. Our perturbation method applies directly, since the projection operator will contain, as an irreducible factor, the implicit equation. The extraneous factor has to be removed by factorization. Distinguishing the implicit equation from the latter is straightforward by using the parametric expressions to generate points on the implicit surface. The rest of the section illustrates our perturbation in the case of base points.

Example 6.1. Let us consider the de-homogenized version of a system defined in [2]:
$$p_0 = t_1^2, \ p_1 = t_1^3, \ p_2 = t_2^2, \ q = t_1^3 + t_2^3.$$

It has one base point, namely $(0,0)$, of multiplicity 4. The toric resultant here does not vanish, so it yields the implicit equation
$$x_2^3 x_1^2 - x_0^3 x_1^2 + 2x_0^3 x_1 - x_0^3.$$

But under the change of variable $t_2 \to t_2 - 1$ the new system has zero toric resultant. The determinant of the perturbed 27×27 resultant matrix has a trailing coefficient which is precisely the implicit equation. The degree of the trailing term is 4, which equals in this case, the number of base points in the toric variety counted with multiplicity.

Example 6.2. The parametric surface is given by
$$x_i = \prod_{j=1}^{7} \left(b_{0j} + \sum_{k=1}^{2} t_k b_{kj} \right)^{b_{ij}}, \quad i = 0, 1, 2,$$

where $B = (b_{ij})$ for $i = 0, \ldots, 2$, $j = 1, \ldots, 7$ is as follows:

$$B = \begin{bmatrix} 1 & 0 & -1 & 0 & 2 & -1 & -1 \\ 0 & 1 & -1 & 2 & 0 & -1 & -1 \\ 1 & 1 & -2 & 1 & 0 & -1 & 0 \end{bmatrix}.$$

There are base points forming components of codimension 2, including a single affine base point $(1, -1)$. Our algorithm constructs a 33×33 matrix, whose perturbed determinant has a trailing term of degree 3. The corresponding coefficient has total degree 14 in x_0, x_1, x_2. When factorized, it yields the precise implicit equation, which is of degree 9.

References

1. C. Burnikel, K. Mehlhorn, and S. Schirra. How to compute the Voronoi diagram of line segments: Theoretical and experimental results. In *Proc. ESA-94, LNCS 855*, pages 227–237, 1994.
2. L. Busé. Residual Resultant over the Projective Plane and the Implicitization Problem. In *Proc. ACM Intern. Symp. on Symbolic & Algebraic Comput.*, pages 48–55, 2001.
3. J. Canny. Generalised characteristic polynomials. *J. Symbolic Computation*, 9:241–250, 1990.
4. J.F. Canny and I.Z. Emiris. A subdivision-based algorithm for the sparse resultant. *J. ACM*, 47(3):417–451, May 2000.
5. D. Cox, J. Little, and D. O'Shea. *Using Algebraic Geometry*. Number 185 in Graduate Texts in Mathematics. Springer-Verlag, New York, 1998.
6. C. D'Andrea. Macaulay-style formulas for the sparse resultant. *Trans. of the AMS*, 354:2595–2629, 2002.
7. C. D'Andrea and I.Z. Emiris. Computing sparse projection operators. In *Symbolic Computation: Solving Equations in Algebra, Geometry, and Engineering*, volume 286 of *Contemporary Mathematics*, pages 121–139, Providence, Rhode Island, 2001. AMS.
8. I.Z. Emiris. Discrete geometry for algebraic elimination. In *this volume*, pages 77–91.
9. I.Z. Emiris and J.F. Canny. A general approach to removing degeneracies. *SIAM J. Computing*, 24(3):650–664, 1995.
10. I.Z. Emiris, J.F. Canny, and R. Seidel. Efficient perturbations for handling geometric degeneracies. *Algorithmica, Special Issue on Computational Geometry in Manufacturing*, 19(1/2):219–242, Sep./Oct. 1997.
11. I.Z. Emiris and V.Y. Pan. Symbolic and numeric methods for exploiting structure in constructing resultant matrices. *J. Symbolic Computation*, 33:393–413, 2002.
12. L. González-Vega. Implicitization of Parametric Curves and Surfaces by Using Multidimensional Newton Formulae. *J. Symbolic Comput.*, 23:137–152, 1997.
13. H. Hirukawa and Y. Papegay. Motion planning of objects in contact by the silhouette algorithm. In *Proc. IEEE Intern. Conf. Robotics & Automation*, San Francisco, April 2000. IEEE Robotics and Automation Society.

14. C.M. Hoffmann, J.R. Sendra, and F. Winkler. *Parametric Algebraic Curves and Applications*, volume 23 of *J. Symbolic Comput.*. Academic Press, 1997.
15. E. Kaltofen and G. Villard. On the Complexity of Computing Determinants. In *Proc. Fifth Asian Symp. Computer Math.*, K. Shirayanagi and K. Yokoyama, eds., Lect. Notes in Computing, vol. 9, pages 13–27, World Scientific, Singapore, 2001.
16. D. Manocha and J. Canny. The implicit representation of rational parametric surfaces. *J. Symbolic Computation*, 13:485–510, 1992.
17. J.M. Rojas. Solving degenerate sparse polynomial systems faster. *J. Symbolic Computation, Special Issue on Elimination*, 28:155–186, 1999.
18. R. Seidel. The nature and meaning of perturbations in geometric computing. In *Proc. 11th Symp. Theoret. Aspects Computer Science, LNCS 775*, pages 3–17. Springer-Verlag, 1994.
19. T.W. Sederberg, R. Goldman, and H. Du. Implicitizing rational curves by the method of moving algebraic curves. *J. Symbolic Comput., Spec. Issue Parametric Algebraic Curves & Appl.*, 23:153–175, 1997.
20. B. Sturmfels. On the Newton polytope of the resultant. *J. of Algebr. Combinatorics*, 3:207–236, 1994.

Numerical Irreducible Decomposition Using PHCpack*

Andrew J. Sommese[1], Jan Verschelde[2], and Charles W. Wampler[3]

[1] Department of Mathematics, University of Notre Dame,
   Notre Dame, IN 46556-4618, USA.
   Email: sommese@nd.edu. *URL:* http://www.nd.edu/~sommese.
[2] Department of Mathematics, Statistics, and Computer Science,
   University of Illinois at Chicago, 851 South Morgan (M/C 249),
   Chicago, IL 60607-7045, USA.
   Email: jan@math.uic.edu or jan.verschelde@na-net.ornl.gov.
   URL: http://www.math.uic.edu/~jan.
[3] General Motors Research Laboratories, Enterprise Systems Lab,
   Mail Code 480-106-359, 30500 Mound Road, Warren, MI 48090-9055, USA.
   Email: Charles.W.Wampler@gm.com.

Abstract. Homotopy continuation methods have proven to be reliable and efficient to approximate all isolated solutions of polynomial systems. In this paper we show how we can use this capability as a blackbox device to solve systems which have positive dimensional components of solutions. We indicate how the software package PHCpack can be used in conjunction with Maple and programs written in C. We describe a numerically stable algorithm for decomposing positive dimensional solution sets of polynomial systems into irreducible components.
2000 Mathematics Subject Classification. Primary 65H10; Secondary 13P05, 14Q99, 68N30, 68W30.
Key words and phrases. homotopy continuation, interfaces, numerical algebraic geometry, polynomial system, software.

1 Introduction

Using numerical algorithms to solve polynomial systems arising in science and engineering with tools from algebraic geometry is the main activity in "Numerical Algebraic Geometry." This is a new developing field on the crossroads of algebraic geometry, numerical analysis, computer science and engineering. One of the key problems in this area (and also in Computational Algebraic Geometry [11]) is to decompose positive dimensional solution sets into irreducible components. A special instance of this problem is the factoring of polynomials in several variables.

---

* The authors gratefully acknowledge the support of this work by Volkswagen-Stiftung (RiP-program at Oberwolfach), where the writing for this paper began. The first author is supported in part also by NSF Grant DMS - 0105653; and the Duncan Chair of the University of Notre Dame. The second author is supported in part by NSF Grant DMS-0105739.

The notion of an *irreducible variety* is a natural geometric refinement of the notion of connectedness. For example, consider the algebraic set in $\mathbb{C}^2$ defined by $xy = 0$. This set is connected, but it is natural to think of it as the union of the x-axis and the y-axis. A set defined by polynomials is said to be *irreducible* if the nonsingular points of the set are connected. Thus the x-axis and the y-axis are both irreducible. Given an arbitrary algebraic set Z in $\mathbb{C}^N$, the natural candidates for irreducible components of Z are the closures of the connected components of the nonsingular points of Z. For example, the only singular point of the solution set $xy = 0$ is the origin $(0,0)$. The solution set minus $(0,0)$ breaks up into two connected components, i.e., the x-axis minus the origin and the y-axis minus the origin. The closure of the x-axis (respectively, the y-axis) minus the origin is the x-axis (respectively, the y-axis). Thus for the solution set of $xy = 0$, the breakup of $xy = 0$ into the x-axis and the y-axis is the decomposition into irreducible components. This decomposition is very well behaved, e.g., there are finitely many components, which are algebraic sets in their own right.

With the dictionary in Table 1 we show how to translate the key concepts of algebraic geometry to define an irreducible decomposition into data structures used by numerical algorithms. For each irreducible component we find as many generic points as the degree of the component. With this set of generic points we construct filters that provide a probability-one test to decide whether any given point belongs to that component.

We say that an algorithm is a probability-one algorithm if the algorithm depends on a choice of a point in an irreducible variety X and the algorithm works for a Zariski open dense set points $U \subset X$.

For example, consider that we want to check whether a polynomial $p(x)$ on $\mathbb{C}$ is identically zero. We might have as our algorithm: take an explicit random $x_* \in \mathbb{C}$ and check if $p(x_*) = 0$. Then $p(x)$ is identically zero if $p(x_*) = 0$. Here $X := \mathbb{C}$. The algorithm fails precisely when $p(x)$ is not identically zero and $p(x_*) = 0$. Since $p(x)$ is not identically zero, $p^{-1}(0)$ is finite and we need to choose $x_* \in U := \mathbb{C} \setminus p^{-1}(0)$. We are assuming that a random point on $\mathbb{C}$ will not lie in $p^{-1}(0)$. Of course, since we are working with machine numbers, there is an exceedingly small chance we will be wrong, e.g., see [35]. Usually we choose not one but many constants in an algorithm, e.g., the coefficients of the equation of a random hyperplane. In this case, the point will be the vector made up of the coefficients.

The goal of this paper is to elaborate the sentence "all algorithms have been implemented as a separate module of PHCpack" in recent papers [35–40], a sequence that has its origin in [41]. PHCpack [44] is a general purpose numerical solver for polynomials systems. In this paper we describe the extensions to this package, with a special emphasis on interfaces.

Our main tool is homotopy continuation, which has proven to be reliable and efficient to compute approximations to all isolated solutions of polynomials systems. Nowadays [27], this involves the use of computational geometry

Table 1. Dictionary to translate algebraic geometry into numerical analysis.

Algebraic Geometry	Example in 3-space	Numerical Analysis
variety	collection of points, algebraic curves, and algebraic surfaces	polynomial system + union of witness point sets, see below for the definition of a witness point
irreducible variety	a single point, or a single curve, or a single surface	polynomial system + witness point set + probability-one membership test
generic point on an irreducible variety	random point on an algebraic curve or surface	point in witness point set; a witness point is a solution of polynomial system on the variety and on a random slice whose codimension is the dim. of the variety
pure dimensional variety	one or more points, or one or more curves, or one or more surfaces	polynomial system + set of witness point sets of same dim. + probability-one membership tests
irreducible decomposition of a variety	several pieces of different dimensions	polynomial system + array of sets of witness point sets and probability-one membership tests

techniques to compute a homotopy with start points (based on mixed volumes and Newton polytopes) combined with numerical methods to follow the solution paths defined by the homotopies.

After outlining the design of PHCpack, reporting on an interface to fast mixed volume calculations, we describe a simple Maple procedure to call the blackbox solver of PHCpack. Via sampling and projecting we obtain a numerical elimination procedure. The algorithms are numeric-symbolic: with numerical interpolation we construct equations to represent the solution components. In section four we illustrate this sampling on a three dimensional spatial mechanical linkage, using Maple as a plotting tool. Since factoring of polynomials in several variables is a special instance of our general decomposition algorithms, we developed a low level interface to call the Ada routines from C programs. Section six explains the membership problem and our homotopy test to solve it. This test is then used as one of the tools in the decomposition method, illustrated in section seven. In the last section we list some of our major benchmark applications.

2   Toolbox and Blackbox Design of PHCpack

PHCpack [44] was designed to test new homotopy algorithms to solve systems of polynomial equations. In particular, three classes of homotopies [45] have been implemented. For dense polynomial systems, refined versions of Bézout's theorem lead to linear-product start systems [47,49]. Polyhedral homotopies [48,50] are optimal for generic sparse polynomial systems. The third class contains SAGBI [46] and Pieri homotopies [24] implementing a Numerical Schubert Calculus [22].

These three classes of homotopies can be accessed directly when the software is used in toolbox mode. For general purpose polynomial systems, a blackbox solver was designed and tested on a wide variety of systems [44]. Although a blackbox will rarely be optimal and therefore cannot beat the particular and sophisticated uses offered by a toolbox, both a toolbox and a blackbox are needed. The writing of PHCpack reflects the dual use of the software: on the one hand as a *package*, offering a library of specialized homotopy algorithms, and on the other hand as a blackbox, where only the executable version of the program, i.e., *PHC*, is needed, whence the mixed capitalization: PHCpack.

In this paper we focus on recent developments and interfaces. Combining both recent research and the use of other software, we developed a customized interface to the software of T.Y. Li and Li Xing [28] for computing mixed volumes. We report on an experiment using this interface to compute generic points on the two dimensional components of cyclic 9-roots. In the following paragraphs we describe the context of how such polyhedral methods are used in continuation.

To solve a polynomial system with homotopies, we deform a start system with known solutions to the system we wish to solve, applying Newton's method to track the solution paths. If we want to approximate all isolated solutions, then one major problem is the construction of a "good" start system that leads to the least number of paths. As shown in [1], mixed volumes of the tuples of convex hulls of the exponent sets of the polynomials provide a sharp bound for the number of isolated solutions when the coefficients of the polynomials are generic. The theorems of Bernshtein [1] led to polyhedral homotopies [23], [50], and to a renewed interest in resultants, spurred by the field sparse elimination theory [7], [13], [14] and [33]; see also [42].

Mixed volumes for the cyclic n-roots problems were first computed in [13] and [14]. In [15], the cyclic 9-roots problem was reported to be solved with Gröbner bases. This problem has two dimensional components of solutions. Following [35], an embedding of the original polynomial system was constructed.

The first time we used the interface to the software of Li and Li [28] was to find all 18 generic points on the two dimensional components of the cyclic 9-roots problem, see [38]. We also used this type of interface to compute all 184,756 isolated cyclic 11-roots with PHC. The 8,398 generating solutions (with respect to the permutation symmetry) are on display, via the web site of the second author. Note that computing all cyclic 11-roots is numerically "easier" than solving the cyclic 9-roots problem, as all solutions to the cyclic 11-roots problem are isolated.

Besides [28], other recent computational progress is described in [17,18] and [43], with application to cyclic n-roots in [9], for n going up to 13. See the web at [8] for the listing of all cyclic n-roots, for n from 8 to 13. Polyhedral methods led to a computational breakthrough in approximating all isolated solutions. To deal with positive dimensional solution sets we solve an embedded system that has all its roots isolated. Thus the recent activity on polyhedral root counting methods is highly relevant to the general solution method.

3   A Maple Interface to PHCpack

In this section we describe how to use the blackbox solver of PHCpack from within a Maple worksheet. The interface is file oriented, just like PHC was used with OpenXM [29].

The blackbox solver offered by PHCpack can be invoked on the command line as

```
phc -b input output
```

with output the name of the file for the results of the computation. The file input contains a system in the format

```
2
  x*y - 1;
  x**2 - 1;
```

where the "2" at the beginning of the file indicates the number of equations and unknowns. We point out that this blackbox only attempts to approximate the isolated solutions.

To bring the solutions into a Maple worksheet, PHC was extended with an addition option -z. The command

```
phc -z output sols
```

takes the output of the blackbox solver and creates the file sols which contains the list of approximations to all isolated solutions in a format readable by Maple.

The simple Maple procedure listed below has been tested for Maple7. Similar versions run with Maple 6 and MapleV on SUN workstations running Solaris and PCs running Linux and Windows. Besides a list of polynomials, the user should provide a path name for the executable version of PHC.

```
run_phc := proc(phcloc::string,p::list)
  description 'Calls phc from within a Maple 7 session.
The first argument is the file name of the executable version
of phc.  The second argument is a list of polynomials.
On return is a list of approximations to all isolated
roots of the system defined by p.':
  local i,sr,infile,outfile,solfile,sols:
  sr := convert(rand(),string): solfile := sols||sr:
  infile := input||sr: outfile := output||sr:
  fopen(infile,WRITE): fprintf(infile,'%d\n',nops(p)):
  for i from 1 to nops(p) do
    fprintf(infile,'%s;\n',convert(p[i],string)):
  end do;
  fclose(infile):
  ssystem(phcloc||' -b '||infile||' '||outfile):
  ssystem(phcloc||' -z '||outfile||' '||solfile):
  read(solfile): sols := %:
  fremove(infile): fremove(outfile): fremove(solfile):
  RETURN(sols);
end proc:
```

As pointed out earlier, we can only obtain approximations to all isolated solutions with this blackbox. The aim of the remainder of this paper is to sketch the ideas of the algorithms needed in a toolbox to describe also the positive dimensional solution components.

4   Numerical Elimination Methods

Recently we extended the use of homotopies from "just" approximating all isolated roots to describing all solution components of any dimension. In this section we introduce our approach by example. The system

$$\begin{cases} y - x^2 = 0 \\ z - x^3 = 0 \end{cases} \tag{1}$$

defines the so-called "twisted cubic" as the intersection of a quadratic and cubic surface.

For this example, we distinguish three possible orders of elimination:

1. projection onto the (x, y)-plane gives $y - x^2 = 0$;
2. projection onto the (x, z)-plane gives $z - x^3 = 0$;
3. projection onto a random plane gives a cubic curve.

To eliminate we first sample generic points from the curve using the system

$$\begin{cases} y - x^2 = 0 \\ z - x^3 = 0 \\ ax + by + cz + d = 0 \end{cases} \tag{2}$$

where the constants a, b, c, and d are randomly chosen complex numbers. For a general (a, b, c, d) we get exactly three regular solutions which are generic points on the twisted cubic. Moving the last equation of (2) we generate as many samples as desired.

We may perform numerical elimination in two ways. One is to simply project the sampled points in the desired direction and compute an interpolating polynomial. For example, projecting the samples as $(x, y, z) \mapsto (x, y)$ and fitting the points will yield the equation $y - x^2 = 0$ (or a multiple of it). This is fine for low degree hypersurfaces. For higher degree components, the interpolation is numerically undesirable, and for components of codimension greater than one, polynomials must be fitted to projections in several directions to cut out the desired set.

A better approach in general is to compute witness points by slicing the solution set in a special way. In particular, to eliminate z, make the last equation of (2) a plane parallel to the z-axis by setting the coefficient for z to zero. This will give two witness points representing the quadratic $y - x^2 = 0$. Similarly, to eliminate y, the last equation of (2) must be parallel to the y-axis, with a zero coefficient for y. To project in a general direction, the slicing plane is made parallel to that direction, or equivalently, the normal vector to the plane (a, b, c) is made perpendicular to the direction of the projection. This generalizes easily to projections in any dimension.

The next example comes from mechanical design. Suppose N placements of a rigid body in space are given and consider the associated positions, $\mathbf{x}_1, \mathbf{x}_2, \ldots, \mathbf{x}_N$, occupied by a designated point, $\mathbf{x}$, of the body. For four general placements and any general point $\mathbf{x}$ of the body, the associated points $\mathbf{x}_1, \ldots, \mathbf{x}_4$ define a sphere having a center point, say $\mathbf{y}$. However, for five general placements, $\mathbf{x}_1, \ldots, \mathbf{x}_5$ lie on a sphere only if $\mathbf{x}$ is on a certain surface within the body, and for six general placements, $\mathbf{x}_1, \ldots, \mathbf{x}_6$ lie on a sphere only if $\mathbf{x}$ is on a certain curve in the body. Seven general positions determine 20 center-point/sphere-point pairs, a result proven by Schönflies at the end of the nineteenth century [6]. These points are of interest because we may build a linkage to guide the body through the given placements by connecting a rigid link between point $\mathbf{x}$ and its center $\mathbf{y}$. In the following, we consider the center-point/sphere-point curve arising when only six placements are given.

The polynomial system is given by five quadratic equations in six variables: $\mathbf{x} = (x_1, x_2, x_3)$ the coordinates on the sphere and $\mathbf{y} = (y_1, y_2, y_3)$ points on the centerpoint curve. The equations of the system have the following form:

$$\|R_i \mathbf{x} + \mathbf{p}_i - \mathbf{y}\|^2 - \|R_0 \mathbf{x} + \mathbf{p}_0 - \mathbf{y}\|^2 = 0, \quad i = 1, 2, \ldots, 5, \qquad (3)$$

where $\mathbf{p}_i \in \mathbb{R}^3$ are positions of the body and we use $R_i \in \mathbb{R}^{3\times3}$, $R^T R = I$, to denote the rotation matrices for the orientations of the bodies. The problem is to find values for the unknown variables $(\mathbf{x}, \mathbf{y}) \in \mathbb{C}^6$, given the positions and orientations of the body, encoded respectively by $\mathbf{p}_i \in \mathbb{R}^3$ and $R_i \in \mathbb{R}^{3\times3}$.

In our experiment the values for **y** are separated from those for **x** and we construct line segments between the coordinates. Figure 1 shows part of the ruled surface made by Maple.

```
[ > read sbr100x: read sbr100y:   # samples
[ > with(plottools):
[ > a := 1: b := 14:
[ The curve for x appears in dashed lines, the curve for y is drawn in solid lines :
[ > x := curve(xl1[a..b],linestyle=4,thickness=3,color=black):
[ > y := curve(yl1[a..b],thickness=3,color=black):
[ > T1 := plots[textplot3d]([-.5,-.3,.8,`curve
    x`],align=LEFT,color=black):
[ > T2 := plots[textplot3d]([0,0.3,0.2,`curve
    y`],align=RIGHT,color=black):
[ > l := []:
[ > for i from a to b do
[ >   l := [op(l),line(xl1[i],yl1[i],thickness=2)]:
[ > od:
[ > plots[display](x,y,T1,T2,l,axes=BOXED);
```

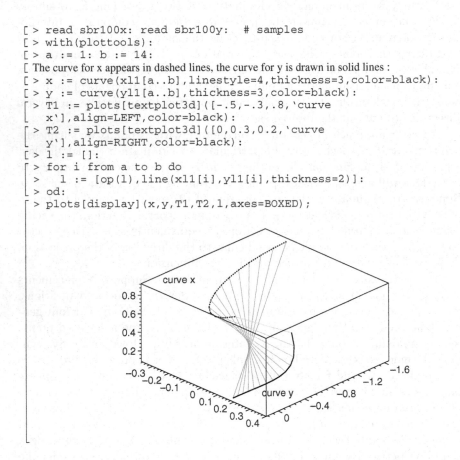

Figure 1. A Maple session to draw a piece of a ruled surface, from samples generated by PHC.

The data for Figure 1 was generated as follows. While the degree of the solution curve is twenty and we are guaranteed to find twenty generic points as isolated solutions when we add one random *complex* hyperplane to the original system, we cannot be sure to find twenty real points when we use a random *real* hyperplane. It is possible for all generic points to have nonzero imaginary parts, but we chose by chance a real hyperplane that gave four real points. Starting at the first real point, PHC sampled 100 points, moving the constant coefficient in the real hyperplane from 0.0 with steps of size 0.1.

Observe that in the Maple session we only use the first 14 samples as the curve started moving too fast for the fixed step size of 0.1.

This last point illustrates that it does not suffice to have a one way communication to export samples for plotting. The plotting program (in our case Maple) must be able to take an active role to control the step size, or equivalently, one should change the sampling for plotting purposes.

Finally, one may wonder about the degrees of the curves for **x** and **y** in Figure 1. To find out, we sliced the curve with a hyperplane involving only the **x** coordinates and obtained only 10 witness points, thus demonstrating that the degree drops from twenty to ten in such a projection. The same drop in degree occurs for **y** as well.

5   Factoring into Irreducible Components

For the spatial Burmester problem above, it is natural to wonder if the solution curve is a single irreducible component or whether it breaks up into several irreducible components.

A special case of the decomposition of positive dimensional solution sets into irreducible components is the problem of decomposing a (multivariate) polynomial into irreducible factors. We emphasize that our algorithms are closer to geometry than to algebra. In particular, we view the input polynomials as Riemann surfaces. See Figure 2 for a visualization of $z^3 - x = 0$ representing the multi-valued cube root function $z = x^{1/3}$, for $x \in \mathbb{C}$. Consider $x = re^{i\theta}$ and the parametrization $x(t) = re^{i\theta+2\pi t}$. As t goes from 0 to 1, $x(t)$ returns at the original value $x(1) = x(0)$, but the order of the roots has been changed, showing that $z^3 - x = 0$ is irreducible. This observation illustrates the use of monodromy to factor polynomials into irreducibles and to decompose solution sets into irreducible components.

The monodromy breakup algorithm has the following specifications:

Input: A set of witness points on a positive dimensional solution set.
Output: A partition of the set; points in same subset of the partition
        belong to the same irreducible component.

To find out which points lie on the same irreducible component, we use homotopies to make the connections. For a system $f(\mathbf{x}) = \mathbf{0}$ we cut out witness points by adding a set of hyperplanes $L(\mathbf{x}) = \mathbf{0}$. The number of hyperplanes in L equals the dimension of the solution set. With the homotopy

$$h_{KL}(\mathbf{x}, t) = \lambda \begin{pmatrix} f \\ K \end{pmatrix} (1 - t) + \begin{pmatrix} f \\ L \end{pmatrix} t = \mathbf{0}, \quad \lambda \in \mathbb{C}, \tag{4}$$

we find new witness points on the hyperplanes $K(\mathbf{x}) = \mathbf{0}$, starting at those witness points satisfying $L(\mathbf{x}) = \mathbf{0}$, letting t move from one to zero.

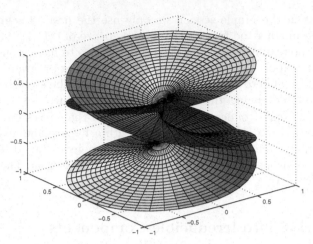

Figure 2. The Riemann surface produced by `cplxroot` of MATLAB plots $\mathrm{Re}(z)$ with z satisfying $z^3 - x = 0$, for $x \in \mathbb{C}$. These solutions form a connected component. Observe that a loop of x around the origin permutes the order of the three solution points.

The role of the random complex constant λ is very important in what follows. Suppose we move back from K to L, using the homotopy

$$h_{LK}(\mathbf{x}, t) = \mu \begin{pmatrix} f \\ L \end{pmatrix} (1 - t) + \begin{pmatrix} f \\ K \end{pmatrix} t = \mathbf{0}, \quad \mu \in \mathbb{C}, \tag{5}$$

using some random constant $\mu \neq \lambda$.

After using $h_{KL}(\mathbf{x}, t) = \mathbf{0}$ and $h_{LK}(\mathbf{x}, t) = \mathbf{0}$, we moved the hyperplanes added to $f(\mathbf{x}) = \mathbf{0}$, from L to K and back from K to L. The set of solutions of $f(\mathbf{x}) = \mathbf{0}$ at L is the same with the important difference that the order of the solutions in the set may have been changed. In particular, we obtain a permutation from the witness points in the set. Points that are permuted (e.g., the first point went to the third, the third to the second, and the second to the first), belong the same irreducible component, as illustrated on Figure 2.

The example above contains the main idea of the monodromy breakup algorithm developed in [37]. While the connections found in this way indicate which points lie on the same irreducible component, we may miss certain loops. Therefore, we recently developed an efficient test [39], based on the linear trace, to certify the partitions obtained by the monodromy algorithm. Besides the efficiency, the test is numerically stable as it only requires the set up of a linear function.

For the spatial Burmester problem, there was a unique irreducible component, which was found to be an irreducible curve of degree 20.

At this stage we wish to show how a low level interface to the numerical factorization routines with PHC is planned. PHCpack is developed in Ada 95, which provides mechanisms to interface with other languages, such as Fortran, C, and Cobol. In our example we wish to call the factorization routines in Ada from a C program. This multi-lingual programming is supported by the gcc compilation system. A polynomial is represented in human readable format like Maple input, e.g. for $x^2 - y^2$, we use the string `"x**2 - y**2;"`, using the semicolon as terminator (also according the conventions in Maple). The sample C program is listed below.

```
#include <stdio.h>
extern char *_ada_phc_factor(int n, char *s);
extern void adainit();
extern void adafinal();
int main() {
    char *f;
    adainit();
    f = _ada_phc_factor(2,"x**2-y**2;");
    adafinal();
    printf("The factors are %s \n",f);
}
```

A more elaborate interface to C programs that allows passing numerical data (such as matrices of doubles), directly from C to Ada has been constructed.

Also in [16] the authors propose monodromy to factor multivariate polynomials numerically.

6   A Membership Test

From the dictionary in Table 1 we see that a membership test figures prominently. Traditional uses of homotopies discard roots as nonisolated when the Jacobian matrix is sufficiently close to being singular. But this approach fails in the presence of isolated roots of multiplicity greater than one.
The membership problem can be formulated as follows:

Given: A solution $\mathbf{x} \in \mathbb{C}^n$ of $f(\mathbf{x}) = \mathbf{0}$ and
             a witness point set for an irreducible solution component V.
Wanted: To determine whether $\mathbf{x}$ lies on V.

In the following example, we present the equations of f in factored form, so we can read off the equations for the solution components. We emphasize that we only take this for the sake of presentation, our method does not require this factorization. Consider,

$$f(x,y,z) = \begin{cases} (y - x^2)(y + z) = 0 \\ (z - x^3)(y - z) = 0 \end{cases} \tag{6}$$

From the factored form, we read off the four solution components of the system $f(x,y,z) = \mathbf{0}$:

1. $V_1 = \{ (x, y, z) \mid y - x^2 = 0, \ z - x^3 = 0 \}$ is the twisted cubic;
2. $V_2 = \{ (x, y, z) \mid y - x^2 = 0, \ y - z = 0 \}$ is a quadratic curve in the plane $y - z = 0$;
3. $V_3 = \{ (x, y, z) \mid y + z = 0, \ z - x^3 = 0 \}$ is a cubic curve in the plane $y + z = 0$;
4. $V_4 = \{ (x, y, z) \mid y + z = 0, \ y - z = 0 \}$ is the x-axis.

While the symbolic solution of the system is given by the factored form of the equations, the numerical solution is given by nine generic points falling on the components according to their degrees: three on V_1, two on V_2, three on V_3, and one on V_4. These generic points are solutions of the system

$$
e(x, y, z) = \begin{cases}
(y - x^2)(y + z) = 0 \\
(z - x^3)(y - z) = 0 \\
c_0 + c_1 x + c_2 y + c_3 z = 0
\end{cases}
\tag{7}
$$

where the constants c_0, c_1, c_2, and c_3 are randomly chosen complex numbers. The partition of the set of nine witness points corresponding to the components is achieved by running the monodromy algorithm on the nine solutions of the system $e(x, y, z) = \mathbf{0}$.

The homotopy membership test consists of the following steps:

1. Adjust the constant term c_0 in (7) to c_0' so that the plane defined by $c_0' + c_1 x + c_2 y + c_3 z = 0$ passes through the test point.
2. Use the homotopy $h(x, y, z, t) = \mathbf{0}$ to track paths with t going from 1 to 0 in the system

$$
h(x, y, z, t) = \begin{cases}
(y - x^2)(y + z) = 0 \\
(z - x^3)(y - z) = 0 \\
c_0 t + c_0'(1 - t) + c_1 x + c_2 y + c_3 z = 0
\end{cases}
\tag{8}
$$

At $t = 1$ we start the path tracking at the witness point set of one of the irreducible components V_i of $f(\mathbf{x}) = \mathbf{0}$ and find their end points at $t = 0$.
3. If the test point lies on the component V_i, then it will be one of the endpoints at $t = 0$.
4. Repeat steps (2) and (3) for each of the components V_i.

This test is numerically stable because the comparison of the test point with the end points at $t = 0$ does not require any extra precision, at least not for components of multiplicity one. For components of higher multiplicity we refer to [40].

This numerical stability is very important for components of high degree. In contrast, in [36], multi-precision arithmetic was needed to compute filtering polynomials [36] to high enough accuracy to identify components. This is because high-degree polynomials are numerically unstable, so unless one knows the test points with a sufficiently high accuracy, the result of the evaluation in the filtering polynomials cannot be certified.

The membership test is a crucial component in treating solution sets of different dimensions. To find generic points on all solution components of all dimensions, we apply a sequence of homotopies, introduced in [35]. The decomposition starts at the top dimension. At each step, the sets of generic points are partitioned, after removing superfluous points, using the membership test. We explain this procedure in the next section.

7 A Numerical Blackbox Decomposer

So far we have discussed the following tools:

1. use of monodromy to partition the sets of generic points into subsets of points that lie on the same irreducible solution component, or in more general terms, to decompose pure dimensional varieties into irreducibles; and

2. a homotopy membership test to separate isolated solutions from nonisolated points on solution components, or in general, to separate generic points on one component from those on higher dimensional components.

We still have to explain how to obtain the sets of generic points as the solutions of the original polynomial equations with additional linear constraints representing the random hyperplanes. As above we explain by example:

$$f(\mathbf{x}) = \begin{cases} (x_1 - 1)(x_2 - x_1^2) = 0 \\ (x_1 - 1)(x_3 - x_1^3) = 0 \\ (x_1^2 - 1)(x_2 - x_1^2) = 0 \end{cases} \tag{9}$$

From its factored form we see that $f(\mathbf{x}) = \mathbf{0}$ has two solution components: the two dimensional plane $x_1 = 1$ and the twisted cubic $\{ (x_1, x_2, x_3) \mid x_2 - x_1^2 = 0,\ x_3 - x_1^3 = 0 \}$.

To describe the solution set of this system, we use a sequence of homotopies, the chart in Figure 3 illustrates the flow of data for this example.

Because the top dimensional component is of dimension two, we add two random hyperplanes to the system and make it square again by adding two slack variables z_1 and z_2:

$$e(\mathbf{x}, z_1, z_2) = \begin{cases} (x_1 - 1)(x_2 - x_1^2) + a_{11}z_1 + a_{12}z_2 = 0 \\ (x_1 - 1)(x_3 - x_1^3) + a_{21}z_1 + a_{22}z_2 = 0 \\ (x_1^2 - 1)(x_2 - x_1^2) + a_{31}z_1 + a_{32}z_2 = 0 \\ c_{10} + c_{11}x_1 + c_{12}x_2 + c_{13}x_3 + z_1 = 0 \\ c_{20} + c_{21}x_1 + c_{22}x_2 + c_{23}x_3 + z_2 = 0 \end{cases} \tag{10}$$

where all constants a_{ij}, $i = 1, 2, 3$, $j = 1, 2$, and c_{kl}, $k = 1, 2$, $l = 0, 1, 2, 3$ are randomly chosen complex numbers. Observe that when $z_1 = 0$ and $z_2 = 0$ the solutions to $e(\mathbf{x}, z_1, z_2) = \mathbf{0}$ satisfy $f(\mathbf{x}) = \mathbf{0}$. So if we solve $e(\mathbf{x}, z_1, z_2) = \mathbf{0}$ we will find a single witness point on the two dimensional solution component

$x_1 = 1$ as a solution with $z_1 = 0$ and $z_2 = 0$. Using polyhedral homotopies, this requires the tracing of six solutions paths.

The embedding was proposed in [35] to find generic points on all positive dimensional solution components with a sequence of homotopies. In [35] it was proven that solutions with slack variables $z_i \neq 0$ are regular and, moreover, that those solutions can be used as start solutions in a homotopy to find witness points on lower dimensional solution components. At each stage of the algorithm, we call solutions with nonzero slack variables *nonsolutions*.

In the solution of $e(\mathbf{x}, z_1, z_2) = \mathbf{0}$, one path ended with $z_1 = 0 = z_2$, the five other paths ended in regular solutions with $z_1 \neq 0$ and $z_2 \neq 0$. These five "nonsolutions" are start solutions for the next stage, which uses the homotopy

$$
h_2(\mathbf{x}, z_1, z_2, t) \\
= \begin{cases}
(x_1 - 1)(x_2 - x_1^2) + a_{11}z_1 + a_{12}z_2 = 0 \\
(x_1 - 1)(x_3 - x_1^3) + a_{21}z_1 + a_{22}z_2 = 0 \\
(x_1^2 - 1)(x_2 - x_1^2) + a_{31}z_1 + a_{32}z_2 = 0 \\
c_{10} + c_{11}x_1 + c_{12}x_2 + c_{13}x_3 + z_1 = 0 \\
z_2(1 - t) + (c_{20} + c_{21}x_1 + c_{22}x_2 + c_{23}x_3 + z_2)t = 0
\end{cases} \quad (11)
$$

where t goes from one to zero, replacing the last hyperplane with $z_2 = 0$. Of the five paths, four of them converge to solutions with $z_1 = 0$. Of those four solutions, one of them is found to lie on the two dimensional solution component $x_1 = 1$, the other three are generic points on the twisted cubic. As there is one solution with $z_1 \neq 0$, we have one candidate left to use as a start point in the final stage, which searches for isolated solutions of $f(\mathbf{x}) = \mathbf{0}$. The homotopy for this stage is

$$
h_1(\mathbf{x}, z_1, t) = \begin{cases}
(x_1 - 1)(x_2 - x_1^2) + a_{11}z_1 = 0 \\
(x_1 - 1)(x_3 - x_1^3) + a_{21}z_1 = 0 \\
(x_1^2 - 1)(x_2 - x_1^2) + a_{31}z_1 = 0 \\
z_1(1 - t) + (c_{10} + c_{11}x_1 + c_{12}x_2 + c_{13}x_3 + z_1)t = 0
\end{cases} \quad (12)
$$

which as t goes from 1 to 0, replaces the last hyperplane $z_1 = 0$. At $t = 0$, the solution is found to lie on the twisted cubic, so there are no isolated solutions.

The calculations are summarized in Figure 3.

8   Benchmark Applications

To measure progress of our algorithms we executed the algorithms on a wide variety of polynomial systems. The second author maintains at http://www.math.uic.edu/~jan/demo a collection of 120 polynomial systems. We focus below on three of our most significant benchmarks. The systems come from relevant application fields and are challenging.

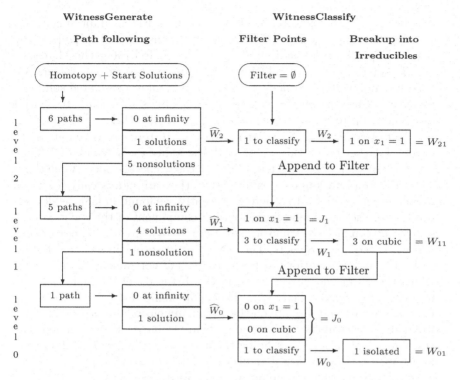

Figure 3. Numerical Irreducible Decomposition of a system whose solutions are the 2-dimensional plane $x_1 = 1$, the twisted cubic, and one isolated point. At level i, for $i = 2, 1, 0$, we filter candidate witness point sets $\widehat{W}_i$ into junk sets J_i and witness point sets W_i. The sets W_i are partitioned into witness point sets W_{ij} for the irreducible components.

cyclic n-roots: This system is probably the most notorious benchmark, popularized by Davenport in [10] and coming from an application involving Fourier transforms [2,3]. Already for $n = 4$, the system

$$\begin{cases} x_1 + x_2 + x_3 + x_4 = 0 \\ x_1 x_2 + x_2 x_3 + x_3 x_4 + x_4 x_1 = 0 \\ x_1 x_2 x_3 + x_2 x_3 x_4 + x_3 x_4 x_1 + x_4 x_1 x_2 = 0 \\ x_1 x_2 x_3 x_4 - 1 = 0 \end{cases} \tag{13}$$

is numerically troublesome when directly fed to traditional homotopy continuation methods because there are no isolated solutions and all paths terminate close to points on the two quadratic solution curves.

Haagerup confirmed [20] a conjecture of Fröberg (mentioned in [31]): for n prime, the number of solutions is finite and equals $\frac{(2n-2)!}{(n-1)!^2}$. Fröberg furthermore conjectured that for a quadratic divisor of n, there are infinitely many solutions.

Progress with computer algebra methods is described for dimensions 7 in [4], 8 in [5] and 9 in [15]. Unpublished (mentioned in [20]) is the result of Björck, who found all distinct isolated 184,756 cyclic 11-roots.

To solve this problem efficiently by numerical homotopy solvers, we need fast polyhedral methods to bound the number of isolated solutions by means of mixed volumes. These mixed volumes were computed for all dimensions up to $n = 11$ by Emiris in [13]; see also [14].

With PHCpack, all 35,840 isolated cyclic 10-roots were computed. As reported earlier in the this paper, with the aid of the software of T.Y. Li and Li Xing [28], PHC also found all 184,756 cyclic 11-roots. While those cases are computationally very intensive, they are numerically "easy" to handle as all solutions are isolated. Recently, we found in [38] the decomposition of the one dimensional solution component of the cyclic 8-roots system into 16 pieces, 8 quadrics and 8 curves of degree 16. There are also 1,152 isolated cyclic 8-roots. The cyclic 9-roots problem has, in addition to 6,642 isolated solutions, a two dimensional component of degree 18, which breaks up into six cubic surfaces. Of the 6,642 isolated solutions, 648 are roots of multiplicity four. The membership test is therefore essential to determine if these are on the higher-dimensional sets or if they are isolated. After confirming that the roots are isolated, we determined that the multiplicity is four by refining the roots using Newton's method with multi-precision arithmetic (32 decimal places) until 13 decimal places were correct and then grouping the roots into clusters.

As the technology for computing mixed volumes advances, (see [17,18] and [43]), we may expect further cases to be solved. In particular, [8] lists all isolated cyclic 12-roots and cyclic 13-roots.

adjacent minors: This application is taken from [12], where ideals are decomposed to study the connectivity of certain graphs arising in a random walks. One particular system in this paper is defined by all adjacent minors of a $2 \times (n + 1)$-matrix. For example, for $n = 3$ we have as matrix and polynomial system:

$$
\begin{bmatrix} x_{11} & x_{12} & x_{13} & x_{14} \\ x_{21} & x_{22} & x_{23} & x_{24} \end{bmatrix} \qquad \begin{cases} x_{11}x_{22} - x_{21}x_{12} = 0 \\ x_{12}x_{23} - x_{22}x_{13} = 0 \\ x_{13}x_{24} - x_{23}x_{14} = 0 \end{cases} \tag{14}
$$

In [12], it is proven that the ideal of adjacent 2×2-minors of a generic $2 \times (n + 1)$-matrix is the intersection of F_n prime ideals (where F_n is the nth Fibonacci number), and that the ideal is radical. All irreducible solution components have the same dimension, and the sum of their degrees equals 2^n.

Applying the monodromy breakup algorithm of [38], we found the breakup of a curve of degree 2,048 into 144 irreducible components, as the solution set to a system defined by all adjacent minors of a general 2×12-matrix.

We have limited our calculations to the case of matrices with 2 rows; see [12] for results on more general matrices. We expect methods exploiting the binomial structure of the equations, like the ones in [21] to be superior over general-purpose solvers.

Griffis-Duffy platform: In mechanical engineering, we study the direct kinematics of a so-called parallel-link robot (also known as a Stewart-Gough platform), consisting of two platforms attached to each other by six extensible legs. The problem is to find all possible positions of the top platform for a given position of the base platform and given lengths of the legs. Numerical continuation methods [32] first established that this problem has 40 isolated solutions. This result was later confirmed analytically in [25], [34] and [51]. These parallel robots are an important area of study, see for example [30].

In [19] a special type of the Stewart-Gough platform was proposed. In [26], it was shown that this platform permits motion, i.e., instead of 40 isolated solutions we now find a solution curve. Following [26], we call this type of platform a Griffis-Duffy platform.

Instead of 40 isolated solutions, we now have a one dimensional solution curve of degree 40. We investigated two cases of this Griffis-Duffy platform, see [38,39]. In both cases the curve of degree 40 contains twelve degenerate lines. For general choices of the parameters we have an irreducible curve of degree 28, while a more special case (also analyzed in [26]) breaks up in several components of lower degree.

9 Conclusions

In this paper we gave a description of our software tools to decompose positive dimensional solution sets of polynomial systems into irreducible components, emphasizing the geometrical and numerical aspects.

We also reported on the first steps to let the software interact with a computer algebra system, such as Maple. This interaction is currently limited to passing polynomials from Maple into PHCpack and passing results from PHCpack (solution vectors or interpolating polynomials) back into Maple. We experienced that efficient visualization requires more advanced communication protocols. These protocols will be built, using the C interface to PHCpack.

Acknowledgment: The authors wish to thank Jean-Charles Faugère for confirming the number of isolated cyclic 9-roots. The authors are grateful for the comments of the referee.

References

1. D.N. Bernshteĭn. The number of roots of a system of equations. *Functional Anal. Appl.* 9(3):183–185, 1975. Translated from *Funktsional. Anal. i Prilozhen.* 9(3):1–4,1975.
2. G. Björck. Functions of modulus one on Z_p whose Fourier transforms have constant modulus. In *Proceedings of the Alfred Haar Memorial Conference, Budapest,* volume 49 of *Colloquia Mathematica Societatis János Bolyai,* 193–197. 1985.
3. G. Björck. Functions of modulus one on Z_n whose Fourier transforms have constant modulus, and "cyclic n-roots". In *Recent Advances in Fourier Analysis and its Applications,* edited by J.S. Byrnes and J.F. Byrnes. Volume 315 of *NATO Adv. Sci. Inst. Ser. C: Math. Phys. Sci.,* 131–140. Kluwer, 1989.
4. G. Björck and R. Fröberg. A faster way to count the solutions of inhomogeneous systems of algebraic equations, with applications to cyclic n-roots. *J. Symbolic Computation* 12(3):329–336, 1991.
5. G. Björck and R. Fröberg. Methods to "divide out" certain solutions from systems of algebraic equations, applied to find all cyclic 8-roots. In *Analysis, Algebra and Computers in Math. research,* edited by M. Gyllenberg and L.E. Persson. Volume 564 of *Lecture Notes in Applied Mathematics,* 57–70. Marcel Dekker, 1994.
6. O. Bottema and B. Roth. *Theoretical Kinematics.* North-Holland, Amsterdam, 1979.
7. J. Canny and J.M. Rojas. An optimal condition for determining the exact number of roots of a polynomial system. In *Proceedings of the 1991 International Symposium on Symbolic and Algebraic Computation,* 96–101. ACM, 1991.
8. Y. Dai, K. Fujisawa, S. Kim, M. Kojima, and A. Takeda. Solution Information on Some Polynomial Systems. Web Site `http://www.is.titech.ac.jp/~kojima/polynomials/index.html`. January 4, 2002.
9. Y. Dai, S. Kim, and M. Kojima. Computing all nonsingular solutions of cyclic-n polynomial using polyhedral homotopy continuation methods. *Journal of Computational and Applied Mathematics,* to appear.
10. J. Davenport. Looking at a set of equations. Technical report 87-06, Bath Computer Science, 1987.
11. W. Decker and F.-O. Schreyer. Computational algebraic geometry today. In *Application of Algebraic Geometry to Coding Theory, Physics and Computation,* edited by C. Ciliberto, F. Hirzebruch, R. Miranda, and M. Teicher, 65–119, 2001. Proceedings of a NATO Conference, February 25 - March 1, 2001, Eilat, Israel. Kluwer Academic Publishers.
12. P. Diaconis, D. Eisenbud, and B. Sturmfels. Lattice walks and primary decomposition. In *Mathematical Essays in Honor of Gian-Carlo Rota,* edited by B.E. Sagan, R.P. Stanley. Volume 161 of *Progress in Mathematics,* 173–193, Birkhäuser, 1998.
13. I.Z. Emiris. *Sparse Elimination and Applications in Kinematics.* PhD thesis, Computer Science Division, Dept. of Electrical Engineering and Computer Science, University of California, Berkeley, 1994. Available at `http://www-sop.inria.fr/galaad/emiris`.
14. I.Z. Emiris and J.F. Canny. Efficient incremental algorithms for the sparse resultant and the mixed volume. *J. Symbolic Computation* 20(2):117–149, 1995. Software available at `http://www-sop.inria.fr/galaad/emiris`.

15. J.C. Faugère. A new efficient algorithm for computing Gröbner bases (F_4). *Journal of Pure and Applied Algebra* 139(1-3):61–88, 1999. Proceedings of MEGA'98, 22–27 June 1998, Saint-Malo, France.

16. A. Galligo, I.S. Kotsireas, R.M. Corless, and S.M. Watt. A geometric numeric algorithm for multivariate factorization. Poster presented at ISSAC'01.

17. T. Gao and T.Y. Li. Mixed volume computation via linear programming. *Taiwan J. of Math.* 4, 599–619, 2000.

18. T. Gao and T.Y. Li. Mixed volume computation for semi-mixed systems. *Discrete Comput. Geom.*, to appear.

19. M. Griffis and J. Duffy. Method and apparatus for controlling geometrically simple parallel mechanisms with distinctive connections. US Patent #5,179,525, 1993

20. U. Haagerup. Orthogonal maximal abelian *-algebras of the n x n matrices and cyclic n-roots. In *Operator Algebras and Quantum Field Theory*, 296–322. International Press, Cambridge, MA, 1996.

21. S. Hosten and J. Shapiro. Primary decomposition of lattice basis ideals. *Journal of Symbolic Computation* 29(4&5):625–639, 2000. Special Issue on Symbolic Computation in Algebra, Analysis and Geometry. Edited by E. Cattani and R.C. Laubenbacher.

22. B. Huber, F. Sottile, and B. Sturmfels. Numerical Schubert calculus. *J. of Symbolic Computation* 26(6):767–788, 1998.

23. B. Huber and B. Sturmfels. A polyhedral method for solving sparse polynomial systems. *Math. Comp.* 64(212):1541–1555, 1995.

24. B. Huber and J. Verschelde. Pieri homotopies for problems in enumerative geometry applied to pole placement in linear systems control. *SIAM J. Control Optim.* 38(4): 1265–1287, 2000.

25. M.L. Husty. An algorithm for solving the direct kinematics of general Stewart-Gough Platforms. *Mech. Mach. Theory*, 31(4):365–380, 1996.

26. M.L. Husty and A. Karger. Self-motions of Griffis-Duffy type parallel manipulators *Proc. 2000 IEEE Int. Conf. Robotics and Automation*, CDROM, San Francisco, CA, April 24–28, 2000.

27. T.Y. Li. Numerical solution of multivariate polynomial systems by homotopy continuation methods. *Acta Numerica* 6:399–436, 1997.

28. T.Y. Li and X. Li. Finding mixed cells in the mixed volume computation. *Found. Comput. Math.* 1(2): 161–181, 2001. Software available at http://www.math.msu.edu/~li.

29. M. Maekawa, M. Noro, K. Ohara, Y. Okutani, N. Takayama, and Y. Tamura. OpenXM – an open system to integrate mathematical softwares. Available at http://www.OpenXM.org/.

30. J-P. Merlet. Parallel robots. Kluwer, Dordrecht, 2000. For example pictures, see http://www-sop.inria.fr/coprin/equipe/merlet/merlet.html.

31. H.M. Möller. Gröbner bases and numerical analysis. In *Gröbner Bases and Applications*, edited by B. buchberger and F. Winkler. Volume 251 of *London Mathematical Lecture Note Series*, 159–178. Cambridge University Press, 1998.

32. M. Raghavan. The Stewart platform of general geometry has 40 configurations. *ASME J. Mech. Design* 115:277–282, 1993.

33. J.M. Rojas. A convex geometric approach to counting the roots of a polynomial system. *Theoret. Comput. Sci.* 133(1):105–140, 1994.

34. F. Ronga and T. Vust. Stewart platforms without computer? In *Real Analytic and Algebraic Geometry*. Proceedings of the International Conference, (Trento, 1992), 196–212. Walter de Gruyter, 1995.
35. A.J. Sommese and J. Verschelde. Numerical homotopies to compute generic points on positive dimensional algebraic sets. *Journal of Complexity* 16(3):572–602, 2000.
36. A.J. Sommese, J. Verschelde, and C.W. Wampler. Numerical decomposition of the solution sets of polynomial systems into irreducible components. *SIAM J. Numer. Anal.* 38(6):2022–2046, 2001.
37. A.J. Sommese, J. Verschelde, and C.W. Wampler. Numerical irreducible decomposition using projections from points on the components. In *Symbolic Computation: Solving Equations in Algebra, Geometry, and Engineering*, volume 286 of *Contemporary Mathematics*, edited by E.L. Green, S. Hoşten, R.C. Laubenbacher, and V. Powers, pages 37–51. AMS 2001.
38. A.J. Sommese, J. Verschelde, and C.W. Wampler. Using monodromy to decompose solution sets of polynomial systems into irreducible components. In *Application of Algebraic Geometry to Coding Theory, Physics and Computation*, edited by C. Ciliberto, F. Hirzebruch, R. Miranda, and M. Teicher, 297–315, 2001. Proceedings of a NATO Conference, February 25 - March 1, 2001, Eilat, Israel. Kluwer Academic Publishers.
39. A.J. Sommese, J. Verschelde, and C.W. Wampler. Symmetric functions applied to decomposing solution sets of polynomial systems. *SIAM J. Numer. Anal.*, to appear.
40. A.J. Sommese, J. Verschelde, and C.W. Wampler. A method for tracking singular paths with application to the numerical irreducible decomposition. In *Algebraic Geometry, a Volume in Memory of Paolo Francia*, edited by M.C. Beltrametti, F. Catanese, C. Ciliberto, A. Lanteri, and C. Pedrini, pages 329–345. W. de Gruyter, 2002.
41. A.J. Sommese and C.W. Wampler. Numerical algebraic geometry. In *The Mathematics of Numerical Analysis*, volume 32 of *Lectures in Applied Mathematics*, edited by J. Renegar, M. Shub, and S. Smale, 749–763, 1996. Proceedings of the AMS-SIAM Summer Seminar in Applied Mathematics, Park City, Utah, July 17-August 11, 1995, Park City, Utah.
42. B. Sturmfels. Polynomial equations and convex polytopes. *Amer. Math. Monthly* 105(10):907–922, 1998.
43. A. Takeda, M. Kojima, and K. Fujisawa. Enumeration of all solutions of a combinatorial linear inequality system arising from the polyhedral homotopy continuation method. To appear in *J. of Operations Society of Japan*.
44. J. Verschelde. Algorithm 795: PHCpack: A general-purpose solver for polynomial systems by homotopy continuation. *ACM Transactions on Mathematical Software* 25(2): 251–276, 1999. Software available at http://www.math.uic.edu/~jan.
45. J. Verschelde. Polynomial homotopies for dense, sparse and determinantal systems. MSRI Preprint #1999-041. Available at http://www.math.uic.edu/~jan.
46. J. Verschelde. Numerical evidence for a conjecture in real algebraic geometry. *Experimental Mathematics* 9(2): 183–196, 2000.
47. J. Verschelde and R. Cools. Symbolic homotopy construction. *Applicable Algebra in Engineering, Communication and Computing* 4(3):169–183, 1993.

48. J. Verschelde, K. Gatermann, and R. Cools. Mixed-volume computation by dynamic lifting applied to polynomial system solving. *Discrete Comput. Geom.* 16(1):69–112, 1996.
49. J. Verschelde and A. Haegemans. The *GBQ*-Algorithm for constructing start systems of homotopies for polynomial systems. *SIAM J. Numer. Anal.* 30(2):583–594, 1993.
50. J. Verschelde, P. Verlinden, and R. Cools. Homotopies exploiting Newton polytopes for solving sparse polynomial systems. *SIAM J. Numer. Anal.* 31(3):915–930, 1994.
51. C.W. Wampler. Forward displacement analysis of general six-in-parallel SPS (Stewart) platform manipulators using soma coordinates. *Mech. Mach. Theory* 31(3): 331–337, 1996.

Generating Kummer Type Formulas for Hypergeometric Functions

Nobuki Takayama

Kobe University, Rokko, Kobe 657-8501, Japan,
takayama@math.kobe-u.ac.jp

Abstract. Kummer type formulas are identities of hypergeometric series. A symmetry by the permutations of n-letters yields these formulas. We will present an algorithmic method to derive known and new Kummer type formulas. The algorithm utilizes several algorithms in algebra and geometry for generating Kummer type formulas.

1 Introduction

The Gauss hypergeometric function is defined by the series

$$F(a, b, c; x) = \sum_{n=0}^{\infty} \frac{(a)_n (b)_n}{(1)_n (c)_n} x^n, \quad (a)_n = a(a+1)\cdots(a+n-1) = \Gamma(a+n)/\Gamma(a)$$

where a, b, c are complex parameters and we assume $c \notin \mathbb{Z}_{<0}$. The series in the right hand side converges in the unit disk and $F(a, b, c; x)$ can be analytically continued to $\mathbb{C} \setminus \{0, 1\}$ as a multi-valued analytic function. The Gauss hypergeometric function satisfies several attractive formulas. Among them, we will consider Kummer's formula

$$F(a, b, c; x) = (1 - x)^{-a} F\left(a, c - b, c; \frac{x}{x - 1}\right)$$

and its generalizations in this article.

There are several methods to prove it. Among them, we here refer to a method by the Chu-Vandermonde formula of binomial coefficients [20]. Although this proof presents an interesting interplay between the famous identity in combinatorics and Kummer's formula, it seems hard to generalize this approach.

I.M. Gel'fand suggested an idea to derive and prove Kummer's formula in his series of lectures at Kyoto in 1989. The method is a natural generalization of the method to derive $24 = 4!$ solutions of the Gauss hypergeometric equation by Kummer. The point of his idea is that the method can be also applied to hypergeometric functions associated to the product of simplices, which have the symmetry of permutations of n-letters.

Since his presentation of the idea, there have been only a few tries to study systematically Kummer type formulas (see, e.g., [16]). The author thinks that the reason for it is the lack of integrations of algebra and geometry software systems. As readers will see in the body of this paper, the systematic study of Kummer type formulas requires several kinds of explicit constructions in algebra and geometry including Gröbner basis computation, triangulations, and constructions of fans, for which algorithms and systems have been intensively studied in the last 10 years.

In this article, we will present an algorithm to derive Kummer type formulas for hypergeometric functions associated to the product of simplices developing the original idea by I.M. Gel'fand. We can experience an interplay between algebra, geometry and software systems through deriving Kummer type formulas of hypergeometric functions.

2 Hypergeometric Function Associated to $\Delta_{k-1} \times \Delta_{n-k-1}$

Let us recall the definition of hypergeometric functions associated to a product of linear forms [1], [9]. We fix two numbers k and n satisfying $n \geq 2k \geq 4$. Let α_j be parameters satisfying $\sum_{j=1}^{n} \alpha_j = -k$. *Hypergeometric function* of type $E_{k,n}$ is defined by the integral

$$\Phi(\alpha; z) = \int_C \prod_{j=1}^{n} (\sum_{i=1}^{k} z_{ij} s_i)^{\alpha_j} \, ds_2 \cdots ds_k$$

where we put $s_1 = 1$ and $z \in M_{k,n}$ = the space of $k \times n$ matrices and C is a bounded $(k-1)$-cell in the hyperplane arrangement defined by

$$\prod_{j=1}^{n} \sum_{i=1}^{k} z_{ij} s_i = 0$$

in the $(s_2, \ldots, s_k)$-space.

When $k = 2$, $n = 4$, it is written as

$$\Phi(\alpha; z) = \int_C (z_{11} + z_{21} s_2)^{\alpha_1} \cdots (z_{14} + z_{24} s_2)^{\alpha_4} ds_2.$$

The integral $\Phi(z)$ agrees with the Gauss hypergeometric integral when

$$z = \begin{pmatrix} 1 & 0 & 1 & 1 \\ 0 & 1 & -1 & -x \end{pmatrix}.$$

The hypergeometric function of type $E_{k,n}$ is quasi-invariant under the action of complex torus $(\mathbb{C}^*)^n$ and the general linear group $GL(k) = GL(k, \mathbb{C})$. In fact, we have for

$$h = \begin{pmatrix} h_1 & 0 & & 0 \\ 0 & h_2 & & 0 \\ 0 & 0 & \cdots & 0 \\ \cdot & \cdot & \cdots & \cdot \\ 0 & 0 & \cdots & h_n \end{pmatrix} \in (\mathbb{C}^*)^n$$

and $g \in GL(k)$,

$$\Phi(\alpha; zh) = \left(\prod_j h_j^{\alpha_j}\right) \Phi(\alpha; z), \tag{1}$$

$$\Phi(\alpha; gz) = |g|^{-1}\Phi(\alpha; z). \tag{2}$$

It follows from the quasi-invariant property and the integral representation that the function $\Phi(\alpha; z)$ satisfies a system of first order equations and a system of second order equations respectively.

Theorem 2.1. (Gel'fand [9]) *The function $\Phi(\alpha; z)$ satisfies*

$$\left(\sum_{i=1}^{k} z_{ip}\frac{\partial}{\partial z_{ip}} - \alpha_p\right) f = 0, \ p = 1, \dots, n$$

$$\left(\sum_{p=1}^{n} z_{ip}\frac{\partial}{\partial z_{jp}} + \delta_{ij}\right) f = 0, \ i, j = 1, \dots, k$$

$$\left(\frac{\partial^2}{\partial z_{ip}\partial z_{jq}} - \frac{\partial^2}{\partial z_{iq}\partial z_{jp}}\right) f = 0, \ i, j = 1, \dots, k, p, q = 1, \dots, n$$

We call this system of equations $E_{k,n}$.

Let $A = (a_{ij})$ be an integer $d \times N$-matrix of rank d and β a vector over complex numbers $(\beta_1, \dots, \beta_d)^T$. The $\mathcal{A}$-*hypergeometric* (GKZ hypergeometric) system $H_A(\beta)$ is the following system of linear partial differential equations for the indeterminate function $f(x_1, \dots, x_N)$:

$$\left(\sum_{j=1}^{N} a_{ij}x_j\partial_j - \beta_i\right) f = 0, \quad \text{for } i = 1, \dots, d$$

$$(\partial^u - \partial^v)f = 0 \quad \text{for all } u, v \in \mathbb{N}_0^N \text{ with } Au = Av.$$

Here, we use the multi-index notation $\partial^u = \prod_{i=1}^{N} \partial_i^{u_i}$.

The $\mathcal{A}$-hypergeometric system was first introduced and studied by Gel'fand, Kapranov and Zelevinsky [10].

If we restrict the system of differential equations in Theorem 2.1 to the affine chart

$$\begin{pmatrix} 1 \cdots 0 \ x_{11} \ \cdots \ x_{1\ell} \\ 0 \cdots 0 \ x_{21} \ \cdots \ x_{2\ell} \\ 0 \cdots 0 \quad \cdots \\ 0 \cdots 1 \ x_{k1} \ \cdots \ x_{k\ell} \end{pmatrix}, \quad \ell = m - k$$

of $GL(k) \backslash M_{k,n}$, we obtain the $\mathcal{A}$-hypergeometric system associated to the product of simplices $\Delta_{k-1} \times \Delta_{n-k-1}$. More precisely, put

$$
a_1 = \begin{pmatrix} 1 \\ 0 \\ \cdot \\ \cdot \\ \cdot \\ 0 \end{pmatrix}, \quad \ldots, \quad a_{n-k} = \begin{pmatrix} 0 \\ 0 \\ \cdot \\ \cdot \\ \cdot \\ 1 \end{pmatrix} \in \mathbb{Z}^{n-k},
$$

$$
A_{k,n} = \begin{pmatrix}
1 & \cdots & 1 & 0 & \cdots & 0 & & 0 & \cdots & 0 \\
0 & \cdots & 0 & 1 & \cdots & 1 & & 0 & \cdots & 0 \\
0 & \cdots & 0 & 0 & \cdots & 0 & & 0 & \cdots & 0 \\
\cdot & \cdots & \cdot & \cdot & \cdots & \cdot & & \cdot & \cdots & \cdot \\
\cdot & \cdots & \cdot & \cdot & \cdots & \cdot & & \cdot & \cdots & \cdot \\
\cdot & \cdots & \cdot & \cdot & \cdots & \cdot & & \cdot & \cdots & \cdot \\
0 & \cdots & 0 & 0 & \cdots & 0 & & 1 & \cdots & 1 \\
a_1 & \cdots & a_{n-k} & a_1 & \cdots & a_{n-k} & & a_1 & \cdots & a_{n-k}
\end{pmatrix},
$$

and $\beta = (-\alpha_1 - 1, \ldots, -\alpha_k - 1, \alpha_{k+1}, \ldots, \alpha_n)^T$. Then, the restricted system of $E_{k,n}$ is the $\mathcal{A}$-hypergeometric system $H_{A_{k,n}}(\beta)$ with $x_1 = x_{11}, x_2 = x_{12}, \ldots, x_\ell = x_{1\ell}, x_{\ell+1} = x_{21}, \ldots, x_N = x_{k\ell}$. Note that, by using the $GL(k)$-quasi-invariant property (2), solutions of $E_{k,n}$ can be expressed in terms of solutions of $H_{A_{k,n}}(\beta)$. We will present later some explicit constructions of solutions of $E_{2,4}$ and $E_{2,5}$.

3   Configuration Space

We denote by $X_{k,n}$ the quotient space

$$
GL(k) \backslash M_{k,n}^* / (\mathbb{C}^*)^n.
$$

Here, $M_{k,n}^*$ denotes the set of the $k \times n$ matrices of which any $k \times k$ minor does not vanish. The space $X_{k,n}$ is called the configuration space of n-points in the $(k-1)$-dimensional project space. It is a $(k-1) \times (n-k-1)$ dimensional affine variety. The group of the permutations of n-columns of $M_{k,n}$ acts on $X_{k,n}$. We denote the group by S_n.

Regular functions on $M_{k,n}^*$ which are invariant under the action of $GL(k)$ and $(\mathbb{C}^*)^n$ are regular functions on $X_{k,n}$. Let us denote by

$$
y_{i_1,\cdots,i_k}
$$

the determinant of $(i_1, \ldots, i_k)$-th columns of the $k \times n$ matrix $z = (z_{ij})$, which is called the *Plücker coordinate*. Then, the affine chart x_{ij} of $GL(k) \backslash M_{k,n}^*$ is expressed as

$$
x_{ij} = \frac{y_{1,2,\cdots,i-1,k+j,i+1,\cdots,k}}{y_{1,2,\cdots,k}}, \tag{3}
$$

which is invariant under the action of $GL(k)$.

Let us take a vector

$$p = (p_{ij}) \in \operatorname{Ker} \mathbb{Z}^{k \times (n-k)} \xrightarrow{A_{k,n}} \mathbb{Z}^n.$$

Then, it is easy to see that $x^p = \prod x_{ij}^{p_{ij}}$ is invariant as a function of $y = (y_{i_1,\ldots,i_k})$ by the actions of $GL(k)$ and $(\mathbb{C}^*)^n$. Hence, x^p is a regular function on the configuration space $X_{k,n}$.

Let σ be an element of S_n. The action of σ on x^p is defined as

$$\sigma \bullet x^p = \prod_{i,j} \left(\frac{y_{\sigma(1),\sigma(2),\cdots,\sigma(i-1),\sigma(k+j),\sigma(i+1),\cdots,\sigma(k)}}{y_{\sigma(1),\sigma(2),\cdots,\sigma(k)}} \right)^{p_{ij}} \tag{4}$$

The action induces a biholomorphic transformation on $X_{k,n}$.

Example 3.1. $k = 2, n = 4$

The configuration space $X_{2,4}$ is isomorphic to $\mathbf{P}^1 \setminus \{0, 1, \infty\} = \mathbb{C} \setminus \{0, 1\}$. An isomorphism is given by

$$X_{2,4} \ni (z_{ij}) \mapsto \frac{y_{4,2}y_{1,3}}{y_{3,2}y_{1,4}} = x \in \mathbb{C} \setminus \{0, 1\}.$$

We denote the Plücker coordinate y_{ij} by $[ij]$. Then, the action of S_4 on $\mathbb{C}^1 \setminus \{0, 1\}$ is summarized as follows.

$$x := \frac{[42][13]}{[32][14]}$$

$$(12) \bullet x = \frac{[41][23]}{[31][24]} = \frac{1}{x}$$

$$(23) \bullet x = \frac{[43][12]}{[23][14]} = \frac{[43][13]}{[32][14]} - 1 = x - 1$$

$$([42][13] - [41][23] + [21][43] = 0, \quad \text{Plücker's relation})$$

$$(34) \bullet x = \frac{1}{x}$$

Here, for example, (23) means $\begin{pmatrix} 1\,2\,3\,4 \\ 1\,3\,2\,4 \end{pmatrix} \in S_4$.

Example 3.2. $k = 2, n = 4$

The configuration space $X_{2,5}$ is isomorphic to

$$X = \mathbb{C}^2 \setminus \{(x, y) \mid xy(x - 1)(y - 1)(x - y) = 0\}$$

by the isomorphism

$$X_{2,5} \ni (z_{ij}) \mapsto \left(\frac{[42][13]}{[32][14]}, \frac{[14][52]}{[42][15]} \right) = (x, y) \in X.$$

where $[ij] = y_{i,j}$. The action of S_5 is as follows.

$$(12) \bullet x = \frac{1}{x}, \quad (12) \bullet y = \frac{1}{y}$$

$$(23) \bullet x = 1 - x, \quad (23) \bullet y = \frac{1 - xy}{1 - x}$$

$$(34) \bullet x = \frac{1}{x}, \quad (34) \bullet y = xy$$

$$(45) \bullet x = xy, \quad (45) \bullet y = \frac{1}{y}$$

where Plücker relations

$$\sum_{k=0}^{2} (-1)^k y_{i_1, j_k} y_{j_0, \cdots, \hat{j}_k, \cdots, j_2} = 0$$

are used to reduce expressions in y_{ij} into expressions in x and y. The index sets $\{i_1\}$ and $\{j_0, j_1, j_2\}$ run over the subsets of $\{1, 2, 3, 4, 5\}$. For instance, when $i_1 = 1$ and $j_0 = 3, j_1 = 4, j_2 = 5$, the relation is $y_{1,3} y_{4,5} - y_{1,4} y_{3,5} + y_{1,5} y_{3,4} = 0$.

We are interested in deriving explicit expressions of the action of S_n on the configuration space $X_{k,n}$ as in the two examples. It can be done by an elimination. Put $m = (k-1)(n-k-1)$, which is the dimension of the configuration space $X_{k,n}$.

Algorithm A: Action of S_n on the configuration space

Data   : $x_1 = \frac{y^{p(1)}}{y^{q(1)}}, \ldots, x_m = \frac{y^{p(m)}}{y^{q(m)}}$: a system of local coordinates of a compactification of $X_{k,n}$, and $\sigma \in S_n$.

Result : Polynomials f_j and g_j such that $\sigma \bullet x_j = \frac{f_j(x_1, \ldots, x_m)}{g_j(x_1, \ldots, x_m)}$

begin

 (1) Put $P = \{$the Plücker relations,
     $y_{i_1, \cdots, i_k} y^*_{i_1, \cdots, i_k} - 1 \mid i_1, \ldots, i_k \subset \{1, \ldots, n\}\}$.
 Define $I = \langle P, y^{q(i)} x_i - y^{p(i)} \ (i = 1, \ldots, m)$,
    $(\sigma \bullet y^{q(j)}) w_j - (\sigma \bullet y^{p(j)}) \rangle$ as an ideal in
  $\mathbb{Q}[y_{1,2,\ldots,k}, \cdots, y_{n-k+1,\ldots,n},$
   $y^*_{1,2,\ldots,k}, \cdots, y^*_{n-k+1,\ldots,n}, x_1, \ldots, x_m, w_j]$.
 (2) Find the ideal intersection
  $J = \mathbb{Q}(x_1, \ldots, x_m) I \cap \mathbb{Q}(x_1, \ldots, x_m)[w_j]$.
 It is generated by an element of the form $w_j - \frac{f_j(x)}{g_j(x)}$.
 (3) Output $\frac{f_j}{g_j}$.

end

The correctness follows from that the ideal generated by the Plücker relations is prime and hence radical. Here is a sketch of a proof of the primeness suggested by B. Sturmfels. "Establish the SAGBI basis property of the $k \times k$-minors and then to infer that the Plücker ideal is precisely the ideal of all algebraic relations on the $k \times k$-minors. This implies primeness, since the kernel of a map from a domain to a domain is always prime." Gröbner basis of Plücker relations and the SAGBI argument are described in the chapter three of the book by B. Sturmfels on invariant theory [18].

Elimination can be done by the characteristic set method or the Gröbner basis method. Readers are invited to articles by M. Noro [11], H. Schönemann [15], and D.M. Wang [21] in this volume for algorithms, implementations and related topics in this area.

We close this section with a short remark on efficiency. Although the method above is simple, it seems that it is not the most efficient method to explicitly derive the action of S_n on a coordinate system of the configuration space. It seems more efficient to use a method based on transforming a given $k \times n$ matrix to the normal form

$$\begin{pmatrix} 1 & 0 & \cdots & 0 & 1 & 1 & \cdots & 1 \\ 0 & 1 & \cdots & 0 & 1 & * & \cdots & * \\ & & \cdots & 1 & * & \cdots & * \\ & & \cdots & 1 & * & \cdots & * \\ 0 & 0 & \cdots & 1 & 1 & * & \cdots & * \end{pmatrix}$$

by the left $GL(k)$ and the right $(\mathbb{C}^*)^n$ action.

4   Series Solutions

Several methods to construct series solutions of $\mathcal{A}$-hypergeometric system have been known. Among them, we would like to mention here the following two constructions.

Theorem 4.1. [10]

1. *There is a correspondence between a regular triangulation T of A and a basis of series solutions of $H_A(\beta)$ for generic β.*
2. *A translation of the secondary cone of a regular triangulation T of A is the domain of convergence of the basis of series solutions associated to T.*

Theorem 4.2. [13] *There is a construction algorithm of series solution basis of $H_A(\beta)$ for any β. The series solutions are constructed by extending the solutions of the initial ideal $\mathrm{in}_{(-w,w)}(H_A(\beta))$ with respect to a given weight $(-w, w)$.*

Details on these methods are explained in [10] and section 3.4 and chapter two of the book [13]. Triangulations play an important role to construct series solutions. Readers are invited to consult the article by J. Pfeifle and J. Rambau [12] on algorithms and implementations for triangulations.

We will add one more note here. Series solutions can be analytically continued to the complement of the *singular locus* of $H_{A_{k,n}}(\beta)$, of which defining equation is expressed in terms of resultants. In case of $A_{2,4}$, the defining equation is

$$x_{11}x_{12}x_{21}x_{22}(x_{11}x_{22} - x_{12}x_{21}).$$

Readers are invited to the article by I.Z. Emiris [7] on sparse resultants in this book.

Let $\phi(\alpha; x)$ be a series solution of $A_{k,n}$-hypergeometric system. Then, by the quasi-invariant relation (2), the function

$$\Phi(\alpha; y) = \frac{1}{y_{1,2,\dots,k}}\phi(\alpha; x(y)) \tag{5}$$

is a solution of $E_{k,n}$ where $x(y)$ is a coordinate system defined in (3).

The invariant property of the system of equations $E_{k,n}$ under the action of S_n yields the following theorem.

Theorem 4.3. *If the function $\Phi(\alpha; y)$ is a solution of $E_{k,n}$, then the function $\sigma \bullet \Phi(\alpha; y) := \Phi(\sigma \bullet \alpha; \sigma \bullet y)$ is also a solution of $E_{k,n}$ for any $\sigma \in S_n$.*

Example 4.4. $k = 2, n = 4$.
The matrix $A_{2,4}$ is

$$
\begin{array}{cccc}
x_{11} & x_{12} & x_{21} & x_{22}
\end{array}
$$
$$
\begin{pmatrix}
1 & 1 & 0 & 0 \\
0 & 0 & 1 & 1 \\
1 & 0 & 1 & 0 \\
0 & 1 & 0 & 1
\end{pmatrix}.
$$

The hypergeometric system $H_{A_{2,4}}(\beta)$ is generated by

$$x_{11}\partial_{11} + x_{12}\partial_{12} - (-\alpha_1 - 1), \ x_{21}\partial_{21} + x_{22}\partial_{22} - (-\alpha_2 - 1),$$

$$x_{11}\partial_{11} + x_{21}\partial_{21} - \alpha_3, \ x_{12}\partial_{12} + x_{22}\partial_{22} - \alpha_3, \ \partial_{11}\partial_{22} - \partial_{12}\partial_{21}$$

where $\partial_{ij} = \frac{\partial}{\partial x_{ij}}$ and $\alpha_1 + \alpha_2 + \alpha_3 + \alpha_4 = -2$. Take the triangulation $T = \{124, 134\}$ or the Gröbner deformation with respect to the weight $w = (0, 1, 1, 0) = \left(\begin{smallmatrix} 0 & 1 \\ 1 & 0 \end{smallmatrix}\right)$. Then the starting terms of series solutions are x^p and x^q where $p = \left(\begin{smallmatrix} p_{11} & p_{12} \\ p_{21} & p_{22} \end{smallmatrix}\right) = \left(\begin{smallmatrix} \alpha_3 & -\alpha_1-\alpha_3-1 \\ 0 & -\alpha_2-1 \end{smallmatrix}\right)$ and $q = \left(\begin{smallmatrix} -\alpha_1-1 & 0 \\ -\alpha_2-\alpha_4-1 & \alpha_4 \end{smallmatrix}\right)$. Note that

$$
\begin{pmatrix} z_{11} & z_{12} \\ z_{21} & z_{22} \end{pmatrix}^{-1} (z_{ij}) = \begin{pmatrix} 1 & 0 & \frac{[32]}{[12]} & \frac{[42]}{[12]} \\ 0 & 1 & \frac{[13]}{[12]} & \frac{[14]}{[12]} \end{pmatrix}.
$$

Then, we put $x_{11} = \frac{[32]}{[12]}, x_{12} = \frac{[42]}{[12]}, x_{21} = \frac{[13]}{[12]}, x_{22} = \frac{[14]}{[12]}$. By extending the starting terms to series solutions, we obtain solutions of $H_{A_{2,4}}(\beta)$ and consequently those of E_{25}:

$$\eta_1 F(-\alpha_3, \alpha_2 + 1, -\alpha_1 - \alpha_3; s)$$
$$\eta_2 F(-\alpha_4, \alpha_1 + 1, -\alpha_2 - \alpha_4; s)$$

where $s = \frac{x_{12}x_{21}}{x_{11}x_{22}} = \frac{[42][13]}{[32][14]}$ and

$$\eta_1 = \frac{1}{[12]} x_{12}^{-\alpha_1} x_{22}^{-\alpha_2} \left(\frac{x_{11}}{x_{12}}\right)^{\alpha_3} \frac{1}{x_{12}x_{22}}$$

$$\eta_2 = \frac{1}{[12]} \left(\frac{x_{21}}{x_{11}}\right)^{\alpha_1} x_{21}^{\alpha_3} x_{22}^{\alpha_4} \frac{x_{21}}{x_{11}},$$

$\sum_{i=1}^{4} \alpha_i = -2$. F is the Gauss hypergeometric function. We obtain $|S_4| = 24$ fundamental sets of solutions by applying the Theorem 4.3.

5 Deriving Kummer Type Formulas

Let $\eta_1(y)\phi_1(\alpha; x(y)), \eta_2(y)\phi_2(\alpha; x(y)), \ldots, \eta_r(y)\phi_r(\alpha; x(y))$ be a basis of solutions of $E_{k,n}$. It follows from the quasi-invariant property (1) and (2) that the quotient function

$$(\sigma \bullet \eta_i(y))\phi_i(\sigma \bullet \alpha; \sigma \bullet x(y))/\eta_1(y)\phi_1(\alpha; x(y))$$

is a multi-valued analytic function on $X_{k,n}$ for $\sigma \in S_n$. Since, the function $\phi_1(\alpha; x(y))$ is a multi-valued analytic function in $X_{k,n}$, the quotient function

$$(\sigma \bullet \eta_i(y))\phi_i(\sigma \bullet \alpha; \sigma \bullet x(y))/\eta_1(y)$$

is also a multi-valued analytic function on the configuration space $X_{k,n}$.

We take a coordinate system

$$s_1 = x(y)^{p(1)}, s_2 = x(y)^{p(2)}, \ldots, s_m = x(y)^{p(m)} \tag{6}$$

of a compactification of the configuration space $X_{k,n}$ which satisfies the condition: all $\phi_i(\alpha; x(y))$ can be expressed in a convergent power series of $s_1, \ldots, s_m$.

If we take a *compatible basis* $\{p(1), \ldots, p(m)\}$ [10], it satisfies the condition above. And, we need a "small" basis in a suitable sense. However, the author does not know an efficient method to find a compatible basis and does not know what compatible basis is "small" for our purpose. They are questions of geometry.

Now, we are ready to state our method to derive Kummer type formulas of hypergeometric series associated to $E_{k,n}$, see Algorithm B.

Since the formula (7) contains complex power functions, we have to be careful about choices of branches. As to this topics, readers are invited to the article of J. Davenport [5] in this book. The problem of choices of branches

Algorithm B: Deriving Kummer type formulas

Data   : $\eta_i(y)$, $\phi_i(\alpha; x(y))$, $s_j = x(y)^{p(j)}$
Result : Kummer Type Formulas
Suppose that $\eta_i(y)/\eta_1(y), i \geq 2$ are not holomorphic at $s_1 = \cdots = s_m$.

begin

   (1) Let G be the isotropy group in S_n that fixes the point $s_1 = \cdots = s_m$.

   (2) **for** $\sigma \in G$ **do**

   consider functions $\{\frac{\sigma \bullet \eta_i}{\eta_1} \mid i = 1, \ldots, r\}$. If the function $\frac{\sigma \bullet \eta_i}{\eta_1}$ is holomorphic at $s_1 = \cdots = s_m = 0$ and other functions $\frac{\sigma \bullet \eta_j}{\eta_1}$, $j \neq i$ are
   not holomorphic, then output

   $$c \frac{\sigma \bullet \eta_i}{\eta_1} \phi_i(\sigma \bullet \alpha; \sigma \bullet x(y)) = \phi_1(\alpha; x(y)). \tag{7}$$

   Here, c is a suitable constant. Note that the both sides are functions
   in $s_1, \ldots, s_m$.

end

in formulas on hypergeometric functions has a long history. For example,
É. Goursat gave connection formulas of the Gauss hypergeometric function
on the upper half plane and the lower half plane in his long paper in 1881 [8].
This approach of giving connection formulas on simply connected domains
is nice, however, most editors of formula books do not take his original nice
rigorous approach and list connection formulas without the side conditions
on branches. In the several variable case, the problem of finding a nice decomposition into simply connected domains is more interesting and difficult.
See, for example, [23, Chapter V] and [14].

Example 5.1. $k = 2, n = 4$.
By a calculation, we have $\eta_2/\eta_1 = s^{\alpha_1 + \alpha_3 + 1}$, which is not holomorphic at
$s = 0$. Put $\sigma = (13) = [3, 2, 1, 4, 5]$. Then, we have

$$\frac{\sigma \bullet \eta_1}{\eta_1} = \left(\frac{[32][14]}{[34][12]}\right)^{\alpha_2 + 1}$$

and

$$\sigma \bullet F(-\alpha_3, \alpha_2 + 1, -\alpha_1 - \alpha_3; s) = F\left(-\alpha_1, \alpha_2 + 1, -\alpha_1 - \alpha_3; \frac{[42][31]}{[12][34]}\right).$$

Since $[12][34] - [13][24] + [23][14] = 0$ we have $\frac{[42][31]}{[12][34]} = \frac{s}{s-1}$. Similarly, we have $\frac{[32][14]}{[12][34]} = 1 - \frac{s}{s-1}$. Therefore, we obtain

$$F(-\alpha_3, \alpha_2 + 1, -\alpha_1 - \alpha_3; s)$$
$$= (1-s)^{-\alpha_2-1} F\left(-\alpha_1, \alpha_2 + 1, -\alpha_1 - \alpha_3; \frac{s}{s-1}\right),$$

which is well-known Kummer's formula for the Gauss hypergeometric function.

Example 5.2. $k = 2, n = 5$. Put $x_{11} = \frac{[32]}{[12]}, x_{12} = \frac{[42]}{[12]}, x_{13} = \frac{[52]}{[12]}, x_{21} = \frac{[13]}{[12]}, x_{22} = \frac{[14]}{[12]}, x_{23} = \frac{[15]}{[12]}$,

$$\phi_1(x(y)) =$$
$$F_1\left(\alpha_1 + 1, -\alpha_4, -\alpha_5, \alpha_3 + \alpha_1 + 2; \left(\frac{x_{21}x_{12}}{x_{11}x_{22}}\right), \left(\frac{x_{21}x_{13}}{x_{11}x_{23}}\right)\right),$$

$$\phi_2(x(y)) =$$
$$G_2\left(-\alpha_3, -\alpha_5, \alpha_3 + \alpha_1 + 1, \alpha_5 + \alpha_2 + 1; \left(-\frac{x_{21}x_{12}}{x_{11}x_{22}}\right), \left(-\frac{x_{22}x_{13}}{x_{12}x_{23}}\right)\right),$$

$$\phi_3(x(y)) =$$
$$F_1\left(\alpha_2 + 1, -\alpha_3, -\alpha_4, \alpha_5 + \alpha_2 + 2; \left(\frac{x_{21}x_{13}}{x_{11}x_{23}}\right), \left(\frac{x_{22}x_{13}}{x_{12}x_{23}}\right)\right),$$

and

$$\eta_1 = [12]^{-1} x_{11}^{-\alpha_1-1} x_{21}^{\alpha_3+\alpha_1+1} x_{22}^{\alpha_4} x_{23}^{\alpha_5}$$
$$\eta_2 = [12]^{-1} x_{11}^{\alpha_3} x_{12}^{-\alpha_1-1-\alpha_3} x_{22}^{-\alpha_2-1-\alpha_5} x_{23}^{\alpha_5}$$
$$\eta_3 = [12]^{-1} x_{11}^{\alpha_3} x_{12}^{\alpha_4} x_{13}^{\alpha_5+\alpha_2+1} x_{23}^{-\alpha_2-1}.$$

Here

$$F_1(a, b, b', c; x_1, x_2) := \sum_{m,n=0}^{\infty} \frac{(a)_{m+n}(b)_m(b')_n}{(c)_{m+n}(1)_m(1)_n} x_1^m x_2^n$$

is the Appell hypergeometric function and G_2 is the hypergeometric function G_2 in Horn's list. Then, the functions $\eta_i \phi_i$ are solutions of $E_{k,n}$. Put

$$s_1 = \frac{x_{21}x_{12}}{x_{11}x_{22}} = \frac{[42][13]}{[32][14]}, \quad s_2 = \frac{x_{22}x_{13}}{x_{12}x_{23}} = \frac{[52][14]}{[42][15]}.$$

The functions $\frac{\eta_i}{\eta_1}\phi_i(x(y))$ are expressed in terms of s_1 and s_2 as follows.

$$F_1(s_1, s_1 s_2)$$

$$s_1^{-1} s_1^{-\alpha_1} s_1^{-\alpha_3} G_2(-s_1, -s_2)$$

$$(s_1 s_2)^{-1} (s_1 s_2)^{-\alpha_1} (s_1 s_2)^{-\alpha_3} s_2^{-\alpha_4} F_1(s_1 s_2, s_2)$$

These functions converge in $|s_1|, |s_2| < 1$ and the first function $\phi_1(s_1, s_1 s_2)$ is holomorphic at $s_1 = s_2 = 0$.

The isotropy group G at $s_1 = s_2 = 0$ consists of eight elements. The group G is the dihedral group. Here, $[i_1, i_2, i_3, i_4, i_5]$ denotes the permutation

$(\sigma \bullet s_1, \sigma \bullet s_2)$	σ
$(\frac{(-s_1+1)s_2}{s_2-1}, \frac{-s_1 s_2 + s_1}{s_1-1})$	$[5, 3, 2, 4, 1]$
$(\frac{s_2}{s_2-1}, \frac{s_1 s_2 - s_1}{s_1 s_2 - 1})$	$[5, 1, 2, 4, 3]$
$(\frac{-s_1 s_2 + s_1}{s_1-1}, \frac{(-s_1+1)s_2}{s_2-1})$	$[3, 5, 1, 4, 2]$
$(\frac{s_1}{s_1-1}, \frac{(s_1-1)s_2}{s_1 s_2 - 1})$	$[3, 2, 1, 4, 5]$
$(\frac{(s_1-1)s_2}{s_1 s_2 - 1}, \frac{s_1}{s_1-1})$	$[2, 3, 5, 4, 1]$
(s_2, s_1)	$[2, 1, 5, 4, 3]$
$(\frac{s_1 s_2 - s_1}{s_1 s_2 - 1}, \frac{s_2}{s_2-1})$	$[1, 5, 3, 4, 2]$
(s_1, s_2)	$[1, 2, 3, 4, 5]$

$$\begin{pmatrix} 1 & 2 & 3 & 4 & 5 \\ i_1 & i_2 & i_3 & i_4 & i_5 \end{pmatrix} \in S_5.$$

Here is a list of

$$\left[\frac{\sigma \bullet \eta_1}{\eta_1}, \frac{\sigma \bullet \eta_2}{\eta_1}, \frac{\sigma \bullet \eta_2}{\eta_1}, \sigma \right]$$

generated by Algorithm A in this article. For example, the first line means that

$$(\sigma \bullet \eta_1)/\eta_1 = \frac{-s_1 s_2 + 1}{s_1 s_2} \left(\frac{s_2-1}{s_2} \right)^{\alpha_4} \left(\frac{-1}{s_1 s_2} \right)^{\alpha_1} (-s_1 s_2 + 1)^{\alpha_5} \left(\frac{s_1 s_2 - 1}{s_1 s_2} \right)^{\alpha_3}$$

for $\sigma = [5, 3, 2, 4, 1]$.

```
[[0,(-s1*s2+1)/(s1*s2)],[4,(s2-1)/(s2)],[1,(-1)/(s1*s2)],[5,-s1*s2+1],
 [3,(s1*s2-1)/(s1*s2)]]
[[0,((s1^2-s1)*s2-s1+1)/(s1*s2-s1)],[4,-s1+1],[1,(s1-1)/(s1*s2-s1)],
 [5,-s1*s2+1],[3,((-s1^2+s1)*s2+s1-1)/(s1*s2-s1)]]
[[0,-s1*s2+1],[4,-s1+1],[1,-1],[5,-s1*s2+1],[3,s1*s2-1]]
[5,3,2,4,1]

[[0,(1)/(s1*s2)],[4,(s2-1)/(s2)],[3,(s1*s2-1)/(s1*s2)],[1,(-1)/(s1*s2)]]
[[0,(1)/(s1*s2-s1)],[3,(s1*s2-1)/(s1*s2-s1)],[1,(-1)/(s1*s2-s1)]]
[[0,(1)/(s1*s2-1)],[1,(-1)/(s1*s2-1)]]
[5,1,2,4,3]
```

```
[[0,s1*s2-1],[1,-1],[4,-s1+1],[3,s1*s2-1],[5,-s1*s2+1]]
[[0,((-s1^2+s1)*s2+s1-1)/(s1*s2-s1)],[1,(s1-1)/(s1*s2-s1)],[4,-s1+1],
 [3,((-s1^2+s1)*s2+s1-1)/(s1*s2-s1)],[5,-s1*s2+1]]
[[0,(s1*s2-1)/(s1*s2)],[1,(-1)/(s1*s2)],[4,(s2-1)/(s2)],
 [3,(s1*s2-1)/(s1*s2)],[5,-s1*s2+1]]
[3,5,1,4,2]

[[0,-1],[1,-1],[4,-s1+1],[5,-s1*s2+1],[3,-1]]
[[0,(-s1+1)/(s1)],[1,(-s1+1)/(s1)],[4,-s1+1],[5,-s1*s2+1],
 [3,(-s1+1)/(s1)]]
[[0,(-s1*s2+1)/(s1*s2)],[1,(-s1*s2+1)/(s1*s2)],[4,(-s1*s2+1)/(s2)],
 [5,-s1*s2+1],[3,(-s1*s2+1)/(s1*s2)]]
[3,2,1,4,5]

[[0,(s1*s2-1)/(s1*s2)],[5,-s1*s2+1],[4,(-s1*s2+1)/(s2)],
 [1,(-s1*s2+1)/(s1*s2)],[3,(-s1*s2+1)/(s1*s2)]]
[[0,(s1-1)/(s1)],[5,-s1*s2+1],[4,-s1+1],[1,(-s1+1)/(s1)],
 [3,(-s1+1)/(s1)]]
[[5,-s1*s2+1],[4,-s1+1],[1,-1],[3,-1]]
[2,3,5,4,1]

[[0,(-1)/(s1*s2)],[4,(1)/(s2)],[3,(1)/(s1*s2)],[1,(1)/(s1*s2)]]
[[0,(-1)/(s1)],[3,(1)/(s1)],[1,(1)/(s1)]]
[[0,-1]]
[2,1,5,4,3]

[[0,(-1)/(s1*s2-1)],[1,(-1)/(s1*s2-1)]]
[[0,(-1)/(s1*s2-s1)],[3,(s1*s2-1)/(s1*s2-s1)],[1,(-1)/(s1*s2-s1)]]
[[0,(-1)/(s1*s2)],[3,(s1*s2-1)/(s1*s2)],[4,(s2-1)/(s2)],[1,(-1)/(s1*s2)]]
[1,5,3,4,2]

[]
[[0,(1)/(s1)],[3,(1)/(s1)],[1,(1)/(s1)]]
[[0,(1)/(s1*s2)],[3,(1)/(s1*s2)],[4,(1)/(s2)],[1,(1)/(s1*s2)]]
[1,2,3,4,5]
```

From this table, we obtain one of Kummer type formulas for the Appell function F_1 firstly studied by Vavasseure [2]

$$(1 - s_1 s_2)(1 - s_1 s_2)^{\alpha_3}(1 - s_1)^{\alpha_4}(1 - s_1 s_2)^{\alpha_5}$$
$$\times F_1\left(\alpha_3 + 1, -\alpha_2, -\alpha_4, \alpha_1 + \alpha_3 + 2; s_1 s_2, \frac{s_1(1 - s_2)}{s_1 - 1}\right)$$
$$= F_1\left(\alpha_1 + 1, -\alpha_4, -\alpha_5, \alpha_3 + \alpha_1 + 2; s_1, s_1 s_2\right)$$

standing for $\sigma = [5, 3, 2, 4, 1]$ and $(\sigma \bullet \eta_3)/\eta_1$.

We have sketched a method to generate Kummer type formulas. However, it is not the end of the story and the following problems are arising.

1. Improve efficiency. How far can we derive Kummer type formulas of $E_{k,n}$ for large k and n? Can we find an interesting structure?

2. We need an efficient method to find a "small" compatible basis.
3. We will obtain a lot of Kummer type identities. However, we have no method to classify them by a suitable mathematical meaning.
4. Building a system to study and generate formulas of hypergeometric functions by integrating algebra and geometry systems (http://www.openxm. org).
5. How to store generated formulas and proofs in an electronic or digital formula book on hypergeometric functions[6], [22], [19]?

As to the last problem, OpenMath approaches seem to be promising. Readers are invited to articles by A. Cohen, H. Cuypers, E. Reinaldo Barreiro, H. Sterk [3], and A. Solomon [17] in this book on OpenMath activities.

References

1. K. Aomoto and M. Kita, *Theory of Hypergeometric Functions*, (in Japanese) Springer-Tokyo, 1994.
2. P. Appell, J. Kampé de Fériet, *Fonctions Hypergéometrique et Hypersphériques – Polynomes d'Hermite*. Gauthier-Villars, 1926, Paris.
3. A. Cohen, H. Cuypers, E. Reinaldo Barreiro, and H. Sterk, Interactive Mathematical Documents on Web, this volume, pages 289–307.
4. C. D'Andrea and I.Z. Emiris, Sparse Resultant Perturbations, this volume, pages 93–107.
5. J. Davenport, The Geometry of $\mathbb{C}^n$ is Important for the Algebra of Elementary Functions, this volume, pages 207–224.
6. Digital Library of Mathematical Functions (Digital Abramowitz and Stegun) http://dlmf.nist.gov/
7. I.Z. Emiris, Discrete Geometry for Algebraic Elimination, this volume, pages 77–91.
8. É. Goursat, Sur L'Équation Différentielle Linéaire qui Admet pour Intégrale la Série Hypergéométrique, Annales Scientifique l'École Normal Supérieure, Series II, Vol. 10, (1881) 3–142.
9. I.M. Gel'fand, General theory of hypergeometric functions. Soviet Mathematics Doklady **33**, (1986), 573–577.
10. I.M. Gel'fand, M.M. Kapranov, and A.V. Zelevinsky, Hypergeometric functions and toral manifolds. Functional Analysis and its Applications **23**, (1989), 94–106.
11. M. Noro, A Computer Algebra System Risa/Asir, this volume, pages 147–162.
12. J. Pfeifle and J. Rambau, Computing Triangulations Using Oriented Matroids, this volume, pages 49–75.
13. M. Saito, B. Sturmfels, and N.Takayama, *Gröbner Deformations of Hypergeometric Differential Equations*, Algorithms and Computation in Mathematics **6**, Springer, 2000.
14. M. Saito and N. Takayama, Restrictions of A-hypergeometric systems and connection formulas of the $\Delta_1 \times \Delta_{n-1}$-hypergeometric function, International Journal of Mathematics **5** (1994), 537–560.
15. H. Schönemann, SINGULAR in a Framework for Polynomial Computations, this volume, pages 163–176.

16. J. Sekiguchi and N. Takayama, Compactifications of the configuration space of 6 points of the projective plane and fundamental solutions of the hypergeometric system of type $(3, 6)$, Tohoku Mathematical Journal **49** (1997), 379–413.

17. A. Solomon, Distributed Computing for Conglomerate Mathematical Systems, this volume, pages 309–325.

18. B. Sturmfels, *Algorithms in Invariant Theory*, Texts and Monographs in Symbolic Computation. Springer-Verlag, Vienna, 1993.

19. Y. Tamura, The design and implementation of an interactive formula book for generalized hypergeometric functions, Thesis, Kobe University, 2002, in preparation.

20. K. Ueno, Hypergeometric series formulas generated by the Chu-Vandermonde convolution. Memoirs of the Faculty of Science. Kyushu University. Series A. Mathematics 44 (1990), 11–26.

21. D.M. Wang, Automated Generation of Diagrams with Maple and Java, this volume, pages 277–288.

22. Wolfram Research's Mathematical Functions, http://functions.wolfram.com

23. M. Yoshida, *Hypergeometric functions, my love*. Modular interpretations of configuration spaces. Aspects of Mathematics, Vieweg, 1997

A Computer Algebra System: Risa/Asir

Masayuki Noro

Kobe University, Rokko, Kobe 657-8501, Japan
noro@math.kobe-u.ac.jp

Abstract. Risa/Asir [20] is a computer algebra system developed for efficient algebraic computation. It is also designed to be a platform for parallel distributed computation under the OpenXM (Open Message eXchange for Mathematics) protocol [11]. OpenXM defines client-server communication between mathematical software systems and Risa/Asir is one of main components in the OpenXM package [21]. The source code of Risa/Asir is completely open and it is easy to modify the source code or to add new functions. In the present paper, we explain an overview of Risa/Asir, its functions and implemented algorithms with their performances, the OpenXM API and the way to add built-in functions.

1 What is Risa/Asir?

Risa/Asir [20] is a tool for performing various computations in mathematics and engineering applications. The development of Risa/Asir started in 1989 at FUJITSU. Binaries have been freely available since 1994, and now the source code is also available at no cost. Currently, Kobe distribution is the most active branch in the development of Risa/Asir. Risa/Asir is characterized as follows:

1. An environment for large scale and efficient polynomial computation.
   Risa/Asir consists of the Risa engine for performing operations on mathematical objects and an interpreter for programs written in the Asir user language. In Risa/Asir, polynomials are represented in two different internal forms: the recursive representation and the distributed representation. Polynomial factorization and GCD are based on the former representation, and computations related to the Gröbner basis are based on the latter representation. The field of rationals, algebraic number fields and finite fields are available as ground fields of polynomial rings.
2. A platform for parallel and distributed computation.
   In order to combine mathematical software systems, we previously proposed the OpenXM (Open Message eXchange for Mathematics) protocol [11]. Risa/Asir acts as both an OpenXM client and an OpenXM server. The Risa/Asir OpenXM client API provides a way to call functions in external OpenXM servers. Using multiple Risa/Asir OpenXM servers, one can perform parallel and distributed computation in an attempt to achieve linear speedup. Conversely, other OpenXM clients can call functions in the Risa/Asir server. In addition, all of the functions of the

Risa/Asir server are available by linking Risa/Asir subroutine library. We have made various mathematical software systems OpenXM servers or OpenXM clients, which are distributed as the OpenXM package [21].

3. Open source software.

The source code of Risa/Asir is completely open, and algorithms and implementations can be verified if necessary. The addition of new codes is simple. Japanese version of manual describing the internal structures is found in the OpenXM package.

In the present paper, we explain the structure of Risa/Asir, functions and algorithms implemented in Risa/Asir, and how to add functions to Risa/Asir. We refer to the related source files in the source tree of Risa/Asir when necessary. Pathnames are relative to the root directory `asir2000`. The information presented herein is based on the version 20020720 of Risa/Asir.

2   Risa Objects

Risa/Asir is composed of the Risa engine and the Asir parser/interpreter. The former deals with internal Risa objects and the latter converts human readable forms into Risa objects. Risa objects represent various mathematical objects and there is a hierarchy among them according to their mathematical structures.

2.1   Risa/Asir Object Hierarchy

The source code of the Risa engine is contained in the directory `engine`. Various mathematical objects are defined as C structures in `include/ca.h` and these objects have the common header:

```
typedef struct oObj {
  short id;
  short pad;
} *Obj;
```

An integer referred to as the object identifier is assigned to each object category and the field `id` indicates the object identifier. Table 1 shows examples of object identifiers. Each object category is associated with a set of fundamental arithmetic operations. These are addition, subtraction, multiplication, division, power, change of sign and comparison, and these arithmetic operations are registered to a function table in `parse/arith.c`. Arithmetic functions accept multiple types of objects as arguments, and the correct operations are determined according to the identifiers of the arguments. In order to reduce the cost of classification of arguments, we apply the following principles.

Table 1. Object identifiers

Category	Id
zero	0
number	1
recursive polynomial	2
rational function	3
list	4
vector	5
matrix	6
string	7
structure	8
distributed polynomial	9

1. We assume number $\subset$ recursive polynomial $\subset$ rational function and we treat these objects as scalar objects. In addition, we assume that the unique zero is a member of all object categories containing zero. A scalar operation belonging to a category should accept scalar objects belonging to lower categories.

2. If the result of an operation belongs to a lower category, then the result should be an object belonging to that category. For example, if a sum of two polynomials is a number, then a number object is generated and returned.

3. Binary operations in categories other than scalar only must accept arguments in that category. If scalar multiplication is admissible in a category, then the multiplication function should accept a scalar as an argument.

The number itself is divided into several sub categories: the field of rationals based on bignum, the field of real numbers based on floating point numbers, the complex number field, algebraic number fields and finite fields having several different representations. A number has its number identifier in the field **pad**. Table 1 shows the correspondence, whereby each number category is associated with a set of fundamental arithmetic operations, which are registered to a function table in `engine/init.c`. A function belonging to a category may accept an argument in another number category. For example, if the denominator of a rational number is not divisible by a prime p, then the rational number is regarded as a member of $GF(p)$ in the canonical way, and operations dealing with finite fields naturally accept such a rational number as an argument. This feature is implementation-dependent and the source code must be consulted in order to determine the exact behavior of an arithmetic function.

Table 2. Number identifiers

Category	Id
rational	0
alg. num.	1
real	2
bigfloat	3
complex	4
finite field	5
large finite field	6

2.2 Memory Management

Due to difficulties in memory management, non-developers often experience difficulty in modifying software written in C. Risa/Asir employs Boehm's conservative GC (Garbage Collection) package [2] as the memory allocator/manager. The source code of GC is in `gc5.3`. Under this GC, memory can be allocated by calling `GC_malloc()`. Memory need not be freed explicitly because memory is reclaimed automatically. Although such a method may come at the price of a certain degree of overhead, memory leaks and serious bugs due to inappropriate deallocation are eliminated, and the source code can be modified easily and safely by non-developers.

3   Functions in Risa/Asir

Risa/Asir provides various functions on algebraic computation. Among them, we explain the implementations of two main functions: polynomial factorization and Gröbner basis computation over various ground fields. We present several timing data to show their performances. The measurement were made on a PC with PentiumIII 1GHz and 2 GB of main memory. Times are given in seconds.

3.1   Operations over Fields of Characteristic 0

Polynomial arithmetics (`engine/P.c`), polynomial GCD and factorization (`engine/{F.c, H.c, D.c, NEZ.c, E.c}`) are implemented over our own implementation of integer and rational number arithmetics (`engine/{N.c,Q.c}`, `asm/ddN.c`). Most algorithms for multivariate polynomial factorization are recursive, that is, in these algorithms a polynomial is factorized using information on a polynomial having fewer variables, obtained by an evaluation at some point. The recursive representation of a polynomial is suitable for such algorithms.

```
typedef struct oP {   /* polynomial */
   short id;             /* object id */
   short pad;
   V v;                /* main variable */
   struct oDCP *dc; /* leading term */
} *P;

typedef struct oDCP {        /* term */
   Q d;                      /* degree */
   P c;                    /* coefficient */
   struct oDCP *next;  /* next term */
} *DCP;
```

The actual implementation of multivariate factorization is very complicated because square-free factorization, GCD and factorization are called recursively from each other. One key to efficient implementation of multivariate factorization is the estimation of the leading coefficients of factors. If the estimate overshoots, then the number of steps of Hensel lifting becomes large, which is known as the leading coefficient problem. In [18], Wang proposed a method in which an attempt is made to determine completely the leading coefficient of a factor via an integer substitution. Although we implemented a simplified version of this method, which estimates the upper bound of the multiplicity of each irreducible factor in the leading coefficient of a factor, the leading coefficient problem can be avoided in many cases. Table 3 lists the times required to factorize polynomials which tend to cause the leading coefficient problem. Here, $W_{i,j,k}$ is equal to $Wang[i] \cdot Wang[j] \cdot Wang[k]$, where $Wang[i]$ is defined in `asir2000/lib/fctrdata`. Risa/Asir clearly shows better performance than well-known systems such as Maple and Mathematica.

	$W_{1,2,3}$	$W_{4,5,6}$	$W_{7,8,9}$	$W_{10,11,12}$	$W_{13,14,15}$
Asir	0.2	4.7	14	17	0.4
Mathematica 4	0.2	16	23	36	1.1
Maple 7	0.5	18	967	48	1.3

Table 3. Multivariate factorization (in seconds)

Risa/Asir provides a univariate factorization over algebraic number fields (`lib/sp`). An algebraic number is a rational expression of a root of an irreducible univariate polynomial $f(t)$. Here, the coefficients of $f(t)$ themselves may be algebraic numbers. That is, an algebraic number field is in general represented by a successive extension. As an application, the splitting field computation of a univariate polynomial is implemented. The following example shows a computation of the splitting field of a polynomial.

```
[0] load("sp")$
[151] sp(x^6+6*x^4+2*x^3+9*x^2+6*x-4);
[[x+(-#2),x+(#2+#0),
  20*x+(((-2*#0^5-8*#0^3-2*#0^2-6*#0-2)*#1^2+(#0^5+4*#0^3+6*#0^2
  +3*#0+6)*#1-4*#0^5-16*#0^3-4*#0^2-12*#0-4)*#2+(2*#0^4+#0^3
  +6*#0^2+5*#0-4)*#1^2+(-#0^4+2*#0^3-3*#0^2+12)*#1+4*#0^4+2*#0^3
  +12*#0^2+10*#0-8),
  20*x+(((2*#0^5+8*#0^3+2*#0^2+6*#0+2)*#1^2+(-#0^5-4*#0^3-6*#0^2
  -3*#0-6)*#1+4*#0^5+16*#0^3+4*#0^2+12*#0+4)*#2+(-2*#0^4-#0^3
  -6*#0^2-5*#0+4)*#1^2+(#0^4-2*#0^3+3*#0^2+8)*#1-4*#0^4-2*#0^3
  -12*#0^2-10*#0+8),
  x+(-#1),x+(-#0)],
 [[(#2),t#2^2+t#0*t#2+t#0^2+3],
  [(#1),t#1^3+3*t#1+t#0^3+3*t#0+2],
  [(#0),t#0^6+6*t#0^4+2*t#0^3+9*t#0^2+6*t#0-4]]]
```

The result is a list consisting of two elements. The first is a list of linear factors of the input polynomial over an algebraic number field. The second represents the field as a successive extension, which is interpreted as follows:

$$F_0 = \mathbb{Q}(\alpha_0), \quad \alpha_0^6 + 6\alpha_0^4 + 2\alpha_0^3 + 9\alpha_0^2 + 6\alpha_0 - 4 = 0$$
$$F_1 = F_0(\alpha_1), \quad \alpha_1^3 + 3\alpha_1 + \alpha_0^3 + 3\alpha_0 + 2 = 0$$
$$F_2 = F_1(\alpha_2), \quad \alpha_2^2 + \alpha_0\alpha_2 + \alpha_0^2 + 3 = 0$$

This algorithm is an improvement [1,9] of the algorithm proposed by Trager [16]. In the present algorithm, we utilize non square-free norms to obtain partial (not necessarily irreducible) factorization. The algorithm finally requires univariate factorization of a polynomial over the rationals, however this factorization is often difficult because the polynomial contains many spurious modular factors. At present, Risa/Asir is not optimized to factor polynomials of this type. However, in this case, a factorization in the PARI library, which implements the knapsack factorization algorithm [7], can be used, and the splitting field can be computed efficiently even in such cases.

3.2   Computation over Finite Fields

Finite fields are useful in avoiding unnecessary coefficient swells in various computations over the rationals. Computation over finite fields is essential in cryptography, coding theory and algebraic geometry over fields of positive characteristic. In Risa/Asir finite fields are represented in several ways.

1. $GF(p)$ (p: a prime $p < 2^{29}$; engine/mi.c)
   Elements are represented by machine integers.
2. $GF(p)$ (p: an arbitrary prime; engine/lmi.c)
   Elements are represented by bignums.

3. $GF(2^n)$ (n: small; `engine/gf2n.c`)

Elements are represented by bit sequences, where each bit represents a coefficient of a univariate polynomial over $GF(2)$. Arithmetics are performed modulo an irreducible polynomial of degree n.

4. $GF(p^n)$ (p: an arbitrary prime, n: small integer; `engine/gfpn.c`)

Elements are represented by a univariate polynomial over $GF(p)$ and arithmetics are performed modulo an irreducible polynomial of degree n.

5. $GF(p^s)$ (p: a small prime, $p^s < 2^{16}$; `engine/gfs.c`)

Elements are represented by a primitive root. That is, $GF(p^s)^{\times}$ is regarded as $\{1, \alpha, \ldots, \alpha^{p^s-2}\}$ for some α, and α^i is represented by its index i. Addition and subtraction are performed using a table.

The type 1 representation is used internally in various modular algorithms. The type 2, 3 and 4 representations are general purpose representations. The univariate factorizer and a function to count rational points on elliptic curves operate under these representations. Multiplication in $GF(2^n)$ is implemented using Karatsuba algorithm extensively (`engine/up2.c`). The type 5 representation has been introduced for efficient implementation of multivariate factorization over finite fields of small characteristic. When we attempt to factor a polynomial over such a field, we often have insufficient evaluation points for reducing the factorization to a case of fewer variables. In this case we must extend the ground field. If the ground field is represented by a type 1 or 2 representation, the efficiency of field arithmetics is decreased by the field extension. However, if the ground field and its extension are both represented by a type 5 representation, we can expect that field arithmetics are performed efficiently in both fields. In [13], a new polynomial time algorithm to factor bivariate polynomials over a finite field is presented. A combination of the Berlekamp-Hensel method and the new algorithm under a type 5 representation enables us to efficiently factor a certain class of polynomials, including hard-to-factor polynomials.

3.3   Gröbner Basis Computation

We started to develop functions related to Gröbner basis in 1993. In order to achieve practically efficient implementation, we have incorporated various new algorithms. For the Buchberger algorithm, we have implemented criteria by which to detect unnecessary pairs as reported by Gebauer-Moeller [5], the sugar strategy [6], trace algorithms by modular computation [17], stabilization of strategy by combining homogenization and the trace algorithm, and efficient content reduction during normal form computation [10]. Furthermore, we implemented several types of change of ordering algorithms. Among these, `tolex()` implements a modular FGLM algorithm (`lib/gr`), which avoids intermediate coefficient swells during the ordinary FGLM algorithm [3] and realizes efficient computation of lexicographic Gröbner bases for zero-dimensional ideals [12]. In Gröbner basis computation, a polynomial is

regarded as a sum of monomials. Toward that end, an internal representation
called distributed representation is defined.

```
typedef struct oMP {    /* monomial */
  struct oDL *dl;        /*exponent */
  P c;                   /* coefficient */
  struct oMP *next;  /* next monom. */
} *MP;


typedef struct oDL {    /* exponent */
  int td;                /* total degree */
  int d[1];              /* exponent vector */
} *DL;
```

Table 4. Gröbner basis computation over $\mathbb{Q}$ (in seconds)

	C_6	C_7	C_8	K_7	K_8	McKay
Risa/Asir	1.6	389	54000	35	351	34950
Maple7	629	—	—	> 3h	—	—
Mathematica4	3917	—	—	> 3h	—	—
Singular 2.0	0.8	15247	—	7.6	79	> 20h

Table 4 shows the timing data for Gröbner basis computation. C_n is the
cyclic n system, and K_n is the Katsura n system, both of which are famous
bench mark problems [19]. McKay [10] is a system for which the Gröbner basis
is difficult to compute over $\mathbb{Q}$. The symbol '—' indicates it was not measured.
Risa/Asir is faster than Maple and Mathematica, and is often faster than
Singular, which is an optimized system for Gröbner basis computation.

When we execute the Buchberger algorithm via Risa/Asir, several days
are often required in order to obtain a Gröbner basis. If a system is down or
memory exhaustion occurs during execution, or if the machine must simply
be shutdown for any reason, then all intermediate results must be abandoned.
In such cases, all intermediate basis elements can be placed in a directory.

```
% mkdir nfdir
% asir
...
[0] dp_gr_flags(["Demand","./nfdir/","Print",1])$
[1] F = [...]$
[2] V = [...]$
[3] G=dp_gr_main(F,V,0,1,0)$
... /* The first n elements has been computed */
interrupt ?(q/t/c/d/u/w/?) q
```

Abort this session? (y or n) y
% asir
...
```
[0] dp_gr_flags(["Demand","./nfdir/","Print",1])$
[1] F = [...]$ V = [...]$ /* the same inputs as before */
[3] G=dp_gr_main(F,V,0,1,0)$
... /* The first n elements are read from a disk */
... /* The (n+1)-th elements are actually computed */
```

If we supply an option Demand with a directory name as its argument, generated basis elements are placed in the directory. Although this requires additional cost in order to read the basis elements required for normal form computations, the total amount of memory is smaller than the case in which all basis elements are placed in memory, which may enable a very large computation, generating numerous intermediate basis elements. When a computation is interrupted, the saved basis elements remain in the directory. If the computation is restarted using the same arguments and a trace algorithm, then the n-th normal form is first computed over a finite field. If the modular normal form is not equal to zero and there exists a file n in the directory, then this file is regarded as the n-th normal form over the rationals, without actually performing computation. This reduces the cost of re-computation required when restarting.

As an application of Gröbner basis computation and polynomial factorization, we implemented primary ideal decomposition and prime decomposition of the radical of an ideal [15]. These functions can be used to decompose solutions of systems of algebraic equations into irreducible components. The following illustrates prime ideal decomposition of the radical of an ideal generated by a polynomial set F.

```
[0] load("primdec")$
[176] F = [(2*t*y-2)*x+z*y^2-z,-z*x^3+(4*t*y+4)*x^2+(4*z*y^2+4*z)
*x+2*t*y^3-10*y^2-10*t*y+2,(t^2-1)*x+2*t*z*y-2*z,
(-z^3+(4*t^2+4)*z)*x+(4*t*z^2+2*t^3-10*t)*y+4*z^2-10*t^2+2]$
[177] V = [x,y,z,t]$
[178] primedec(F,V);
[[t+1,z,y-1,x],[t+1,y+1,x-z,z^4-16*z^2+16],[t+1,z-2,y+1,x+2],
[t+1,y+1,x+z,z^2+4],[t+1,z+2,y+1,x-2],[t-1,z,y+1,x],
[t-1,y-1,x-z,z^4-16*z^2+16],[t-1,z-2,y-1,x+2],
[t-1,y-1,x+z,z^2+4],[t-1,z+2,y-1,x-2],[3*t^2+1,z+2*t,y+t,x-2*t],
[3*t^2+1,z-2*t,y+t,x+2*t],[t^2+3,z+2,y+t,x+2],
[t^2+3,z-2,y+t,x-2],[t^4-10*t^2+1,z,y-t,x],[t^2+1,z,y+t,x]]
```

Recently, we implemented fundamental arithmetics, the Buchberger algorithm and the minimal polynomial computation over Weyl algebra, the ring of differential operators having polynomial coefficients. Using these arithmetics we implemented an efficient algorithm for computing the b-function of a polynomial (lib/bfct). Details can be found in [14].

4 Parallel and Distributed Computation

Risa/Asir is designed to be a platform for parallel distributed computation. To realize such computation, Risa/Asir provides OpenXM client and server API as well as C language API.

4.1 OpenXM

Risa/Asir provides OpenXM API for executing funtions on other mathematical software. In OpenXM, mathematical software systems are wrapped as server stack machines. As an OpenXM client, Risa/Asir dispatches a request to a server and receives the result. Inputs and outputs are represented as CMO (Common Mathematical Object format) objects. The set of stack machine commands also contains various control operations, such as server invocation, termination, interruption and restarting. Usually, each mathematical software system has its own user language. OpenXM provides stack machine commands for executing a command string and receiving a result as a human readable character string, thus wrapping a mathematical software system as an OpenXM server is relatively easy. OpenXM protocol is completely open, and any program can implement the protocol. OpenXM defines interfaces for fundamental mathematical data types, and users are not required to have knowledge of the details of encoding and decoding data. Furthermore, the byte order for communication is defined dynamically at negotiation, which improves the efficiency when the amount of data to be communicated is large [11].

When executing a program for parallel and distributed computation, a server may often freeze due to a bug. In such cases, the server must at least be reset. OpenXM specifies a procedure for robust interruption and restarting of execution, which enables clients to be reset safely from any state [11].

Mathlink provides similar functions for communication. Table 5 lists various specifications of OpenXM and Mathlink.

Table 5. OpenXM and Mathlink

	OpenXM	Mathlink
protocol	open	not open
generality	no restriction	at least one endpoint is Mathematica
data types	various mathematical data	only primitive data (machine integer etc.)
byte order	optimized	should be maintained by programs
resetting	supported	not supported

4.2 OpenXM C Language API in `libasir.a`

A typical OpenXM communication channel is the TCP/IP stream and the OpenXM asir server `ox_asir` is accessible via TCP/IP. However, communication between an OpenXM server and a client can be realized by other methods: files, MPI, PVM, RPC and linking a subroutine library. The Risa/Asir subroutine library `libasir.a` contains functions simulating the stack machine commands supported in `ox_asir`, the OpenXM asir server. By linking `libasir.a`, an application can use the functions as in `ox_asir` without accessing the server via TCP/IP. In order to make full use of this interface, conversion functions between CMO and the data structures appropriate to the application itself must be prepared. However, if the application linking `libasir.a` can also parse human readable outputs, the function `asir_ox_pop_string()` will be sufficient for the reception of results unless efficiency is required. The OpenXM package contains the file `OpenXM/doc/oxlib/test.c`, which is an example of the application of these library functions. In the program, `asir_ox_execute_string()` executes an Asir command line string and the result is pushed onto the stack as a CMO data, then `asir_ox_pop_string()` pops the result from the stack and converts the result to a human readable form.

4.3 Integration of Risa/Asir and Other OpenXM Servers

Asir OpenXM contrib (asir-contrib) is a collection of wrappers of functions in external OpenXM servers. Using asir-contrib, an external function can be called from Asir without knowing that the function is located in an external server. Currently, the following functions are provided in OpenXM: set, combinatorial and matrix operations, pretty printing of formula, Gröbner basis computation, elimination, computation of the Hilbert function, OpenMath CD compliant functions, and interfaces to gunuplot, Mathematica, PHC, sm1, and TiGERS. Sm1 provides functions for computing polynomial or series solutions of systems of partial differential equations as well as general polynomial operations. Through asir-contrib, we are attempting to assign common names to standard mathematical functions. By calling a standard function using a common name, an appropriate server is called and the details of the specifications of actual functions in the servers need not be known.

5 Extending Risa/Asir

The simplest way to add a function in Risa/Asir is to write a program in Asir, which is the user language for Risa. However, programs written in Asir are executed by the interpreter and are less efficient than C-coded functions. Risa/Asir is open source and built-in functions are easy to add. Since we are free from explicit memory management, we need only know the internal data formats of objects, how to generate objects, possible operations on objects, and how to wrap written functions as built-in functions.

158   Masayuki Noro

5.1   Definitions, Generations and Operations of Objects

Each mathematical object is defined as a C structure in `include/ca.h`. The
header file also defines C macros for generating objects, each of which is
conventionally defined as NEW+the structure name. Fundamental operations
on such objects are found in `parse/arith.c`. For example, the following
function computes the partial derivative of a multivariate polynomial with
respect to a variable.

```
void partial_derivative(VL vl,P p,V v,P *r)
{
  P t;
  DCP dc,dcr,dcr0;

  if ( !p || NUM(p) ) *r = 0;
  else if ( v == p->v ) {
    for ( dc = p->dc, dcr0 = 0; dc && dc->d; dc = dc->next ) {
      mulp(vl,dc->c,(P)dc->d,&t);
      if ( t ) {
        NEXTDC(dcr0,dcr); subq(dc->d,ONE,&dcr->d); dcr->c = t;
      }
    }
    if ( !dcr0 ) *r = 0;
    else { dcr->next = 0; MKP(v,dcr0,*r); }
  } else {
    for ( dc = p->dc, dcr0 = 0; dc; dc = dc->next ) {
      partial_derivative(vl,dc->c,v,&t);
      if ( t ) { NEXTDC(dcr0,dcr); dcr->d = dc->d; dcr->c = t; }
    }
    if ( !dcr0 ) *r = 0;
    else { dcr->next = 0; MKP(p->v,dcr0,*r); }
  }
}
```

The inputs of `partial_derivative` are a recursive polynomial p, a variable
v and a list of variables vl, which should be compatible with p. The function
computes the partial derivative and writes the result at the position indicated by r. If a polynomial is passed by the interpreter, a global variable
CO (the current variable order list) is compatible with the polynomial, thus
we can supply CO as the variable list vl. In the program, `mulp()` computes
the product of two polynomials and `subq()` computes the difference of two
rational numbers. Table 6 explains macro operations. By using macros such
as NEXTDC and MKP, the programmer need not allocate memory explicitly.

5.2   Adding Built-in Functions

Built-in functions are defined as follows:

Table 6. Macro operations

NUM(p)	returns 1 if p is a number
NEXTDC(dc0,dc)	appends a nil term to a term list dc0
ONE	1
MKP(v,dc,p)	generates a polynomial p with the variable v and the term list dc

```
typedef struct oNODE {
  pointer body;
  struct oNODE *next;
} *NODE;

void a_builtin_function(NODE arg, Obj *resultp)
{
  P arg0,arg1,result;
  arg0 = (P)arg->body; arg1 = (P)arg->next->body;
  ...
  NEWP(result);
  ...
  *resultp = (Obj)result;
}
```

Here, NODE is a data structure for representing lists and is used as a list of arguments in the above function. The pointers of arguments are held in the arugment list, and resultp points to the location in which the result pointer is to be written. By registering a builtin function to a function table, the function can be executed in Asir. Function tables are placed in files in builtin. For example builtin/poly.c contains the following function table:

```
void Phomogeneous(), Phomogeneous_part(),  Preorder();

struct ftab poly_tab[] = {
  {"homogeneous_deg",Phomogeneous_deg,-2},
  {"homogeneous_part",Phomogeneous_part,-3},
  {"reorder",Preorder,3},
  ...
```

Each entry has the form {function_name,function_pointer,n_args}. For instance {"reorder",Preorder,3} indicates that Preorder() can be called with three arguments by the name reorder in Asir. A negative n_args indicates that the number of arguments may vary from 0 to -n_args. In order for users to add builtin functions, the file builtin/user.c is prepared. Built-in functions can be added by adding necessary functions, prototype declarations and a function table entry to the file. The following example shows how to add partial_derivative as a built-in function.

```
void Ppartial_derivative();
void partial_derivative(VL vl,P p,V v,P *r);
struct ftab user_tab[] = {
  {"partial_derivative",Ppartial_derivative,2},
  ...
};
void Ppartial_derivative(NODE arg,P *rp)
{
  asir_assert(ARG0(arg),O_P,"partial_derivative");
  asir_assert(ARG1(arg),O_P,"partial_derivative");
  partial_derivative(CO,(P)ARG0(arg),((P)ARG1(arg))->v,rp);
}
```

6 Future Work

We have described the current status of the arts of Risa/Asir. Toward a better practically efficient computer algebra system, further improvements should be done for the following:

- Number arithmetics

  The number arithmetics often dominate the practical performance in, for example, solving systems of linear equations appearing in polynomial computation. Our implementation of integer arithmetics is not optimal. For instance our experiment shows that integer multiplication in GNU MP Library [22] is faster (1.7 times for 10^3 bits integer, 7 times for 10^6 bits integer) than that in Risa/Asir.

- Gröbner basis computation

  A new generation algorithm F_4 [4] has been implemented experimentally (builtin/gr.c). The performance of our implementation is slower than the announced performance data by Faugère. However, we believe that the F_4 is a promising algorithm; for instance, the system of algebraic equations McKay can be solved very efficiently by our F_4 implementation compared with the standard Buchberger algorithm with modular techniques. We are still attempting to improve our own implementation, but we are stuck. We hope that the details of his implementation will be published in the future.

- Inter-server communication

  The current specification of OpenXM does not provide collective operations such as MPI_Bcast and MPI_Reduce in MPI [23]. To realize efficient collective operations, inter-server communication is necessary. We already implemented it experimentally over MPI and obtained a preferable result [11]. The specification for inter-server communication will be proposed as OpenXM-RFC 102, which will enable us to gain a parallel speedup for wider range of applications.

References

1. Anai, H., Noro., M., Yokoyama, K. (1996): Computation of the splitting fields and the Galois group of polynomials, Progress in Mathematics 143, Birkhäuser Verlag, 1-15.
2. Boehm, H.,Weiser, M. (1988): Garbage Collection in an Uncooperative Environment, *Software Practice & Experience*, 807-820. http://www.hpl.hp.com/personal/Hans_Boehm/gc/.
3. Faugère, J.-C., Gianni, P., Lazard, D., Mora, T. (1993): Efficient computation of zero-dimensional Gröbner bases by change of ordering, *J. Symb. Comp.* **16**, 329-344.
4. Faugère, J.C. (1999): A new efficient algorithm for computing Gröbner bases (F_4), *Journal of Pure and Applied Algebra* **139** 1-3, 61-88.
5. Gebauer, R., Möller, H.M. (1989): On an installation of Buchberger's algorithm, *J. Symb. Comp.* **6/2/3**, 275-286.
6. Giovini, A., Mora, T., Niesi, G., Robbiano, L., Traverso, C. (1991): "One sugar cube, please" OR Selection strategies in the Buchberger algorithm, Proc. ISSAC'91, ACM Press, 49-54.
7. van Hoeij, M. (2000): Factoring polynomials and the knapsack problem, to appear in *Journal of Number Theory*.
8. Izu, T., Kogure, J., Noro, M., Yokoyama, K. (1998): Efficient implementation of Schoof's algorithm, ASIACRYPT'98, LNCS **1514**, Springer-Verlag, 66-79.
9. Noro, M., Takeshima, T. (1992): Risa/Asir – a computer algebra system, Proc. ISSAC'92, ACM Press, 387-396.
10. Noro, M., McKay, J. (1997): Computation of replicable functions on Risa/Asir, Proc. PASCO'97, ACM Press, 130-138.
11. Maekawa, M., Noro, M., Ohara, K., Takayama, N., Tamura, Y. (2001): The Design and Implementation of OpenXM-RFC 100 and 101, Computer Mathematics, Proc. the Fifth Asian Symposium (ASCM 2001), World Scientific, 102-111.
12. Noro, M., Yokoyama, K. (1999): A Modular Method to Compute the Rational Univariate Representation of Zero-Dimensional Ideals, *J. Symb. Comp.* **28/1**, 243-263.
13. Noro, M., Yokoyama, K. (2002): Yet Another Practical Implementation of Polynomial Factorization over Finite Fields, ISSAC 2002, ACM Press, 200-206.
14. Noro, M. (2002): An Efficient Modular Algorithm for Computing the Global b-function, Mathematical Software, Proc. International Congress of Mathematical Software (ICMS 2002), World Scientific, 147-157.
15. Shimoyama, T., Yokoyama, K. (1996): Localization and Primary Decomposition of Polynomial Ideals, J. Symb. Comp. **22**, 247-277.
16. Trager, B.M. (1976): Algebraic Factoring and Rational Function Integration, Proc. SYMSAC 76, 219-226.
17. Traverso, C. (1988): Gröbner trace algorithms. Proc. ISSAC'88, LNCS **358**, Springer-Verlag, 125-138.
18. Wang, P.S. (1978): An Improved Multivariate Polynomial Factoring Algorithm. Math. Comp. **32**, 1215-1231.
19. http://www.math.uic.edu/~jan/demo.html.
20. A computer algebra system Risa/Asir, http://www.math.kobe-u.ac.jp/Asir/asir.html.

21. OpenXM package, http://www.openxm.org/.
22. The GNU MP Library, http://swox.com/gmp/.
23. Message Passing Interface, http://cs.utk.edu/netsolve.

SINGULAR in a Framework for Polynomial Computations

Hans Schönemann

University of Kaiserslautern, Centre for Computer Algebra,
67653 Kaiserslautern, Germany

Abstract. There are obvious advantages in having fast, specialized tools in a framework for polynomial computations and flexible communication links between them. Techniques to achieve the first goal are presented with SINGULAR as test bed: several implementation techniques used in SINGULAR are discussed. Approaches to fast and flexible communication links are proposed in the second part.

1 Introduction

SINGULAR is a specialized computer algebra system for polynomial computations. The kernel implements a variety of Gröbner base–type algorithms (generalized Buchberger's algorithm, standard basis in local rings, in rings with mixed order, syzygy computations, ...), algorithms to compute free resolutions of ideals, combinatorial algorithms for computations of invariants from standard bases (vector space dimensions, –bases, Hilbert function, ...) and algorithms for numerical solving of polynomial systems.

All others task have to use external tools, which include C/C++–libraries and external programs.

In order to be able to compute non–trivial examples we need an efficient implementation of the Gröbner base–type algorithms as well as efficient communication links between independent packages. The method of Gröbner bases (GB) is undoubtedly one of the most important and prominent success stories of the field of computer algebra. The heart of the GB method are computations of Gröbner or standard bases with Gröbner base–type algorithms. Unfortunately, these algorithms have a worst case exponential time and space complexity and tend to tremendously long running times and consumption of huge amounts of memory

While algorithmic improvements led to more successful algorithms for many classes of problems, an efficient implementation is necessary for all classes of problems: analyzing GB computations from an implementation point of view leads to many interesting questions. For example: how should polynomials and monomial be represented and their operations implemented? What is the best way to implement coefficients? How should the memory management be realized?

2   Some Historical Remarks

Gröbner base–type algorithms are an important tool in many areas of computational mathematics. It is quite difficult to find out historical details in the area of computer algebra, so I apologize for missing important systems. Also it is often difficult to give a date for such an evolving target like a software system — I took the dates of corresponding articles.

Buchberger [7], and independently Hironaka [10], presented the algorithm in the 60's. First implementations arose in the 80's: Buchberger (1985) in MuLISP, Möller/Mora in REDUCE (1986) [11]. The underlying systems provided all the low level stuff like memory management, polynomial representations, coefficient arithmetics and so on. Algorithmic improvements at that time were the elimination of useless pairs, etc. First direct implementations (CoCoA, 1988; Macaulay, 1987 [5]; Singular, 1989, [13]) were restricted to coefficients in $\mathbb{Z}/p$ and did not care much about actual data representations.

In 1994, the system Macaulay 3.0 (maybe also earlier versions) used to enumerate all possible monomials by an integer, according to the monomial ordering. Monomial comparison was very fast, but the maximal possible degree was limited. The leading terms had additionally the exponent vector available for fast divisibility tests (for a detailed discussion see [2]).

PoSSo, a scientific research and development project 1993 – 1995, introduced a monomial representation which contained both, the exponent vector and the exponent vector multiplied by the order matrix (see below). That provided fast monomial operations, but used a "lot" of memory for each monomial.

Faugère's algorithm F_4 (1999) reused the idea of enumerating the monomials — instead of enumerate all he gave numbers only to the occurring ones (and for each degree independently).

SINGULAR 1.4 used C data types `char`, `short`, `int`, `long` to get an smaller monomial representation and to allow vectorized monomial operations (see below).

The article [15] gives interviews with developers of Gröbner engines and contains interesting stories on early developments of Gröbner engines - unfortunately it's written in Japanese.

3   Basic Polynomial Operations and Representations

As a first question, one might wonder which kind of polynomial (and, consequently, monomial) representation is best suited for GB computations. Although there are several alternatives to choose from (e.g., distributive/recursive, sparse/dense), it is generally agreed that efficiency considerations lead to only one real choice: a dense distributive representation. That is, a polynomial is stored as a sequence of terms where each term consists of a coefficient and an array of exponents which encodes the respective monomial.

With such a representation, one has not only very efficient access to the leading monomial of a polynomial (which is the most frequently needed part of a polynomial during GB computations), but also to the single (exponent) values of a monomial.

Now, what type should the exponent values have? Efficiency considerations lead again to only one realistic choice, namely to fixed–length integers whose size is smaller or equal to the size of a machine word. While assuring the most generality, operations on and representations of arbitrary–length exponent values usually incur an intolerable slow–down of GB computations. Of course, a fixed–length exponent size restricts the range of representable exponent values. However, exponent bound restrictions are usually not very critical for GB computations: on the one hand, the (worst case) complexity of GB computations grows exponentially with the degree of the input polynomials, i.e., large exponents usually make a problem practically uncomputable. On the other hand, checks on bounds of the exponent values can be realized by degree bound checks in the outer loops of the GB computation (e.g., during the S–pair selection) which makes exponent value checks in subsequent monomial operations unnecessary.

3.1   Basic Monomial Operations and Representations

Given an integer $n > 0$ we define the set of exponent vectors M_n by $\{\alpha = (\alpha_1,\ldots,\alpha_n)|\alpha \in \mathbb{N}^n\}$. Notice that monomials usually denote terms of the form $c\,x_1^{\alpha_1}\ldots x_n^{\alpha_n}$. However, in this section we do only consider the exponent vector of a monomial and shall, therefore, use the words exponent vector and monomial interchangeably (i.e., we identify a monomial with its exponent vector).

Monomials play a central role in GB computations. In this section, we describe the basic monomial operations and discuss basic facts about monomial (resp. polynomial) representations for GB computations.

Monomial Operations   The basic monomial operations within GB computations are:

1. computations of the degree (resp. weighted degree):
   the *degree* (resp. *weighted degree*) of a monomial α is the sum of the exponents $\deg(\alpha) := \sum_{i=1}^{n} \alpha_i$ (resp. the weighted sum with respect to a weight vector w: $\deg(\alpha) := \sum_{i=1}^{n} \alpha_i\, w_i$);
2. test for divisibility:
   $\alpha|\beta \Leftrightarrow \forall i \in \{1\ldots n\} : \alpha_i \leq \beta_i$;
3. addition of two monomials:
   $\gamma := \alpha + \beta$ with $\forall i \in \{1\ldots n\} : \gamma_i = \alpha_i + \beta_i$;
4. comparison of two monomials with respect to a monomial ordering.

A *monomial ordering* > (term ordering) on the set of monomials M_n is a total ordering on M_n which is compatible with the natural semigroup structure, i.e., $\alpha > \beta$ implies $\gamma + \alpha > \gamma + \beta$ for any $\gamma \in M_n$. A monomial ordering is a well–ordering if $(0, \ldots, 0)$ is the smallest monomial. We furthermore call an ordering negative if $(0, \ldots, 0)$ is the largest monomial.

Robbiano (cf. [14]) proved that any monomial ordering > can be defined by a matrix $A \in GL(n, \mathbb{R})$: $\alpha > \beta \Leftrightarrow A\alpha >_{lex} A\beta$. Matrix–based descriptions of monomial orderings are very general, but have the disadvantage that their realization in an actual implementation is usually rather time–consuming. Therefore, they are not very widely used in practice.

Instead, the most frequently used descriptions of orderings have at most two defining conditions: a (possibly weighted) degree and a (normal or reverse) lexicographical comparison. We call such orderings *simple orderings*.

Most orderings can be described as block orderings: order subsets of the variables by simple orderings.

For monomials $\alpha, \beta \in M_n$ let

$$\text{lex}(\alpha, \beta) = \begin{cases} 1, & \text{if } \exists i : \alpha_1 = \beta_1, \ldots, \alpha_{i-1} = \beta_{i-1}, \alpha_i > \beta_i \\ 0, & \text{if } \alpha = \beta \\ -1, & \text{otherwise,} \end{cases}$$

$$\text{rlex}(\alpha, \beta) = \begin{cases} 1, & \text{if } \exists i : \alpha_n = \beta_n, \ldots, \alpha_{i-1} = \beta_{i-1}, \alpha_i < \beta_i \\ 0, & \text{if } \alpha = \beta \\ -1, & \text{otherwise,} \end{cases}$$

$$\text{Deg}(\alpha, \beta) = \begin{cases} 1, & \text{if } \deg(\alpha) > \deg(\beta) \\ 0, & \text{if } \deg(\alpha) = \deg(\beta) \\ -1, & \text{otherwise.} \end{cases}$$

Then we can define $\alpha > \beta$ for simple monomial orderings by lex, rlex and Deg.

We furthermore call a monomial ordering > a *degree based monomial ordering* if $\forall \alpha, \beta : \deg(\alpha) > \deg(\beta) \Rightarrow \alpha > \beta$ (e.g., Dp and dp and their weighted relatives are degree based orderings).

Due to the nature of the GB algorithm, monomial operations are by far the most frequently used primitive operations. For example, monomial comparisons are performed much more often than, and monomial additions at least as often as, arithmetic operations over the coefficient field. The number of divisibility tests depends very much on the given input ideal, but is usually very large, as well.

Nevertheless, whether or not monomial operations dominate the running time of a GB computation depends on the coefficient field of the underlying polynomial ring: monomial operations are certainly run–time dominating for finite fields with a small[1] characteristic (e.g., integers modulo a small prime

---

[1] say, smaller than the square root of the maximal representable machine integer, i.e. smaller than $\sqrt{INT\_MAX}$

number), since an arithmetic operation over these fields can usually be realized much faster than a monomial operation. However, for fields of characteristic 0 (like the rational numbers), GB computations are usually dominated by the arithmetic operations over these fields, since the time needed for these operations is proportional to the size of the coefficients which tend to grow rapidly during a GB computation.

Therefore, improvements in the efficiency of monomial operations will have less of an impact on GB computations over fields of characteristic 0.

Monomial Representations As an illustration, and for later reference, we show below SINGULAR's internal `Term_t` data structure:

```
struct  Term_t
{
  Term_t*      next;
  void*        coef;
  long         exp[1];
};
```

Following the arguments outlined above, a SINGULAR polynomial is represented as a linked list of terms, where each term consists of a coefficient (implemented as a hidden type: could be a pointer or a `long`) and a monomial. A monomial is represented by a generalized exponent vector, which may contain degree fields (corresponding to a Deg–conditions) and contains the exponents in an order given by the monomial ordering. The size of the `Term_t` structure is dynamically set at run–time, depending on the monomial ordering and the number of variables in the current polynomial ring.

Based on a monomial representation like SINGULAR's, the basic monomial operations are implemented by *vectorized* realizations of their definitions.

Vectorized Monomial Operations The main idea behind what we call *vectorized monomial operations* is the following: provided that the size of the machine word is a multiple (say, m) of the size of one exponent, we perform monomial operations on machine words, instead of directly on exponents. Into this vector we include also weights (if the monomial order requires it). By doing so, we process a vector of m exponents with word operations only, thereby reducing the length of the inner loops of the monomial operations and avoiding non–aligned accesses of exponents.

The structure of this vector is determined by the monomial ordering:

1. calculate a bound on the exponents to determine the needed space: each exponent is smaller than 2^e;
2. decompose the ordering into blocks of simple orderings;
3. decompose each block into at most 2 defining conditions from lex, rlex, Deg;
4. allocate in the generalized exponent vector for each condition:

- ○ Deg: a `long` for the weighted degree,
- ○ lex: k `long` representing $\alpha_1, \ldots, \alpha_n$ where $k := [(ne + w - 1)/w]$ and w is the size of a `long` in bits,
- ○ rlex: k `long` representing $\alpha_n, \ldots, \alpha_1$ where $k := [(ne + w - 1)/w]$ and w is the size of a `long` in bits,

5. allocate all not yet covered exponents at the end.

The monomial operations on generalized exponent vectors can be used to add and subtract them efficiently. [2] shows that, assuming a stricter bound, even divisibility tests can be vectorized, but we use a pre–test (see below).

In comparison to [2] this approach

- ○ can handle any order represented by blocks of simple orderings, not only pure simple orderings;
- ○ the size of the generalized exponent vector is a multiple of the machine word size;
- ○ no distinction between big endian and little endian machine is necessary;
- ○ the position of a single variable depends on the monomial ordering: access is difficult and time consuming.

Some source code fragments can probably explain it best: Figure 1 shows (somewhat simplified) versions of our implementation of the vectorized monomial operations. Some explanatory remarks are in order:

`n_w` is a global variable denoting the length of the exponent vectors in machine words.

`LexSgn` (used in `MonComp`) is a global variable which is used to appropriately manipulate the return value of the comparison routine and whose value is set during the construction of the monomial ordering from the basic conditions Deg, lex and rlex.

Notice that `MonAdd` works on three monomials and it is most often used as a "hidden" initializer (or, assignment), since monomial additions are the "natural source" of most new monomials.

Our actual implementation contains various, tedious to describe, but more or less obvious, optimizations (like loop unrolling, use of pointer arithmetic, replacement of multiplications by bit operations, etc). We apply, furthermore, the idea of "vectorized operations" to monomial assignments and equality tests, too. However, we shall not describe here the details of these routines, since their implementation is more or less obvious and they have less of an impact on the running time than the basic monomial operations.

The test for divisibility is a little bit different: while a vectorized version is possible (see [2]), it assumes a stronger bound on the exponent size, so we use another approach: most of the divisibility test will give a negative result, a simple bit operation on the bit–support of a monomial avoids most of the divisibility tests (this technique is used in CoCoA [6]: the bit–support is a `long` variable with set bits for each non–zero coefficient in the monomial). If the number of variables is larger than the bit size of `long`, we omit the last

```
// r->VarOffset encodes the position in p->exp
// r->BitOffset encodes the  number of bits to shift to the right
inline int GetExp(poly p, int v, ring r)
{
 return (p->exp[r->VarOffset[v]] >> (r->BitOffset[v]))
         & r->bitmask;
}
inline void SetExp(poly p, int v, int e, ring r)
{
   int shift = r->VarOffset[v];
   unsigned long ee = ((unsigned long)e) << shift;
   int offset = r->BitOffset[v];
   p->exp[offset]  &= ~( r->bitmask << shift );
   p->exp[ offset ] |= ee;
}
inline long MonComp(Term_t* a, Term_t* b)
{
   // return = 0, if a = b
   //        > 0, if a > b
   //        < 0, if a < b

   for (long i = 0; i<n_w; i++)
   {
     d=a->exp[i]-b->exp[i];
     if (d) return d*LexSgn[block];
   }
   return 0;
}

inline void MonAdd(Term_t* c, Term_t* a, Term_t* b)
{
   // Set c = a + b
   for (long i=0; i<n_w; i++)
     c->exp[i] = a->exp[i] + b->exp[i];
}
```

Figure 1. Vectorized monomial operations in SINGULAR

variables, if we have enough bits we use two (or more) for each variable: this block is zero for a zero exponent, has one set bit for exponent one, two set bits for exponent two and so on; and finally all bits of the block set for a larger exponent.

So much for the theory, now let us look at some actual timings: Table 1 shows various timings illustrating the effects of the vectorized monomial operations described in this section. In the first column, we list the used examples — details about these can be found at http://www.symbolicdata.

Table 1. Timings for vectorized monomial operations: SINGULAR in polynomial rings over coefficient field $\mathbb{Z}/32003$

Example	Singular 2.0 (s)	Singular 1.0 (s)	improvement
Ellipsoid-2	99.8	> 1000	> 10
Noon-8	123	> 1000	> 8.1
f855	236	> 1000	> 4.2
Ecyclic-7	48.3	527	11
DiscrC2	14.6	174	12

org. All GB computations were done using the degree reverse lexicographical ordering (dp) over the coefficient field $\mathbb{Z}/32003$.

The dominance of basic monomial operations and experiences vectorizing them are given in [2]. This also applies to the generalized approach described in this paper. This approach is more flexible, in particular we can use any number of bits per exponent (as long as the bounds allow it). As we would expect, the more exponents are encoded into one machine word, the faster the GB computation is accomplished. This has two main reasons: first, more exponents are handled by one machine operation; and second, less memory is used, and therefore, the memory performance increased (e.g., number of cache misses is reduced). However, we also need to keep in mind that the more exponents are encoded into one word, the smaller are the upper bounds on the value of a single exponent. But we can "dynamically" switch from one exponent size to the next larger one, whenever it becomes necessary — it is only necessary to copy the data to the representation.

We should like to point out that the monomial operations for the different orderings are identical — the different ordering results in a different encoding only. Therefore, we can use one inlined routine instead of function pointers (and, therefore, each monomial operation requires a function call). Already the in–place realization of monomial operations results in a considerable efficiency gain which by far compensates the additional overhead incurred by the additional indirection for accessing one (single) exponent value (i.e., the overhead incurred by the `GetExp` macro shown above).

Last, but not least, the combination of all these factors with the new memory management leads to a rather significant improvement over the "old" SINGULAR version 1.0.

4   Arithmetic in Fields

Computer algebra systems use a variety of numbers from fields as one basic data type. Numerical algorithms require floating point numbers (with several levels of accuracy, possibly adjustable), while other algorithms from computer algebra use exact arithmetic.

4.1 Arithmetic in $\mathbb{Z}/p$ for Small Primes p

Representation by a representant from $[0 \ldots p-1]$ fits into a machine word. Addition and subtraction are implemented in the usual way, multiplication and division depend on the properties of the CPU used. The two possibilities are

1. compute the inverse via Euclidean algorithm and keep a table of already computed inverses;
2. use a log–based representation for multiplication and division and keep tables for converting from one representation to the other.

We found that the relative speed of these two methods depends on the CPU type (and its memory band width): on PC processors there is nearly no difference while workstations (SUN, HP) prefer the second possibility.

4.2 Rational Numbers

For the arithmetic in $\mathbb{Z}$ or $\mathbb{Q}$ we rely on the GMP library ([9]), with the following two additions:

- during GB computations we try to avoid rationals: so we have only one large number per monomial;
- small integers may be inlined as LISP interpreters do: they are represented by an immediate integer handle, containing the value instead of pointing to it, which has the following form: (guard bit, sign bit, bit 27, ..., bit 0, tag 0, tag 1). Immediate integer handles carry the tag '01' i.e. the last bit is 1. This distinguishes immediate integers from other handles which point to structures aligned on four byte boundaries and, therefore, have last bit zero (the second bit is reserved as tag to allow extensions of this scheme). Using immediates as pointers and dereferencing them gives address errors. To aid overflow check the most significant two bits must always be equal, that is to say that the sign bit of immediate integers has a guard bit, (see [16]).

5 Arithmetics for Polynomials: Refinement: Bucket Addition

In situations with a lot of additions, where only little knowledge about the intermediate results is required, the *geobucket* method provides good results — it avoids the $O(n^2)$–complexity in additions (merge of sorted lists) by postponing them. It is used during computation of a Gröbner basis: the reduction routine will be replaced by one which will handle long polynomials more efficiently using geobuckets, which accommodate the terms in buckets of geometrically increasing length (see [17]). This method was first used successfully by Thomas Yan, at that time a graduate student in CS at Cornell. Other algorithms which profit from geobuckets are the evaluation of polynomials and the Bareiss algorithm [4].

6 Memory Management

Most of SINGULAR's computations boil down to primitive polynomial operations such as copying, deleting, adding, and multiplying of polynomials. For example, standard bases computations over finite fields spent (on average) 90 % of their time realizing the operation p - m*q where m is a monomial, and p,q are polynomials.

SINGULAR uses linked lists of monomials as data structure for representing polynomials. A monomial is represented by a memory block which stores a coefficient and an exponent vector. The size of this memory block varies: its minimum size is three machine words (one word for each, the pointer to the "next" monomial, the coefficient and the exponent vector), its maximal size is (almost) unlimited, and we can assume that its average size is four to six machine words (i.e., the exponent vector is two to four words long).

From a low–level point of view, polynomial operations are operations on linked lists which go as follows:

1. do something with the exponent vector (e.g., copy, compare, add), and possibly do something with the coefficient (e.g., add, multiply, compare with zero) of a monomial;
2. advance to the next monomial and/or allocate a new or free an existing monomial.

Assuming that coefficient operations are (inlined) operations on a single machine word (as they are, for example, for coefficients from most finite fields), and observing that the exponent vector operations are linear w.r.t. the length of the exponent vector, we come to the following major conclusion:

For most computations, a traversal to the next memory block or an allocation/deallocation of a memory block is, on average, required after every four to six machine word operations.

The major requirements of a memory manager for SINGULAR become immediately clear from this conclusion:

(1) allocation/deallocation of (small) memory blocks must be extremely fast
    If the allocation/deallocation of a memory blocks requires more than a couple of machine instructions it would become the bottleneck of the computations. In particular, even the overhead of a function call for a memory operation is already too expensive;
(2) consecutive memory blocks in linked lists must have a high locality of reference. Compared with the performed operations on list elements, cache misses (or, even worse, page misses) of memory blocks are extremely expensive: we estimate that one cache miss of a memory block costs at least ten or more list element operations. Hence, if the addresses of consecutive list elements are not physically closed to each other (i.e., if the linked lists have a low locality of reference), then resolving cache (or, page) misses would become the bottleneck of computations.

Furthermore, there are the following more or less obvious requirements on a memory manager for Singular:

(3) the size overhead to maintain small blocks of memory must be small. If managing one memory block requires one or more words (e.g., to store its size) then the total overhead of the memory manager could sum up to 25 % of the overall used memory, which is an unacceptable waste of memory;

(4) the memory manager must have a clean API and it must support debugging. The API of the memory manager must be similar to the standard API's of memory managers, otherwise its usability is greatly limited. Furthermore, it needs to provide error checking and debugging support (to detect overwriting, twice freed or not–freed memory blocks, etc.) since errors in the usage of the memory manager are otherwise very hard to find and correct;

(5) the memory manager must be customizable, tunable, extensible and portable. The memory manager should support customizations (such as "what to do if no more memory is available"); its parameters (e.g., allocation policies) should be tunable, it should be extensible to allow easy implementations of furthergoing functionality (like overloading of C++ constructors and destructors, etc), and it needs to be portable to all available operating systems.

To the best of our knowledge, there is currently no memory manager available which satisfies these (admittedly in part extreme) requirements. Therefore, we designed and implemented `omalloc`. ([1]).

`omalloc` manages small blocks of memory on a per–page basis. That is, each used page is split up into a page header and equally sized memory blocks. The page header has a size of six words (i.e., 24 bytes on a 32 bit machine), and stores (among others) a pointer to the free–list and a counter of the used memory blocks of this page.

On *memory allocation*, an appropriate page (i.e., one which has a non–empty free list of the appropriate block size) is determined, based on the used memory allocation mechanism and its arguments (see below). The counter of the page is incremented, and the provided memory block is dequeued from the free–list of the page.

On *memory deallocation*, the address of the memory block is used to determine the page (header) which manages the memory block to be freed. The counter of the page is decremented, and the memory block is enqueued into the free–list. If decrementing the counter yields zero, i.e., if there are no more used memory blocks in this page, then this page is dequeued from its bin and put back into a global pool of unused pages.

This design results in

○ allocation/deallocation of small memory blocks requires (on average) only a couple (say, 5-10) assembler instructions;

o `omalloc` provides an extremely high locality of reference (especially for consecutive elements of linked-list like structures). To realize this, recall that consecutively allocated memory blocks are physically almost always from the same memory page (the current memory page of a bin is only changed when it becomes full or empty);

o small maintenance size overhead: Since `omalloc` manages small memory blocks on a per-page basis, there is no maintenance-overhead on a per-block basis, but only on a per-page basis. This overhead amounts to 24 Bytes per memory page, i.e., we only need 24 Bytes maintenance overhead per 4096 Bytes, or 0.6 %.

For larger memory blocks, this overhead grows since there might be "leftovers" from the division of a memory page into equally sized chunks of memory blocks. Furthermore, there is the additional (but rather small) overhead of keeping Bins and other "global" maintenance structures.

7   A Proposal for Distributing Polynomials

Computing in a distributed fashion is coming of age, driven in equal parts by advances in technology and by the desire to simply and efficiently access the functionality of a growing collection of specialized, stand–alone packages from within a single framework. The challenge of providing connectivity is to produce homogeneity in a heterogeneous environment, overcoming differences at several levels (machine, application, language, etc.). This is complicated by the desire to also make the connection efficient. We have explored this problem within the context of symbolic computation and, in particular, with respect to systems specially designed for polynomial computations.

We tested our ideas with the mechanisms found in the Multi Protocol: dictionaries, prototypes, and annotations. A detailed description of the implementation is given in [3].

A polynomial is more than just an expression: the system usually has a default representation (dense–distributive in the case of SINGULAR, ordered by the monomial ordering). Furthermore, the number of variables, the coefficient field are also known in advance and, usually, constant for all arguments of polynomial operations or structures.

The polynomial objects consist of several parts: the description of the structure (which may be shared with other objects of the same type), the actual raw data, and a list of properties. Understanding the structure is necessary for each client: they would not be able to decode the data otherwise. On the other side, the properties should be considered as hints: a typesetting programme does not need to understand the property of an ideal to be a Gröbner basis.

The following list of requirements should be fulfilled by a communication protocol:

- o the encoding of polynomial data should closely follow the internal representation of the sender;
- o several possibilities for the encoding of polynomial data must be provided:
  corresponding to several possibilities for data representation (distributive,
  recursive, as expression tree);
- o the description of the structure should be shared between objects;
- o possibility to provide additional information which may or may not be
  used;
- o flexibility in the set of provided operations and properties.

While MP provides the first requirements, it fails to fulfill the last one: it is
difficult to extend and only at compile time.

Currently there are several protocols in use which communicate mathematical expressions. OpenMath ([12]) seems to be the most promising in
terms of generality, but it has still to prove that it provides also fast links —
but, as I think, it may become the standard we need.

References

1. Bachmann, O.: An Overview of omalloc, Preprint 2001.
2. Bachmann, O.; Schönemann, H.: Monomial Representations for Gröbner Bases Computations, ISSAC 1998, 309-316.
3. Bachmann, O.; Gray, S.; Schönemann, H.: MPP: a Framework for Distributed Polynomial Computations, ISSAC 1996, 103-112.
4. Bareiss, R.H.; Sylvester's identity and multistep integer-preserving Gaussian elimination, Math. Comput. 22(1968), 565-578.
5. Bayer, D.; Stillman, M.: A Theorem on Refining Division Orders by the Reverse Lexicographic Order, Duke Math. J. 55 (1987), 321-328.
6. Bigatti, A.M.: Computation of Hilbert-Poincaré Series, J. Pure and Appl. Algebra 119, 237–253, 1997.
7. Buchberger, B.: Ein Algorthmus zum Auffinden der Basiselemente des Restklassenringes nach einem nulldimensionalen Ideal, Thesis, Univ. Innsbruck, 1965.
8. Faugère, J.-C.: A new efficient algorithm for computing Grobner bases (F4). Journal of Pure and Applied Algebra 139, 1-3, 61-88 (1999).
9. Granlund, T.: The GNU Multiple Precision Arithmetic Library, http://www.swox.com/gmp/
10. Hironaka, H.: Resolution of singularities of an algebraic variety over a field of characteristic zero. I, II, Ann. of Math. (2) 79, 109-203 (1964), 79, 205-326, (1964).
11. Möller, H.M., Mora, T.: New Constructive Methods in Classical Ideal Theory. Journal of Algebra 100, 138-178 (1986)
12. OpenMath, http://www.openmath.org/
13. Pfister, G.; Schönemann, H.: Singularities with exact Poincaré complex but not quasihomogeneous, Revista Matematica 2 (1989)
14. Robbiano, L.: Term Orderings on the Polynomial Ring, Proceedings of EUROCAL 85, Lecture Notes in Computer Science 204
15. Takayama, N.: Developpers of Gröbner Engines, Sugaku no Tanoshimi Vol 11, 35–48,(1999)
16. Schönert, M. et al.: AP – Groups, Algorithms, and Programming. Lehrstuhl D für Mathematik, Rheinisch Westfälische Technische Hochschule, Aachen, Germany, fifth edition, 1995.
17. Yan, T.: The Geobucket Data Structure for Polynomials, Journal of Symbolic Computation 25, 3 (1998), 285-294.

Computing Simplicial Homology Based on Efficient Smith Normal Form Algorithms

Jean-Guillaume Dumas[1], Frank Heckenbach[2], David Saunders[3], and Volkmar Welker[4]

[1] Laboratoire de Modélisation et Calcul, IMAG-B. P. 53,
   38041 Grenoble, France
[2] Universität Erlangen-Nürnberg, Mathematisches Institut, Bismarckstr. 1 1/2,
   91054 Erlangen, Germany
[3] University of Delaware, Department of Computer and Information Sciences,
   Newark, DE 19716, USA
[4] Philipps-Universität Marburg, Fachbereich Mathematik und Informatik,
   35032 Marburg, Germany

Abstract. We recall that the calculation of homology with integer coefficients of a simplicial complex reduces to the calculation of the Smith Normal Form of the boundary matrices which in general are sparse. We provide a review of several algorithms for the calculation of Smith Normal Form of sparse matrices and compare their running times for actual boundary matrices. Then we describe alternative approaches to the calculation of simplicial homology. The last section then describes motivating examples and actual experiments with the GAP package that was implemented by the authors. These examples also include as an example of other homology theories some calculations of Lie algebra homology.

1 Introduction

Geometric properties of topological spaces are often conveniently expressed by algebraic invariants of the space. Notably the fundamental group, the higher homotopy groups, the homology groups and the cohomology algebra play a prominent role in this respect. This paper focuses on methods for the computer calculation of the homology of finite simplicial complexes and its applications (see comments on the other invariants in Section 7).

We assume that the topological space, whose homology we want to calculate, is given as an abstract simplicial complex. We recall now the basic notions and refer the reader to Munkres' books [33] and [32] for further details. A finite simplicial complex Δ over the non-empty ground set Ω is a non-empty subset of the power set 2^Ω such that $A \subseteq B \in \Delta$ implies $A \in \Delta$. Without loss of generality we will always assume that $\Omega = [n] := \{1, \ldots, n\}$. An element of Δ is called a face. An inclusion-wise maximal element of Δ is called a facet. The dimension $\dim A$ of a face $A \in \Delta$ is given by $\#A - 1$. The dimension $\dim \Delta$ of Δ is the maximal dimension of one of its faces. The definitions imply that every simplicial complex contains the empty set as its unique face of dimension -1. By $f_i = f_i(\Delta)$ we denote the number of

i-dimensional faces of Δ and call $(f_{-1}, \ldots, f_d)$, $d = \dim \Delta$, the f-vector of Δ. The ith chain group $C_i(\Delta, R)$ of Δ with coefficients in the ring R is the free R-module of rank f_i with basis e_A indexed by the i-dimensional faces $A \in \Delta$. The ith differential $\partial_i : C_i(\Delta, R) \to C_{i-1}(\Delta, R)$ is the R-module homomorphism defined by $\partial_i e_A = \sum_{j=0}^{i} (-1)^j e_{A \setminus \{a_j\}}$, where $A = \{a_0 < \cdots < a_i\}$. It is easily checked that $\partial_i \circ \partial_{i+1} = 0$ and hence $\mathrm{Im}\partial_{i+1} \subseteq \mathrm{Ker}\partial_i$. This leads to the definition of the ith (reduced) homology group $\widetilde{H}_i(\Delta; R)$ as the R-module $\mathrm{Ker}\partial_i / \mathrm{Im}\partial_{i+1}$. In this paper we will only be concerned with the case when $R = k$ is a field and $R = \mathbb{Z}$ is the ring of integers.

The definition of the reduced homology group clearly implies that its computation reduces to linear algebra over a field or $\mathbb{Z}$.

In case $R = k$ is a field the dimension of $\widetilde{H}_i(\Delta; k)$ as a k-vector space is easily calculated as $\beta_i^k = \dim \mathrm{Ker}\partial_i - \dim \mathrm{Im}\partial_{i+1} = f_i - \mathrm{rank}\partial_i - \mathrm{rank}\partial_{i+1}$. Here and whenever convenient we identify the linear map ∂_i with its matrix $\mathrm{Mat}(\partial_i)$ with respect to the canonical bases e_A and denote by rank the rank of a matrix. The definitions imply that β_i^k depends on the characteristic of k only and that β_i^k is constant except for fields k whose characteristic lies in a finite set of primes which of course depend on Δ.

If $R = \mathbb{Z}$, then $\widetilde{H}_i(\Delta; R)$ is a a quotient of subgroups of free $\mathbb{Z}$-modules of finite rank. Hence $\widetilde{H}_i(\Delta; R)$ is a finitely generated abelian group – note that we only consider finite simplicial complexes. In order to compute the structure invariants of $\widetilde{H}_i(\Delta; \mathbb{Z})$, we calculate the Smith Normal Form of the matrix $\mathrm{Mat}(\partial_i)$. Recall that the Smith Normal Form of an integer Matrix A is a diagonal matrix of the same size as A which is unimodularly equivalent to A (for more technical details see Definition 4).

Assume that

$$D_{i+1} = \begin{pmatrix} b_1 & 0 & \cdots & 0 & 0 & \cdots & 0 \\ 0 & b_2 & \cdots & 0 & 0 & \cdots & 0 \\ \cdot & \cdot & \cdots & \cdot & 0 & \cdots & 0 \\ 0 & 0 & \cdots & b_t & 0 & \cdots & 0 \\ 0 & 0 & \cdots & 0 & 0 & \cdots & 0 \\ \cdot & \cdot & \cdots & \cdot & 0 & \cdots & 0 \\ 0 & 0 & \cdots & 0 & 0 & \cdots & 0 \end{pmatrix}$$

is the Smith Normal Form of the matrix $\mathrm{Mat}(\partial_{i+1})$, with natural numbers $b_1 | b_2 | \ldots | b_t \neq 0$. Let $e_1', \ldots, e_s'$ be the basis of $C_i(\Delta)$ which corresponds to the Smith Normal Form D_{i+1}. Then $b_1 e_1', \ldots, b_t e_t'$ is a basis of $\mathrm{Im}\partial_{i+1}$. Since $\partial_i \circ \partial_{i+1} = 0$ and ∂_{i+1} is $\mathbb{Z}$-linear it follows that $e_1', \ldots, e_t' \in \mathrm{Ker}\partial_i$. Now after possible renumbering, we can choose the basis e_i' such that for $r = f_i - \mathrm{rank}\partial_i$ we have $e_1', \ldots, e_r'$ is a basis of $\mathrm{Ker}\partial_i$. Thus

$$\widetilde{H}_i(\Delta; \mathbb{Z}) \cong \mathbb{Z}_{b_1} \oplus \cdots \oplus \mathbb{Z}_{b_t} \oplus \mathbb{Z}^{r-t}.$$

Here we write $\mathbb{Z}_l$ for the abelian group $\mathbb{Z}/l\mathbb{Z}$ of residues modulo l.

Thus the computation of homology groups of simplicial complexes boils down to the following two fundamental problems in exact linear algebra:

(i) Calculate the rank of a matrix over a finite field or $\mathbb{Q}$.
(ii) Calculate the Smith Normal Form of a matrix over $\mathbb{Z}$.

In our paper we will focus on (ii), integer coefficients. But some of the methods relevant for (i) will also be useful in this context. Clearly, for the first problem Gauss' algorithm gives a polynomial time algorithm. But when the size of the matrix grows, more efficient methods are in demand. This is even more so for (ii). Here [23] shows that the problem has an efficient solution. But again the matrices, which arise in our context, request algorithms that take advantage of their special structure in order to have a Smith Normal Form algorithm that terminates within a lifetime. For general information on the complexity of the problem of calculating homology and related problems see [27].

We have implemented our software as a package for the computer algebra system GAP [15]. It can be downloaded from

http://www.cis.udel.edu/~dumas/Homology/

The package is split into two parts, a part which consists of an external binary that implements four algorithms calculating the Smith Normal Form and a part that implements algorithms to create and manipulate simplicial complexes in the GAP [15] language.

We report here the key features of the design, implementation, and use of our Simplicial Homology package for the GAP [15] system.

We will present the basic definitions and relationships involved in simplicial complexes, their homology, and the Smith Normal Forms. We will discuss some projects in geometric combinatorics which led us to design the package.

Secondly we will describe in some detail the methods we use in the package. This includes information on how the boundary matrices are built, and the properties of the several algorithms for computing Smith Normal Forms. In addition we will mention several possible new methods and possible new connections of the package to other mathematical software.

Finally we will illustrate the usefulness of the approach and of the software in addressing a couple of topics concerning simplicial homology and Lie algebra homology.

In a concluding section we comment on other invariants of simplicial complexes.

2   Generating the Boundary Matrices

For our purposes a simplicial complex is most conveniently described by the set of its facets. Clearly, if we know the facets then we know all faces of the simplicial complex, since a set is a face if and only if there is a facet containing it.

Even though a list of facets gives a succinct presentation of a simplicial complex, a linear algebra approach to simplicial homology requires the then expensive task to generate all faces of the simplicial complex from which the matrices of the boundary maps ∂_i are set up. On the other hand it is an interesting theoretical question whether it is indeed necessary to go through these time consuming computations in order to calculate homology.

The following section will summarize the algorithms we have implemented in our package that generate the set of faces from the set of facets and the boundary matrices from the set of faces.

A simplicial complex Δ, when read from the input data, is stored as an array of its facets. The facets, as well as other simplices in the following steps, are stored as arrays of the points they contain where the points are indexed by integers.

In order to generate a boundary matrix $\mathtt{Mat}(\partial_k)$, we first generate the lists of simplices of dimension k and $k-1$ ($\mathtt{GetFaces}$ in file $\mathtt{simplicial.pas}$). Of course when computing the homology in several consecutive dimensions, we can reuse the list of k-simplices for the generation of ∂_{k+1}, so we have only to generate one new simplex list for each new dimension. Generating the $k-$ or $(k-1)-$ faces basically amounts to generating the $(k+1)-$ and $k-$element subsets of each facet.

The main problem here is that many simplices, especially in the low and middle dimensions, are subsets of many facets so they are generated many times when processing all the facets. Storing all the copies in memory first and removing duplicates at the end would therefore require much more memory than is needed to store the distinct simplices. Indeed there are cases where an otherwise feasible computation would fail at this point when keeping the duplicates.

For sorting the simplices we use the lexicographic order on the ordered set of point indices which is a natural choice to allow for fast binary searching in the following steps ($\mathtt{FindFace}$). A merge sort algorithm is implemented to sort the simplices and remove duplicates early. More precisely, we use an array of lists of simplices with the restriction that the ith list must never contain more than 2^{i-1} entries. Newly generated simplices are always inserted into the first list ($\mathtt{NewFace}$). Whenever a list grows too big, it is merged with the following list, and when elements exist in both lists, one copy is removed ($\mathtt{Merge}$). This process is repeated with the subsequent lists until the size of each list is again below its limit. When all facets have been processed, all the lists are merged together ($\mathtt{CollectFaces}$).

With the lists of simplices generated, it is straightforward to generate the boundary matrices, by iterating over the k-simplices, corresponding to the rows of the matrix. For technical reasons we consider the transposed boundary matrix. Clearly, transposition does not change the Smith Normal Form. This way we obtain the matrix row by row rather than column by column. For the elimination algorithm we can take advantage of this procedure. Namely instead of generating the whole matrix at once, we do it one row at a time (GetOneRow) and process each row before the next one is generated. This makes it possible to work with matrices whose total size would exceed memory limits.

Each row is generated by taking the $(k-1)$-subsimplices, obtained by deleting one point from the k-simplex, looking them up via binary search in the list of all $(k-1)$-simplices to get the column index, and alternating the matrix entries between 1 and -1. By processing the subsimplices in the natural lexicographic order, we automatically get the row correctly ordered in the index/value format that we use in the further steps for storing matrix rows and other sparse vectors.

3   The Elimination Algorithm

One of the algorithms for computing the Smith Normal Form is a Gaussian elimination with some specific modifications described in this section.

This algorithm is implemented in two variants, one using machine word size integers, the other one using integers of unlimited size, provided by the GMP library. The first version checks for overflows and aborts when it detects one. The second version initially uses machine integers as well. When it detects an overflow, it switches to GMP integers, but only for the affected values. This is based on the experience that even in situations where machine integers are not sufficient, often only few values really overflow, and for the majority of small integers, computing with machine integers is faster than with GMP integers. Apart from this, the two implementations of the algorithm are identical.

Before the elimination is started, we make an attempt to avoid it entirely. If we have computed $\mathrm{Mat}(\partial_{k-1})$ before, we know an upper bound on the rank of $\mathrm{Mat}(\partial_k)$, namely $\dim C_{k-1} - \mathrm{rank} Mat(\partial_{k-1})$. Therefore, a quick pass is made over the matrix, trying to find as many linearly independent rows as *easily* possible (QuickCheck in file arith.pas). It is worth noting that due to the lexicographic ordering of both the k and $(k-1)$-simplices, the column index of the first non-zero entry, in the following denoted by r_i where i is the row index, is monotonic over the rows. Therefore, it is easy to count the number of rows with distinct r_i which are clearly linearly independent. If this number is as big as the known bound, we are done since the matrix contains only ± 1 entries at this point, so we know that the Smith Normal Form can only have 1 entries. This short-cut typically works for some smaller matrices, not for the largest and most difficult matrices, and it can only work if the homology group $\widetilde{H}_{k-1}$ is in fact 0, but since it does not take much time, it seems worth trying it, anyway.

The actual elimination (SmithNormalForm) processes the matrix row by row (DoOneRow), first trying to find as many rows as possible with distinct r_i

and $A_{i,r_i} = \pm 1$ (where $A_{i,j}$ denotes the matrix entries). Each row is modified using row additions until one of the following situations occurs:

- The row is the 0-vector and can be removed. Clearly, in this case the size of the matrix to deal with is reduced before it is fully generated, as noted in the description of the generation of the boundary matrix.
- r_i is distinct from r_j for all previously considered rows j, and $A_{i,r_i} = \pm 1$. Then the row is kept in this form.
- r_i is distinct from r_j for all previously considered rows j, but $|A_{i,r_i}| > 1$. Such rows are deferred in this pass, to avoid having to deal with non-1 pivots in the subsequent row operations, in the hope that the row can be further modified after more rows have been considered, and that the total number of such problematic rows is small compared to the size of the matrix. This is confirmed by observations.

After all rows have been processed, the rows that were deferred are processed again in the same way, but without deferring them again. This makes it necessary to use more complicated row operations at this point, such as a gcd step when dealing with two rows i and j with $r_i = r_j$ and $|A_{i,r_i}|, |A_{j,r_j}| > 1$ and A_{i,r_i} is not divisible by A_{j,r_j} or vice versa.

At the end of this second pass, we have the matrix in echelon form, with many pivots being ± 1. In the (frequent) case that in fact all pivots are ± 1, we only have to count the remaining rows and be done.

Otherwise, we can now remove the rows with ± 1 pivots, counting them, to vastly reduce the size of the matrix. The empirical observation which makes this algorithm work is that in cases with several hundreds of thousands of rows, the number of remaining rows was rarely found to be more than a few hundred. Since the ± 1 pivots are generally not the leftmost ones, this requires some more row additions.

The remaining matrix is now transformed to Smith Normal Form using a standard algorithm as described (GetTorsion), e.g., in [32, §11], involving both row and column operations. Since the latter are expensive with the sparse vector representation used, they are avoided as far as possible, but in general they are necessary in this step.

A result of Storjohann [37] allows us to do this last step modulo the determinant of a nonsingular maximal rank minor. This is very helpful to avoid coefficient explosion which was often observed in this step without this modulo reduction. Since the remaining matrix is of full row rank, we can choose as the minor the leftmost $n \times n$ submatrix (if the matrix is of size $m \times n$, $m \geq n$), i.e., the smallest matrix containing all the pivots. Since the matrix is in echelon form at this point, the determinant is just the product of all the pivots.

4 Valence Algorithm

In this section we describe a second algorithm for the computation of the Smith Normal Form of an integer matrix. This algorithm has proven very effective on some of the boundary matrices discussed in this paper, though the worst case asymptotic complexity is not better than that of the best currently known [17]. One practically important aspect of our method is that we avoid preconditioners which introduce a log factor in the running time. When the matrix size is near a million, this factor is near 20, and is hard to bear when it applies to a computation which already requires hours or days. In this paper we offer a brief overview of our method. For more detail about the valence algorithm, the reader can refer to [10]. The method is particularly effective when the degree of the minimal polynomial of AA^t is small.

There are two kinds of problems arising when computing the Smith Normal Form via elimination over the integers: one is the fill-in of the sparse matrices, and the other one is coefficient growth of the intermediate values. Both of those can lead to memory thrashing, thus killing the computation (see attempts to avoid these phenomena in Section 3). We want to bypass those problems. For the fill-in, the ideas are to use reordering to reduce fill-in and to use iterative methods to maintain a constant level of memory usage. For the coefficient growth, we work modulo primes, or small powers of primes and reconstruct as integers only the results (not the intermediate values).

We begin with some definitions. The *valence* of a polynomial is its trailing nonzero coefficient. The *characteristic valence* of a matrix is the valence of its characteristic polynomial. The *minimal valence* or simply the *valence* of a matrix A is the valence of its minimal polynomial, $\texttt{minpoly}_A(X)$. The *valuation* of a matrix is the degree of the trailing term of its minimal polynomial. The characteristic and minimal valuations of a matrix are similarly defined. For $1 \leq i \leq \min(m,n)$, the i-th *determinantal divisor* of a matrix A, denoted by $d_i(A)$, is the greatest common divisor of the $i \times i$ minors of A. Define $d_0(A) = 1$. For $1 \leq i \leq \min(m,n)$, the i-th *invariant factor* of A is $s_i(A) = d_i(A)/d_{i-1}(A)$, or 0 if $d_i(A) = 0$. Let $s_0 = 1$. And as before, the *Smith Normal Form* of a A is the diagonal matrix

$$S = \mathrm{diag}(s_1(A),\ldots,s_{\min(m,n)}(A))$$

of the same shape as A. It is well known that for $1 \leq i \leq \min(m,n)$, we have $d_{i-1}|d_i$, $s_{i-1}|s_i$, and that A is unimodularly equivalent to its Smith Normal Form.

For a positive integer q, we define 'rank of $A \bmod q$', to be the greatest i such that q does not divide the i-th invariant factor of A, and we denote this rank by $r_q = \texttt{rank}_q(A)$.

The next algorithm statement is an overview of the valence method. The individual steps can be done in many ways. Afterwards we will discuss the implementation details.

In order to prove the correctness of the method we will need the following theorem.

Theorem 4.1 ([10]). *Let A be a matrix in $\mathbb{Z}^{m \times n}$. Let $(s_1, \ldots, s_r)$ be its nonzero invariant factors. If a prime $p \in \mathbb{Z}$ divides some nonzero s_i, then p^2 divides the characteristic valence of AA^t, and p divides the minimal valence of AA^t. The same is true of the valences of $A^t A$.*

Corollary 4.2. *Algorithm* Valence-Smith-Form *correctly computes the Smith Normal Form.*

Proof. The theorem shows that we consider the relevant primes. It is evident that the integer Smith Normal Form may be composed from the relevant local Smith Normal Forms, since the integer Smith Normal Form *is* the local Smith Normal Form at p up to multiples which are units mod p.

4.1 Minimal valence computation

The first two steps of the valence algorithm have the purpose of determining a small, at any rate finite, set of primes which includes all primes occurring in the Smith Normal Form. The choice in the algorithm between $A^t A$ and AA^t is easily made in view of the following.

Lemma 4.3 ([10]).
  For a matrix A the minimal polynomials of $A^t A$ and AA^t are equal or differ by a factor of x.

Thus the difference of degree has a negligible effect on the run time of the algorithm. It is advantageous to choose the smaller of AA^t and $A^t A$ in the algorithm, to reduce the cost of the inner products involved. Moreover any bound on the coefficients of $\texttt{minpoly}_{A^t A}$ can then be applied to those of $\texttt{minpoly}_{AA^t}$ and vice versa.

Chinese Remaindering. We compute the integer minimal valence, v, of a matrix B (the valence of its minimal polynomial over the integers) by Chinese remaindering valences of its minimal polynomials mod p for various primes p. The algorithm has three steps. First compute the degree of the minimal polynomial by doing a few trials modulo some primes. Then compute a sharp bound on the size of the valence using this degree. End by Chinese remaindering the valences modulo several primes.

The first question is how many primes must be used for the Chinese remaindering. Using Hadamard's inequality [16, Theorem 16.6] would induce a use of $O(n)$ primes. The Hadamard bound may be used, but is too pessimistic an estimate for many sparse matrices. Therefore, we use a bound determined by consideration of Gershgörin disks and ovals of Cassini. This bound is of the form β^d where β is a bound on the eigenvalues and d is the degree of the minimal polynomial.

The i-th Gershgörin disk is centered at $b_{i,i}$ and has for a radius the sum of the absolute values of the other entries on the i-th row. Gershgörin's theorem is that all of the eigenvalues are contained in the union of the Gershgörin disks [4,38,18]. One can then go further and consider the ovals of Cassini [5,6,40], which may produce sharper bounds. For our purposes here it suffices to note that each Cassini oval is a subset of two Gershgörin circles, and that all of the eigenvalues are contained in the union of the ovals. We can then use the following proposition to bound the coefficients of the minimal polynomial.

Proposition 4.4 ([10]). *Let $B \in \mathbb{C}^{n \times n}$ with its spectral radius bounded by β. Let $\mathtt{minpoly}_B(X) = \sum_{k=0}^{d} m_i X^i$. Then the valence of B is no more than β^d in absolute value and $|m_i| \leq \max\{\sqrt{d\beta}; \beta\}^d$, for $i = 0, \ldots, d$.*

For matrices of constant size entries, both β and d are $\mathtt{O}(n)$. However, when d and/or β is small relative to n (especially d) this may be a striking improvement over the Hadamard bound since the length of the latter would be of order $n \log(n)$ rather than $d \log(\beta)$.

Our experiments suggest that this is often the case for homology matrices. Indeed, for those, $B = AA^t$ has very small minimal polynomial degree and has some other useful properties. For instance it is diagonally dominant. But we do not think that these facts are true for all homology matrices. On the other hand, we neither have a proof of this assertion nor its converse.

There remains to compute the bound on the spectral radius. We remark that it is expensive to compute any of the bounds mentioned above while staying strictly in the black box model. It seems to require two matrix vector products (with A) to extract each row or column of B. But, if one has access to the elements of A, a bound for the spectral radius of B can easily be obtained with very few arbitrary precision operations: The diagonal values can be computed first, by a simple dot product: $q_i = \sum_{j \in 1\ldots n} a_{ij}^2$, and serve as centers for the disks and ovals. Then the radii are compute via the absolute value of $A : v = |A||A|^t[1, 1, \ldots, 1]^t$ and set $r_i = v_i - |q_i|$ for $i = 1, \ldots, m$.

Minimal Polynomial over the Integers and Bad Primes. The next question is to actually compute the minimal polynomial of a matrix modulo primes. To perform this, we use Wiedemann's probabilistic algorithm [42]. It is probabilistic in the sense that it returns a polynomial which is deterministically a factor of the minimum polynomial and is likely to be the minimal polynomial. When we know the degree of the minimal polynomial, just by a degree check, we can then be sure that the computed polynomial is correct. Thus in order to complete the valence computation we must be sure of the degree of this polynomial over the integers. To compute this degree, we choose some primes at random. The degree of the integer minimal polynomial will be the maximal degree of the minimal polynomials mod p with high probability. Some primes may give a lower degree minimal polynomial. We call them *bad primes*. We next bound the probability of choosing a bad prime at random, by bounding the size of a minor of the matrix (M_δ) that such a prime must divide.

Let δ be the degree of the integer minimal polynomial and U an upper bound on the number of bad primes. We proved that U can have the following value [10]:

$$|M_\delta| \le \lceil \frac{\delta}{2} \rceil \sqrt{n} \beta^{\frac{\delta^2}{2}} = U.$$

Suppose we choose primes at random from a set P of primes each greater than a lower bound l. There can be no more than $\log_l(U)$ primes greater than l dividing M_δ. It suffices to pick from an adequately large set P to reduce the probability of choosing bad primes. The distribution of primes assures that adequately large P can be constructed containing primes that are not excessively large.

We now give the complete algorithm for the computation of the Valence. The algorithm involves computation of minimal polynomials over Z_p. For the fast probabilistic computation of these we use Wiedemann's method (and probability estimates) [42] with early termination as in [26]. We then construct the integer minimal polynomial using Chinese remaindering.

In the following we will denote by $\mu_l(x)$ a lower bound on the number of distinct primes between l and x; this bound is easily computed using direct bounds on $\pi(x)$, the number of primes lower than x, [11, Theorem 1.10].

The binary cost of the multiplication of two integers of length n, will be denoted by $I_m(n)$: classical multiplication uses $I_m(n) = 2n^2$ bit operations, Karatsuba's method uses $I_m(n) = O(n^{1.59})$ and Schönhage & Strassen's method uses $I_m(n) = O(n \log(n) \log \log(n))$ [16]. For convenience, we will also use "soft-Oh" notation: for any cost functions f and g, we write $f = O^\sim(g)$ if and only if $f = O(g \log^c(g))$ for some constant $c > 0$. For a matrix $A \in \mathbb{C}^{m \times n}$ let e be the number of nonzero entries. Let $E = \max\{m, n, e\}$.

Theorem 4.5 ([10]). *Algorithm* Integer-Minimal-Polynomial *is correct. Let $s = d \max\{\log_2(\beta(A)); \log_2(d)\}$, which bounds the lengths of the minimal polynomial coefficients. The algorithm uses expected time*

$$O(sdE \log(\epsilon^{-1}))$$

for constant size entries. It uses $O(n \log(s) + ds)$ memory if every coefficient of the minimal polynomial is computed, and $O(n \log(s) + s)$ if only the valence is computed. The latter is also $O^\sim(n)$.

In practice the actual number of distinct primes greater than 2^{15} dividing valences of homology matrices is very small (no more than 50, say) and we often picked primes between 2^{15} and 2^{16} where there are 3030 primes. This giving us, at most, only 1.7% of bad primes. With only 10 polynomials this reduces to a probability of failure less than 2×10^{-16}.

4.2   Ranks Computation

Once the valence is computed, we know all the primes involved in the Smith Normal Form. We then have to factor the valence and perform rank computations modulo those primes. Two problems can arise. First we might need to compute the rank modulo a *power* of a prime: we deal with this via an elimination method. Second, the valence can be hard to factor: in this case we consider the big composite factor to be prime. If problems arise in the elimination or in the iterative resolution, in fact a factorization of this number has been achieved. Thus we do not have to pay the high cost of explicit factorization of large numbers with large prime factors. [10]).

Next consider the question of computing the local Smith Normal Form in $\mathbb{Z}_{(p)}$. This is equivalent to a computation of the rank mod p^k for sufficiently many k. Recall that we define the rank mod p^k as the number of nonzero invariant factors mod p^k. In a number of cases, we have had success with an elimination approach, despite the fill-in problem. We first present this elimination method then the iterative method with lower space requirements.

Elimination Method. Due to intermediate expression swell, it is not effective to compute directly in $\mathbb{Z}_{(p)}$, the localization of $\mathbb{Z}$ at the prime ideal (p), so we perform a computation mod p^e, which determines the ranks mod p^k for $k < e$ and hence the powers of p in the Smith Normal Form up to p^{e-1}. Suppose by this means we find that s_r is not zero mod p^e, where r is the previously determined integer rank of A. Then we have determined the Smith Normal Form of A locally at p. If, however, the rank mod p^e is less than r, we can repeat the LRE computation with larger exponent e until s_r is nonzero mod p^e.

Theorem 4.6 ([10]). *For a positive integer e, algorithm* Local Ranks by Elimination *mod p^e is correct and runs using* $O(rmn)$ *arithmetic operations mod p^e, where r is the rank mod p^e, and* $O(emn)$ *memory.*

Wiedemann's algorithm and Diagonal Scaling. For some matrices the elimination approach just described fails due to excessive memory demand (disk thrashing). It is desirable to have a memory efficient method for such cases. Two iterative methods are proposed for use here. The first one is "off the shelf". It is to use Wiedemann's algorithm with the diagonal scaling of [13] to compute the rank mod p. This scaling ensures with high probability that the minimal polynomial is a shift of the characteristic polynomial, in fact that it is of the form $m(x) = xf(x)$ where $f(0) \neq 0$ and the characteristic polynomial is of the form $x^k f(x)$ for some k. It follows that the rank is the degree of f. For a given ϵ, to do this with probability of correctness greater than $1 - \epsilon$ requires computation in a field of size $O(n^2/\epsilon)$ [13]. If p is insufficiently large, an extension field may be used. To avoid the large field requirement, we may use the technique as an heuristic, computing the Wiedemann polynomial over a smaller field. The resulting polynomial, $w(x)$, is guaranteed to be a factor of the true minimal polynomial, so that it suffices to verify that $w(A) = 0$. This may be probabilistically done by choosing a vector v at random and computing $w(A)v$. The probability that $w(A)v$ is zero while $w(A)$ is nonzero is no more than $1/p$, hence repetition of this process $\log_2(\epsilon^{-1})$ times ensures that the rank has been computed correctly with probability no less than $1 - \epsilon$.

This algorithm has much lower memory requirements than elimination, requiring $O(E)$ field elements, for $A \in \mathbb{C}^{m \times n}$ with e the number of nonzero entries and $E = \max\{m, n, e\}$. The algorithm also has better asymptotic time complexity, $O(dE \log_2(\epsilon^{-1}))$ field operations, and it is effective in practice for large sparse matrices over large fields. However it does not give the complete local Smith Normal Form at p. In [10] we propose a p-adic way to compute the last invariant factor of this local Smith Normal Form at p. From this, one may infer the complete structure of the local Smith Normal Form at p in some other cases.

4.3   Experiments with Homology Matrices

In this section, we will talk mainly about three homology matrix classes. The matrices will be denoted by three naming patterns:

▷ **mk$i$.bj** denotes the boundary matrix j from the matching complex (see 6.1) with i vertices.
▷ **ch$i$-k.bj** denotes the boundary matrix j from the i by k chessboard complex (see 6.1).
▷ **nick.bj** denotes the boundary matrix j from the not i-connected graph (see 6.1) with k vertices.

The boundary matrices are sparse matrices with a fixed number k of nonzero elements per row and l per column. If A_i is the boundary map between the dimensions i and $i - 1$ of a simplicial complex then $k = i + 1$. All entries are -1, 0 or 1. Moreover, the Laplacians $A_i A_i^t$ and $A_i^t A_i$ also have -1, 0 and 1 as entries except for the diagonal which is respectively all k and all l. However, as expected, those Laplacians have more than twice as many nonzero elements as A_i. Thus, we did not perform the matrix multiplications to compute $A_i A_i^t v$. We performed two matrix-vector products, $A_i(A_i^t v)$, instead.

The Laplacians have indeed a very low degree minimal polynomial (say up to 25 for the matching and chessboard matrices, close to 200 for the not-connected). This fact was our chief motivation to develop the Valence method.

Our experiments were realized on a cluster of 20 Sun Microsystem Ultra Enterprise 450 each with four 250 MHz Ultra-II processor and 1024 Mb or 512 Mb of memory. We computed ranks of matrices over finite fields $GF(q)$ where q has half word size. The chosen arithmetic used discrete Zech logarithms with precomputed tables, as in [36]. The algorithms were implemented in C++ with the LINBOX[1] library for computer algebra and the ATHAPAS-CAN[2] environment for parallelism.

In Table 1 we report some comparisons between Wiedemann's algorithm and elimination with reordering for computing the rank. We just want to emphasize the fact that for these matrices from homology, as long as enough memory is available, elimination is more efficient. However, for larger matrices, Wiedemann's algorithm is competitive and is sometimes the only solution.

Table 1. Rank mod65521 – Elimination vs. Wiedemann

Matrix	Elimination	Wiedemann
mk9.b3	0.26	2.11
mk13.b5	MT	23 hours
ch7-7.b6	4.67	119.53
ch7-6.b4	49.32	412.42
ch7-7.b5	2179.62	4141.32
ch8-8.b4	19 hours	33 hours
ch8-8.b5	MT	55 hours
n2c6.b6	6.44	64.66
n2c6.b7	3.64	49.46
n4c5.b6	2.73	47.17
n4c6.b12	231.34	4131.06
n4c6.b13	8.92	288.57

In Table 2 we compare timings of our algorithm to some implementations of other methods. We compare here only the results obtained using the version of the Valence Smith Form algorithm in which we use Wiedemann's algorithm to compute the Valence and then elimination modulo small powers of primes

---

[1] Symbolic linear algebra library, http://www.linalg.org
[2] Parallel execution support of the APACHE Project, http://www-id.imag.fr/software

p to compute the invariant factors locally at p. Simplicial Homology [9] is a proposed GAP [15] share package. It computes homology groups of simplicial complexes via the Smith Normal Form of their boundary maps. It features a version of our Valence algorithm as well as the classical elimination method for homology groups. The entry "Hom-Elim-GMP" in this table refers to this elimination-based method using Gnu Multi Precision integers. Fermat [31] is a computer algebra system for Macs and Windows. Its Smith Normal Form routine is an implementation of [2].

Table 2. Fermat vs. Hom-Elim-GMP vs. SFV

Matrix	Fermat	Hom-Elim-GMP	Valence (Eliminations)	
ch6-6.b4	49.4	2.151	27.417	(6)
mk9.b3	2.03	0.211	0.946	(4)
mk10.b3	8.4	0.936	13.971	(7)
mk11.b4	98937.27	2789.707	384.51	(7)
mk12.b3	189.9	26.111	304.22	(7)
mk12.b4	MT	MT	13173.49	(7)

"Hom-Elim-GMP" and "Valence" ran on a 400 MHz sparc SUNW, Ultra-4 processor with 512 Mb, but Fermat is only available on Mac and Windows. We therefore report on experiments with Fermat on a 400 MHz Intel i860 processor with only 512 Mb. First we see that "Fermat" cannot compete with "Hom-Elim-GMP" in any case. The main explanation is that the pivot strategy used by "Hom-Elim-GMP" is very well suited to the homology matrices. We can see also that, as long as no coefficient growth is involved, "Hom-Elim-GMP" is often better than "Valence". Indeed, where "Hom-Elim-GMP" performs only one integer elimination, "Valence" performs an elimination for every prime involved (the number of those eliminations is shown between parenthesis in the column Valence (Eliminations) of the table) - of course in parallel this difference will weaken. But as soon as coefficient growth becomes important "Valence" is winning. Moreover, "Valence" using only memory efficient iterative methods can give some partial results where memory exhaustion due to fill-in prevents any eliminations from running to completion. In Table 1 we can see some of these effects and we present some of those partial results in Table 3. Ffor some matrices we were able to compute ranks modulo some primes, and therefore the occurrence of these primes in the Smith Normal Form, but not the actual powers of these primes. These, however, are the only currently known results about these matrices.

With the Valence approach, we were able to compute the rank modulo primes for matrices with 500,000 or more rows and columns, while elimination was failing for matrices of sizes larger than about 50,000. It remains open how

---

Algorithm A: VSF [Valence-Smith-Form]

Data : a matrix $A \in \mathbb{Z}^{m \times n}$. A may be a "black box" meaning that the only requirement is that left and right matrix-vector products may be computed: $x \longrightarrow Ax$ for $x \in \mathbb{Z}^n$, $y \longrightarrow yA$ for $y^t \in \mathbb{Z}^m$.

Result : $S = \mathtt{diag}(s_1, \ldots, s_{\min(m,n)})$, the integer Smith Normal Form of A.

begin

   (1) [Valence computation]
   if $m < n$ **then**
     | let $B = AA^t$
   else
     L   let $B = A^t A$.

   Let $N = \min(m, n)$
   Compute the (minimal) valence v of B.   [See section 4.1]

   (2) [Integer factorization]
   Factor the valence v.
   Let L be the list of primes which divide v.

   (3) [Rank and first estimate of Smith Normal Form.]
   Choose a prime p not in L (i.e. p does not divide v).
   Compute $r = \mathtt{rank}_p(A)$.   [This is the integer rank.
     The first r invariant factors are nonzero and the last $N - r$ *are* 0's.]
   Set $S = \mathtt{diag}(s_1, \ldots, s_N)$, where $s_i = 1$ for $i \leq r$ and $s_i = 0$ for $i > r$.

   (4) [Nontrivial invariant factors]
   foreach *prime* $p \in L$ **do**
       Compute $S_p = \mathtt{diag}(s_{p,1}, \ldots, s_{p,N})$, the Smith Normal Form of A over the localization $\mathbb{Z}_{(p)}$ of $\mathbb{Z}$ at the prime ideal (p).
                                [See section 4.2]
       Set $S = SS_p$.   [That is, set $s_i = s_i s_{p,i}$ for those $s_{p,i}$ which are
                                       nontrivial powers of p.]

   (5) [Return invariants]
   return $S = \mathtt{diag}(s_1 \ldots s_N) \in \mathbb{Z}^{m \times n}$.
end

Table 3. Valence Smith Normal Form with Black Box techniques

Matrix	Time		Results
mk13.b5	98 hours	Partial	133991 ones & 220 powers of 3
ch7-8.b5	180 hours	Partial	92916 ones, 35 powers of 3 & 8 powers of 2 and 3
ch8-8.b4	264 hours	Complete	100289 ones

Algorithm B: IMP [Integer-Minimal-Polynomial]

Data   : a matrix A in $\mathbb{Z}^{n \times n}$
         an error tolerance ϵ, such that $0 < \epsilon < 1$
         an upper bound m on primes for which computations are fast,
         $m > 2^{15}$.
Result : the integer minimal polynomial of A, correct with probability at
         least $1 - \epsilon$.

begin

   (1) [Initialization, first set of primes]
   set $l = 2^{15}$;
   set $d = 0$; set $F = \emptyset$; set $P = \emptyset$;
   $\beta =$ Ovals of Cassini Bound of A;
   set $M = m$;   [computations will be fast]

   (2) [Compute polynomials $\mathtt{mod}\ p_i$]
   repeat
       Choose a prime p_i, with $l < p_i < M$.   [at least $\mu_l(M)$ of those]
       Compute polynomial w_{A,p_i} by Wiedemann's method.
       [$w_{A,p_i} = \mathtt{minpoly}_{A,p_i}$ with probability at least $1 - \frac{1}{p_i}$]
       correct with probability at least $1 - \frac{1}{p_i}$]
       if $\deg(w_{A,p_i}) > d$ **then**
           set $d = \deg(w_{A,p_i})$; set $F = \{p_i\}$; set $P = \{w_{A,p_i}\}$;
           set $U = \sqrt{n}\lceil\frac{d}{2}\rceil\beta^{\frac{(d+1)^2}{2}}$;
           set $bad = \log_l(U)$;   [at most that many bad primes]
           set $b_i = 2 \times bad(1 + \frac{2}{l-2}) + 3512$;   [3512 primes $< 2^{15}$]
           set $M_i =$ upper bound for p_{b_i} ;   [at least b_i primes are $< M_i$]
           if $(M_i > M)$ **then**
               set $M = M_i$;   [computations will be slower,
                         degree will be correct with probability at least $\frac{1}{2}$]

       else
           if $\deg(\mathtt{minpoly}_{A,p_i}) = d$ **then**
               $F = F \cup \{p_i\}$;
               $P = P \cup \{\mathtt{minpoly}_{A,p_i}\}$;

   until $\prod_{F} p_i \geq \max\{\sqrt{d\beta}; \beta\}^d$ *and* $\epsilon \geq \prod_{F}(\frac{1}{p_i} + \frac{bad}{\mu_l(M)})$
   (4) [Chinese remainders]
   return $\mathtt{minpoly}_A = \sum_{j=1}^{d} \alpha_j X^j$,
   where each $\alpha_j \in \mathbb{Z}$ is built from P and F.

end

Algorithm C: LRE [Local-Ranks-by-Elimination-mod-p^e]

Data   : a matrix $A \in \mathbb{Z}^{(m+1)\times(n+1)}$, with elements (a_{ij}) for $i, j \in 0 \ldots m \times$
             $0 \ldots n$.
             a prime p.
             a positive integer e.
Result : the ranks r_{p^i}, for $1 \leq i \leq e$.

begin
    (1) [Initializations]
    set $k = e$
    set $r = 0$

    (2) [e successive Gauss steps]
    for *exponent* $k = 1$ *to* e **do**
        while $\exists (s,t) \in r \ldots m \times r \ldots n, p \nmid a_{st}$ **do**
            [a_{st} is the pivot]
            Swap rows r, s, and columns r, t
            foreach $(i,j) \in \{r + 1 \ldots m\} \times \{r + 1 \ldots n\}$ **do**
                [elimination, with division, mod p^{e-k+1}]
                set $a_{i,j} = a_{i,j} - a_{r,j} a_{i,r} / a_{r,r} \pmod{p^{e-k+1}}$
            set $r = r + 1$.
        set $r_{p^k} = r$.
        [and invariant factors $s_i = p^{k-1}$ for $r_{p^{k-1}} < i \leq r_{p^k}$.]
        foreach $(i,j) \in \{r \ldots m\} \times \{r \ldots n\}$ **do**
            set $a_{ij} = a_{ij}/p$

    (3) [Return local ranks]
    return r_{p^i}, for $i \in \{1 \ldots e\}$
end

---

to efficiently determine the ranks modulo powers (> 1) of primes while using memory-efficient iterative methods.

5 Other Methods

Algebraic Shifting. It would be desirable to be able to read off the homology of a simplicial complex 'directly' from the list of its maximal faces. For shellable complexes for example a shellability of its maximal faces will allow to determine the homology. But calculating a shelling order probably is hard, even if one knows a priori that the complex is shellable (see [27]). But for a particular class of shellable complexes a simple procedure for determining the homology is known. A simplicial complex Δ over ground set $\Omega = [n]$ is called shifted if for $\sigma \in \Delta$ and number $1 \leq i < j \leq n$ such that $i \notin \sigma$ and $j \in \sigma$ we have $\sigma \setminus \{j\} \cup \{i\} \in \Delta$.

Lemma 5.1. *If Δ is a shifted complex then Δ is shellable and its ith homology group $\tilde{H}_i(\Delta; \mathbb{Z})$ is free and its rank is given by the maximal faces of Δ of cardinality $i + 1$ that do not contain the element 1.*

Clearly, this is a very simple algorithm that allows a linear time determination of the homology groups from the maximal faces. But we must admit that shifted complexes are a very small subclass even of the class of shellable complexes. But there is an algorithmic procedure that determines a shifted complex $\Delta(\Delta)$ from a simplicial complex Δ. This procedure is called algebraic shifting and was discovered by Gil Kalai (see [25] for the original references and details).

Let Δ be a simplicial complex over $[n]$. Let E be the exterior algebra freely generated by $e_1, \ldots, e_n$. We write $e_A = e_{j_1} \wedge \cdots \wedge e_{j_s}$ for ordered subsets $\{j_1 < \cdots < j_s\}$ of $[n]$. By J_Δ be denote the ideal generated by the monomials e_A where $A \notin \Delta$. The ideal J_Δ is called the Stanley-Reisner ideal of Δ in E. Clearly, by virtue of exterior algebra, all monomials in E are of the form e_A for some $A \subseteq [n]$. Thus in order to specify a linear order on the monomials in E it suffices to order the subsets of $[n]$ linearly. Here we use lexicographic order '$\prec_l$', where $A \prec_l B$ if the minimum of $(A - B) \cup (B - A)$ lies in A. Now let us perform a generic linear transformation of the generators $e_1, \ldots, e_n$. Thus we obtain new generators $f_i = \sum_{j=1}^n a_{ij} e_j$. Now we consider the set system:

$$\Delta^{\text{shifted}} = \left\{ A \subseteq [n] \mid f_A \notin \text{span}_k \{f_B | B \prec_l A\} + J_\Delta \right\}.$$

Proposition 5.2 (Kalai; see [25]). *The set system Δ^{shifted} is a shifted simplicial complex.*

Now we state the theorem which allows us to use algebraic shifting for homology calculations.

Theorem 5.3 (Kalai; see [25]). *Let Δ be a simplicial complex and Δ^{shifted} be the result of algebraically shifting Δ over the field k then:*

$$\tilde{H}_i(\Delta; k) = \tilde{H}_i(\Delta^{\text{shifted}}; k).$$

We recall a question by Kalai [25].
Is there a polynomial time algorithm for computing the algebraic shifted complex of a simplicial complex Δ which is given by its maximal faces ?

Minimal Free Resolutions of Stanley-Reisner Ideals. Let Δ be a simplicial complex over ground set $[n]$. Let $S = k[x_1, \ldots, x_n]$ be the polynomial ring in n variables over k. The Stanley-Reisner ideal of Δ in S is, analogous to the situation in the exterior algebra, the ideal I_Δ generated by monomials $x_A = \prod_{i \in A} x_i$, where $A \notin \Delta$. A multigraded module M over S is an S-module which allows as a vectorspace a direct sum decomposition $M = \bigoplus_{\alpha \in \mathbb{N}^n} M_\alpha$, such that $\underline{\mathbf{x}}^\beta \mathbf{M}_\alpha \subseteq \mathbf{M}_{\alpha+\beta}$, where $\underline{x}^\alpha = x_1^{\alpha_1} \cdots x_n^{\alpha_n}$ for $\alpha = (\alpha_1, \ldots, \alpha_n)$. Clearly, S itself and I_Δ are multigraded. We consider multigraded free resolutions of I_Δ as a module over the polynomial ring.

$$\mathcal{F}: \cdots \to \bigoplus_{\alpha \in \mathbb{N}^n} S(-\alpha)^{\beta_i^\alpha} \to \cdots \to \bigoplus_{\alpha \in \mathbb{N}^n} S(-\alpha)^{\beta_0^\alpha} \to I_\Delta \to 0,$$

such that each ∂_i is a homogeneous map of multigraded modules and $\text{Ker}\partial_i = \text{Im}\partial_{i+1}$. By tensoring $\mathcal{F}$ with the multigraded S-module $k = S/(x_1, \ldots, x_n)$ we obtain a complex

$$\mathcal{F} \otimes k: \cdots \to \bigoplus_{\alpha \in \mathbb{N}^n} k(-\alpha)^{\beta_i^\alpha} \to \cdots \to \bigoplus_{\alpha \in \mathbb{N}^n} k(-\alpha)^{\beta_0^\alpha} \to 0$$

of k-vectorspaces, with homogeneous differential $\partial_i \otimes k$, i.e., $\partial_{i-1} \otimes k \circ \partial_i \otimes k = 0$. Its homology groups $H_i(\mathcal{F} \otimes k)$ are known as $\text{Tor}_i^S(I_\Delta, k)$. Since all differentials are multigraded this induces a decomposition

$$\bigoplus_{\alpha \in \mathbb{N}^n} \text{Tor}_i^S(I_\Delta, k)_\alpha.$$

A special case of a result by Hochster says:

Theorem 5.4 (Hochster [20]). *Let Δ be a simplicial complex over $[n]$. Then*

$$\text{Tor}_i^S(I_\Delta, k)_{(1,\ldots,1)} \cong \widetilde{H}_{n-2-i}(\Delta, k).$$

Thus the differential $\partial_i \otimes k$ yields a differential that is in general different from the simplicial differential of Δ but computes the same homology groups with field coefficients. Now the goal is to construct a most economic $\partial_i \otimes k$ (i.e. find a free resolution of I_Δ which has a very small $(1, \ldots, 1)$ strand).

Clearly the most economic resolution in this respects is the minimal free resolution of I_Δ for which the differential assumes it simplest form $\partial_i \otimes k = 0$ for all i. But even though computer algebra packages like Macaulay 2 [19] and Cocoa [7] can compute the minimal free resolution, it involves computing Gröbner bases and hence the feasibility of the approach is most likely very low – we have not performed comparative experiments though.

There are other free resolutions which are much simpler to construct given monomial generators of I_Δ. As an example we would like to mention Taylor's resolution [8].

We consider it an interesting and challenging problem to analyze free resolution of I_Δ with respect to their effectiveness calculating the homology of Δ in their $(1, \ldots, 1)$ strand.

Deducing homology of a simplicial complex from its dual. The algebraic methods described above use ideals that have in their description the minimal non-faces of a simplicial complex. This fundamentally differs from the description of a simplicial complex by facets. Here we discuss how this effects homology computations.

If Δ is a simplicial complex over ground set $[n]$ then we denote by $\Delta^* = \{A \mid [n] - A \notin \Delta\}$ the dual simplicial complex. So A is a minimal non-face of Δ if and only if $[n] - A$ is a maximal face of Δ^*. Now the homology with field coefficients of Δ and Δ^* are related by Alexander duality [32, Cor. 72.4] and the isomorphism of homology and cohomology with field coefficients [32, Cor. 53.6]:

$$\widetilde{H}_i(\Delta; k) \cong \widetilde{H}_{n-i-3}(\Delta^*; k).$$

In particular this shows that for calculating homology it is irrelevant whether the simplicial complex is given by its maximal faces or its minimal non-faces even though in general it seems to be a difficult problem to pass from one representation to the other (we are grateful to Marc Pfetsch for pointing us to [14] or [28]). In particular, the complexity of the algebraic methods – algebraic shifting and finite free resolutions – described above does not depend on whether the simplicial complex is given by its maximal faces or its minimal non-faces.

Deducing integer homology from field coefficient homology. The method of algebraic shifting and the method of free resolutions will only calculate simplicial homology with coefficients in a field k. Clearly there are only finitely many primes, we call them the critical primes, for which a simplicial complex can have p-torsion. By the Universal Coefficient Theorem [32, Thm. 53.5] the sizes of the torsion parts for the various primes for which torsion occurs can be deduced from the homology computations over primes fields for all critical primes and in characteristic 0. Also the set of critical primes can be deduced as it is done within the Valence Algorithm (see Section 4). But then the problem arises how to distinguish p, p^2, ... torsion. We find it a challenging problem to find efficiently computable invariants that help to distinguish the p^i from p^j torsion for $i \neq j$.

6   Sample Applications

6.1   Complexes of Graphs

Complexes of graphs have been the motivating source for the development of our GAP [15] package. Let $G = (V, E)$ be a graph on vertex set V with edge set E. We assume $V = [n]$ and also assume that G contains no loops and no multiple edges. Thus E can be considered as a subset of $\binom{[n]}{2} = \{A \subseteq [n] \mid \#A = 2\}$. Conversely any subset of $\binom{[n]}{2}$ can be considered as the edge set of a graph. Hence from now on we do not distinguish between graphs and subsets of $\binom{[n]}{2}$. We call a simplicial complex Δ on ground set $\binom{[n]}{2}$ a complex of graphs. This concept is very general but usually we are interested in graph complexes that have a high symmetry with respect to the set of vertices.

Complexes of not i-connected graphs. A graph G is called 1-connected if for all $i, j \in V$ there is a sequence of edges that connects i and j – this is what we usually call a connected graph. For $i \geq 2$ a graph G is called i-connected if it is j-connected for all $1 \leq j \leq i - 1$ and whenever $(i - 1)$ vertices together with their adjacent edges are deleted the graph remains 1-connected. Clearly, removal of edges preserves the property of being not i-connected. Thus the set $\Delta_{n,i}$ of all not i-connected graphs on $[n]$ is a simplicial complex. The case $i = 2$ has been the primary motivation and starting point for the development of our GAP [15] package. Indeed calculations of early versions and predecessors of the package led to a conjecture that turned into the following theorem.

Theorem 6.1 ([1] and [39]).

$$\widetilde{H}_j(\Delta_{n,i}; \mathbb{Z}) = \begin{cases} 0 & j \neq 2n - 5 \\ \mathbb{Z}^{(n-2)!} & j = 2n - 4 \end{cases}$$

Indeed matrices coming from this example have been used as timing examples in previous parts of this manuscript.

Next we demonstrate in a sample session the use of some of the functions of the homology package.

First, start GAP and load the package.

```
> gap
```

```
        #########          ######        ##########             ###
      #############        ######        ############          ####
      #############        ########       #############        #####
     ###############       ########       #####  ######        #####
     ######      #        #########       #####   #####       ######
     ######                #########       #####   #####      #######
     #####                 ##### ####       #####   ######     ########
     ####                  #####  #####      #############    ###  ####
     #####     #######      ####   ####      ##########      ####  ####
     #####     #######      #####   #####     ######        ####   ####
     #####     #######      #####   #####     #####        #############
      #####      #####     ###############     #####        #############
     ######      #####     ###############     #####        #############
     ################      #################    #####            ####
      ##############       #####       #####    #####            ####
       #############       #####       #####    #####            ####
        #########          #####       ##### #####               ####
```

```
Information at:  http://www-gap.dcs.st-and.ac.uk/~gap
```

```
? for help. Copyright and authors list by ?copyright, ?authors

Loading the library. Please be patient, this may take a while.
GAP4, Version: 4.1 fix 2 of 27-Aug-1999, sparc-sun-solaris2.6-gcc
Components:  small, small2, small3, id2, id3, trans, prim, tbl,
            tom  installed.
gap> RequirePackage("homology");
true
```

Let us calculate the homology of the complex of not 2-connected graphs on 7 vertices. By Theorem 6.1 it is know that the homology is concentrated in dimension $2 \cdot 7 - 5 = 9$ and is free of dimension $(7 - 2)! = 120$. So let us reconfirm the homology in dimension 9 only.

```
gap> SimplicialHomology(SCNot2ConnectedGraphs(7),9);
#I:D Simplicial complex of dimension 15 with 217 facets
#I:D   homology: Computing homology groups
#I:D     faces: Finding faces of dimension 8
#I:D     faces: Found 247100 faces of dimension 8
#I:D     faces: Finding faces of dimension 9
#I:D     faces: Found 219135 faces of dimension 9
#I:D   homology: Finding rank of boundary map d9
#I:D     elimination: Triangulating matrix by modified Gauss
                                              elimination
#I:D     elimination: current torsion-free rank: 0, current rank:
                                       0, max. rank: 247100
#I:D     elimination: current torsion-free rank: 715, current rank:
                                       715, max. rank: 247100
```

The last line tells us that rows of the matrix of the 9th differential processed so far have a span of rank 715. Using general implications from the fact that the sequence of differential is a complex (i.e., $\partial_{i-1} \circ \partial_i = 0$) the maximal possible rank this matrix can have is 247100. Let us look at the output a few minutes later.

```
#I:D     elimination: current torsion-free rank: 128311,
                        current rank: 128311, max. rank: 247100
#I:D     elimination: current torsion-free rank: 128330,
                        current rank: 128330, max. rank: 247100
#I:D     elimination: Matrix triangulated
#I:R   homology: Rank d9 = 128330
#I:D     faces: Finding faces of dimension 10
#I:D     faces: Found 135765 faces of dimension 10
#I:D   homology: Finding rank of boundary map d10
#I:D     elimination: Triangulating matrix by modified Gauss
                                              elimination
#I:D     elimination: current torsion-free rank: 0,
                        current rank: 0, max. rank: 90805
```

The 9th differential has been brought into Smith Normal form. But for calculation the homology group in homological dimension 9 the program has already started to analyze the matrix of the 10th differential. Another few minutes waiting bring us to:

```
#I:D      elimination: current torsion-free rank: 90344,
                        current rank: 90344, max. rank: 90805
#I:D      elimination: current torsion-free rank: 90685,
                        current rank: 90685, max. rank: 90805
#I:D      elimination: Matrix triangulated
#I:R   homology: Rank d10 = 90685
#I:R   homology: H_9 = 120.
#I:D   homology: Homology groups computed
[ [ 120 ] ]
```

After Smith Normal forms of the 9th and 10th differential are calculated the program determines the homology group we have asked for and indeed it is as expected. Note that in the process of calculating this group we have brought within roughly 15 minutes (on a Linux PC, 700 MHZ) two matrices into Smith Normal Form whose row and column sizes are in the 5 or 6 digit numbers.

For $i = 1$ the homology of $\Delta_{n.i}$ had been determined as a one of the first examples of homology computations in combinatorics (see [1] and [39]). For $i = 3$ a conjecture from [1] also based on computer calculations was later verified by Jakob Jonsson. For $i > 3$ the problem is wide open. But for $i = n - 2$ a simple Alexander duality argument reduces the problem to calculating homology of matching complexes.

Matching Complexes. If G is a graph with vertex set E then a partial matching on G is a subset H of E such that no two edges in G share a vertex. Clearly the set Δ_M^G of all partial matchings on G is a simplicial complex. If G is the complete graph on n vertices then we simply call $\Delta_M^G = \Delta_M^n$ the matching complex and when G is the complete bipartite complex on $2n$ vertices with n vertices in each partition then we call $\Delta_M^G = \Delta_M^{n,n}$ the chessboard complex – the name is derived from the fact that the complex can also defined as the set of non-taking rook positions on an n by n chessboard.

Both Δ_M^n and $\Delta_M^{n,n}$ are ubiquitous in mathematics (see [41]) for an excellent survey). Also both complexes have raised interest from the computational point of view. In contrast to the complexes $\Delta_{n,i}$ for $i \leq 3$ these complexes exhibit torsion. Strikingly, the torsion seems to be 3-torsion only.

Indeed Shareshian and Wachs (see [41]) have confirmed this, but their proof still is based on some computer calculations with our package. We regard this a prototype application of our package: Run examples, Get a Conjecture, Prove the Conjecture. But in this case the computer calculations are not yet out of the the picture. Going along with this philosophy Wachs [41] poses the problem to eliminate the computer calculations from their proof.

6.2 Lie Algebra Homology

In this section we show that the computational tools we have implemented are not restricted to calculating homology of simplicial complexes, but can be applied to computation of the homology of more general algebraic complexes. As an example we here consider the homology of finite dimensional Lie algebras. Let L be a Lie algebra of finite dimension over the field k. Consider the following map:

$$\partial_i : \overset{i}{\bigwedge} L \to \overset{i-1}{\bigwedge} L$$

defined by

$$\partial_i(a_1 \wedge \cdots \wedge a_i) = \sum_{1 \leq l < j \leq i} (-1)^{l+j}[a_j, a_j] \wedge a_1 \wedge \cdots \wedge \widehat{a_l} \wedge \cdots \wedge \wedge a_j \wedge \cdots \wedge a_i.$$

Then $\partial_i \circ \partial_{i+1} = 0$ and $\text{Im}\partial_{i+1} \subseteq \text{Ker}\partial_i$. Therefore, we can define a homology group

$$\widetilde{H}_i(L) = \text{Ker}\partial_i / \Im\partial_{i+1}.$$

The definition makes sense for arbitrary Lie algebras L over a field k. If L is finite-dimensional as a vector-space over k then each $\bigwedge^i L$ is a finite-dimensional vectorspace over k and hence the computation of $\widetilde{H}_i(L)$ reduces to the computation of ranks of matrices.

In this section we report about experiments run on Lie algebras associated with finite partially ordered sets. Let us consider $P = \{p_1, \ldots, p_n\}$, a finite partially ordered set with order relation $\prec$, and k a field. We denote by $N_k(P)$ the set of matrices $N = (n_{ij}) \in k^{n \times n}$ such that $n_{ij} \neq 0$ implies $p_i \prec p_j$ (strict inequality). If we assume that $p_i \prec p_j$ implies $i \leq j$ then $N_k(P)$ is the set of lower triangular nilpotent matrices with 0's in places where "there is no order relation in P." Is is easy to see that $N_k(P)$ with $[A, B] = AB - BA$ is a Lie algebra. The dimension of $\dim N_k(P)$ is given by the number of strict order relations in P. For example if $P = C(n)$ is a chain $C(n) = \{p_1 \prec p_1 \prec \cdots \prec p_n\}$ then $N_k(C(n))$ is the set of all lower triangular nilpotent matrices and $\dim N_k(C(n)) = \binom{n}{2}$. If char $k = 0$ then a well known result by Kostant [30] gives complete information on the homology groups. Here we denote by $\text{Inv}(\sigma)$ the set of inversions of a permutation σ; that is the set of pairs (i, j) where $1 \leq i < j \leq n$ and $\sigma(j) < \sigma(i)$.

Theorem 6.2 (Kostant [30]). *Let k be field of characteristic 0. Then:*

$$\dim_k H_i(N_k(C(n)) = \#\{\sigma \in S_n \mid \#\text{Inv}(\sigma) = i\}.$$

It is also well known that there is the same number of permutations in S_n with i inversions as with $\binom{n}{2} - i$ inversions. This is an incarnation of Poincaré duality for Lie algebra homology (see [29, Ch. VII 9]).

Theorem 6.3. *Let L be a finite dimensional Lie algebra of dimension $d = \dim_k L$. Then $\dim_k H_i(L) = \dim_k H_{d-i}(L)$, $0 \leq i \leq d$.*

Thus in oder to calculate the homology of a finite dimensional Lie algebra it suffices to calculate the homology groups half the dimension up. There are two challenging problems that arise for Lie algebras $N_k(P)$:

▷ Express the dimension $\dim_k H_k(N_k(P))$ in terms of P when k is a field of characteristic 0.
▷ Express the dependency of $\dim_k H_k(N_k(P))$ on the characteristic of the field k.

For the latter question integer coefficients become of interest. If we consider the same differential with coefficients in the integers, calculating homology again amounts to computing Smith Normal Form of the matrices of the differentials. For $P = C(n)$ we observe the following picture:

The ranks of the free parts obey Poincare-duality and as usual with integer coefficients the torsion groups obey Poincare-duality in homological dimension shifted by 1.

We do not know an interpretation for the torsion, but we think that this is an interesting question to construct a combinatorial model determining the torsion coefficients.

There is a result by Dwyer [12] which says that p^l torsion occurs in the homology of $N_{\mathbb{Z}}(C(n))$ only if $p \leq n - 2$. Moreover, for $p \leq n - 2$ there is p-torsion in $N_{\mathbb{Z}}(C(n))$.

Thus for both questions raised above and their analogs over the integers a variant of our software can help to calculate examples and infer conjectures.

As an example we present values for the case when $P = B_n$ is the Boolean algebra of all subsets of $[n]$.

7   Other Invariants of Simplicial Complexes

Algebraic topology has developed many other invariants of topological spaces that can also be looked at in the case of simplicial complexes. We just want to comment on a few of them which we consider the most prominent.

Cohomology. The calculation of the cohomology $\widetilde{H}^i(\Delta; R)$ of a simplicial complex Δ reduces to the same problems as for homology. Note that the boundary matrices for cohomology are, when choosing the canonical bases, the transpose of the boundary matrices for homology.

Explicit cycles and co-cycles. More challenging is the problem of actually calculating explicit generators for the homology and cohomology groups. Thus the goal is to find a set of elements of $C_i(\Delta; R)$ that are representatives of a minimal generating set of $\widetilde{H}_i(\Delta; R)$ (resp. $\widetilde{H}^i(\Delta; R)$). Representatives of elements of $\widetilde{H}_i(\Delta; R)$ (resp. $\widetilde{H}^i(\Delta; R)$) are called cycles (resp. co-cycles). If we

Table 4. Lie Algebra homology for $P = C(n)$

$i \backslash n$	2	3	4	5	6	7
0	$\mathbb{Z}$	$\mathbb{Z}$	$\mathbb{Z}$	$\mathbb{Z}$	$\mathbb{Z}$	$\mathbb{Z}$
1	$\mathbb{Z}$	$\mathbb{Z}^2$	$\mathbb{Z}^3$	$\mathbb{Z}^4$	$\mathbb{Z}^5$	$\mathbb{Z}^6$
2		$\mathbb{Z}^2$	$\mathbb{Z}^5 \oplus \mathbb{Z}_2$	$\mathbb{Z}^9 \oplus \mathbb{Z}_2^2$	$\mathbb{Z}^{14} \oplus \mathbb{Z}_2^3$	$\mathbb{Z}^{20} \oplus \mathbb{Z}_2^4$
3		$\mathbb{Z}$	$\mathbb{Z}^6 \oplus \mathbb{Z}_2$	$\mathbb{Z}^{15} \oplus \mathbb{Z}_2^8 \oplus \mathbb{Z}_3^2$	$\mathbb{Z}^{29} \oplus \mathbb{Z}_2^{20} \oplus \mathbb{Z}_3^4$	$\mathbb{Z}^{49} \oplus \mathbb{Z}_2^{35} \oplus \mathbb{Z}_3^6$
4			$\mathbb{Z}^5$	$\mathbb{Z}^{20} \oplus \mathbb{Z}_2^{10} \oplus \mathbb{Z}_3^3$	$\mathbb{Z}^{49} \oplus \mathbb{Z}_2^{47} \oplus \mathbb{Z}_4^3 \oplus \mathbb{Z}_3^{13}$	$\mathbb{Z}^{98} \oplus \mathbb{Z}_2^{124} \oplus \mathbb{Z}_4^6 \oplus \mathbb{Z}_3^{21}$
5			$\mathbb{Z}^3$	$\mathbb{Z}^{22} \oplus \mathbb{Z}_2^{10} \oplus \mathbb{Z}_3^3$	$\mathbb{Z}^{71} \oplus \mathbb{Z}_2^{79} \oplus \mathbb{Z}_4^9 \oplus \mathbb{Z}_3^{26}$	$\mathbb{Z}^{169} \oplus \mathbb{Z}_2^{303} \oplus \mathbb{Z}_4^{28} \oplus \mathbb{Z}_3^{78} \oplus \mathbb{Z}_5^4$
6			$\mathbb{Z}$	$\mathbb{Z}^{20} \oplus \mathbb{Z}_2^8 \oplus \mathbb{Z}_3^2$	$\mathbb{Z}^{90} \oplus \mathbb{Z}_2^{118} \oplus \mathbb{Z}_4^{12} \oplus \mathbb{Z}_3^{35}$	$\mathbb{Z}^{259} \oplus \mathbb{Z}_2^{635} \oplus \mathbb{Z}_4^{65} \oplus \mathbb{Z}_3^{168} \oplus \mathbb{Z}_5^{17}$
7				$\mathbb{Z}^{15} \oplus \mathbb{Z}_2^2$	$\mathbb{Z}^{101} \oplus \mathbb{Z}_2^{138} \oplus \mathbb{Z}_4^{12} \oplus \mathbb{Z}_3^{36}$	$\mathbb{Z}^{359} \oplus \mathbb{Z}_2^{1122} \oplus \mathbb{Z}_4^{112} \oplus \mathbb{Z}_3^{275} \oplus \mathbb{Z}_5^{39}$
8				$\mathbb{Z}^9$	$\mathbb{Z}^{101} \oplus \mathbb{Z}_2^{118} \oplus \mathbb{Z}_4^{12} \mathbb{Z}_3^{35}$	??
9				$\mathbb{Z}^4$	$\mathbb{Z}^{90} \oplus \mathbb{Z}_2^{79} \oplus \mathbb{Z}_3^{26} \oplus \mathbb{Z}_4^9$	??
10				$\mathbb{Z}$	$\mathbb{Z}^{71} \oplus \mathbb{Z}_2^{47} \oplus \mathbb{Z}_4^3 \oplus \mathbb{Z}_3^{13}$	??
11					$\mathbb{Z}^{49} \oplus \mathbb{Z}_2^{20} \oplus \mathbb{Z}_3^9$	??
12					$\mathbb{Z}^{29} \oplus \mathbb{Z}_2^3$	??
13					$\mathbb{Z}^{14}$	??
14					$\mathbb{Z}^5$	$\mathbb{Z}^{359} \oplus \mathbb{Z}_2^{635} \oplus \mathbb{Z}_4^{65} \oplus \mathbb{Z}_3^{168} \oplus \mathbb{Z}_5^{17}$
15					$\mathbb{Z}$	$\mathbb{Z}^{259} \oplus \mathbb{Z}_2^{303} \oplus \mathbb{Z}_4^{28} \oplus \mathbb{Z}_3^{78} \oplus \mathbb{Z}_5^4$
16						$\mathbb{Z}^{169} \oplus \mathbb{Z}_2^{124} \oplus \mathbb{Z}_4^6 \oplus \mathbb{Z}_3^{27}$
17						$\mathbb{Z}^{98} \oplus \mathbb{Z}_2^{35} \oplus \mathbb{Z}_3^6$
18						$\mathbb{Z}^{49} \oplus \mathbb{Z}_2^4$
19						$\mathbb{Z}^{20}$
20						$\mathbb{Z}^6$
21						$\mathbb{Z}$

Table 5. Lie-Algebra homology for $P = B_n$

$i\backslash n$	1	2	3
0	$\mathbb{Z}$	$\mathbb{Z}$	$\mathbb{Z}$
1		$\mathbb{Z}^4$	$\mathbb{Z}^{12}$
2		$\mathbb{Z}^5$	$\mathbb{Z}^{60}$
3		$\mathbb{Z}^5$	$\mathbb{Z}^{179} \oplus \mathbb{Z}_2$
4		$\mathbb{Z}^4$	$\mathbb{Z}^{486} \oplus \mathbb{Z}_2^{13}$
5		$\mathbb{Z}$	$\mathbb{Z}^{1026} \oplus \mathbb{Z}_2^{81} \oplus \mathbb{Z}_3^4$
6			$\mathbb{Z}^{1701} \oplus \mathbb{Z}_2^{250} \oplus \mathbb{Z}_4^6 \oplus \mathbb{Z}_3^{30} \oplus \mathbb{Z}_5^6$
7			$\mathbb{Z}^{2353} \oplus \mathbb{Z}_2^{643} \oplus \mathbb{Z}_4^{55} \oplus \mathbb{Z}_3^{118}$
8			$\mathbb{Z}^{2701} \oplus \mathbb{Z}_2^{1094} \oplus \mathbb{Z}_4^{38} \oplus \mathbb{Z}_3^{141} \oplus \mathbb{Z}_5^{12}$
9			$\mathbb{Z}^{2773} \oplus \mathbb{Z}_2^{1350} \oplus \mathbb{Z}_4^{48} \oplus \mathbb{Z}_3^{268} \oplus \mathbb{Z}_5^2$
10			$\mathbb{Z}^{2773} \oplus \mathbb{Z}_2^{1094} \oplus \mathbb{Z}_4^{38} \oplus \mathbb{Z}_3^{141} \oplus \mathbb{Z}_5^{12}$
11			$\mathbb{Z}^{2701} \oplus \mathbb{Z}_2^{643} \oplus \mathbb{Z}_4^{55} \oplus \mathbb{Z}_3^{118}$
12			$\mathbb{Z}^{2353} \oplus \mathbb{Z}_2^{250} \oplus \mathbb{Z}_4^6 \oplus \mathbb{Z}_3^{30} \oplus \mathbb{Z}_5^6$
13			$\mathbb{Z}^{1701} \oplus \mathbb{Z}_2^{81} \oplus \mathbb{Z}_3^4$
14			$\mathbb{Z}^{1026} \oplus \mathbb{Z}_2^{13}$
15			$\mathbb{Z}^{486} \oplus \mathbb{Z}_2$
16			$\mathbb{Z}^{179}$
16			$\mathbb{Z}^{60}$
17			$\mathbb{Z}^{12}$
18			$\mathbb{Z}$

analyze our algorithms then we see that elimination in principle can achieve this goal but with an enormous extra amount of memory consumption – we have to store transformations that are applied in order to get Smith Normal Form (see [24] for worst case considerations of the integer linear algebra problems involved in a solution of the problem). Note that the Valence algorithm will not even be able to give us explicit generators.

Cohomology algebra. One can impose on the direct sum of Abelian groups $H^*(\Delta; \mathbb{Z}) = \bigoplus_{i \geq 0} H^i(\Delta; \mathbb{Z})$ naturally the structure of an algebra [32, §48] (note that here we use non-reduced cohomology, which differs from reduced cohomology only by a copy of $\mathbb{Z}$ in dimension 0). Once explicit cocycles are known a presentation of this algebra can be given using linear algebra only. Applying Gröbner base techniques even allows standardized presentations and calculations of algebra invariants. Even though there is not problem in theory the explicit calculations seem feasible only for very small examples. On the other hand the importance of the cohomology algebra makes it a challenging goal to find efficient algorithms for its computation.

Fundamental group. Let us assume in the following that the simplicial complex Δ is connected. The fundamental group of Δ is a finitely presented group $\pi_1(\Delta)$. Conversely it is easily seen that every finitely presented group G can occur as the fundamental group of a simplicial complex Δ (see [33, Exercise 2, p. 445]). For a simplicial complex a presentation of the fundamental group can be efficiently constructed. But results by Boone [3] and Novikov [34] show that the word problem for such groups is not decidable. Thus the fundamental group is an invariant which we can easily give a presentation for but whose properties are in general not computable. Nevertheless, there are results that show that algorithms give results in nice cases [35]. In this paper Rees and Soicher also use the fundamental group to give an algorithm for calculating the first homology group with coefficients in a field of characteristic $p > 0$. They use the well know fact that the Abelianization of $\pi_1(\Delta)$ is isomorphic to $\widetilde{H}_1(\Delta; \mathbb{Z})$.

References

1. Eric Babson, Anders Björner, Svante Linusson, John Shareshian and Volkmar Welker Complexes of not i-connected graphs. *Topology* **38** (1999) 271-299.
2. Achim Bachem and Ravindran Kannan. Polynomial algorithms for computing the Smith and Hermite normal forms of an integer matrix. *SIAM J. Comput.* **8** 499-507 (1979).
3. William W. Boone. Certain simple unsolvable problems in group theory. Indig. Math. **16** 231-237 (1955).
4. Alfred Brauer. Limits for the characteristic roots of a matrix. I. *Duke Math. J.* **13** 387-395 (1946).
5. Alfred Brauer. Limits for the characteristic roots of a matrix. II. *Duke Math. J.* **14** 21-26 (1947).
6. Richard A. Brualdi and Stephen Mellendorf. Regions in the complex plane containing the eigenvalues of a matrix. *American Mathematical Monthly* **101** 975-985 (1994).
7. Antonio Capani, Gianfranco Niesi and Lorenzo Robbiano, *CoCoA, a system for doing Computations in Commutative Algebra*. Available via anonymous ftp from: `cocoa.dima.unige.it`
8. David Eisenbud. *Commutative Algebra with a View Toward Algebraic Geometry*, Springer, 1993.
9. Jean-Guillaume Dumas, Frank Heckenbach, B. David Saunders and Volkmar Welker. *Simplicial Homology, a (proposed) share package for GAP*, March 2000. Manual (`http://www.cis.udel.edu/~dumas/Homology`).
10. Jean-Guillaume Dumas, B. David Saunders and Gilles Villard. On efficient sparse integer matrix Smith normal form computations. *J. Symb. Comp.* **32** 71-99 (2001).
11. Pierre Dusart. *Autour de la fonction qui compte le nombre de nombres premiers.* PhD thesis, Université de Limoges, 1998.
12. William G. Dwyer. Homology of integral upper-triangular matrices. Proc. Amer. Math. Soc. **94** 523-528 (1985).
13. Wayne Eberly and Erich Kaltofen. On randomized Lanczos algorithms. In Wolfgang W. Küchlin, editor, *Proceedings of the 1997 International Symposium on Symbolic and Algebraic Computation, Maui, Hawaii*, pages 176-183. ACM Press, New York, July 1997.

14. Michael L. Fredman and Leonid Khachiyan. On the complexity of dualization of monotone disjunctive normal forms, *J. Algorithms* **21** (1996) 618-628.
15. The GAP Group, GAP — Groups, Algorithms, and Programming, Version 4.2; 2000. (http://www.gap-system.org)
16. Joachim von zur Gathen and Jürgen Gerhard. *Modern Computer Algebra.* Cambridge University Press, New York, NY, USA, 1999.
17. Mark W. Giesbrecht. *Probabilistic Computation of the Smith Normal Form of a Sparse Integer Matrix.* Lecture Notes in Comp. Sci. 1122, 173–186. Springer, 1996.
18. Gene H. Golub and Charles F. Van Loan. *Matrix computations.* Johns Hopkins Studies in the Mathematical Sciences. The Johns Hopkins University Press, Baltimore, MD, USA, third edition, 1996.
19. Daniel R. Grayson and Michael E. Stillman, *Macaulay 2, a software system for research in algebraic geometry,* Available at http://www.math.uiuc.edu/Macaulay2/
20. Melvin Hochster. Cohen-Macaulay rings, combinatorics, and simplicial complexes. *Ring Theory II, Proc. 2nd Okla. Conf. 1975.* Lecture Notes in Pure and Applied Mathematics. Vol. 26. New York. Marcel Dekker. 171-223 (1977).
21. Phil Hanlon. A survey of combinatorial problems in Lie algebra homology. Billera, Louis J. (ed.) et al., *Formal power series and algebraic combinatorics.* Providence, RI: American Mathematical Society. DIMACS, Ser. Discrete Math. Theor. Comput. Sci. 24, 89-113, 1996.
22. Iztok Hozo, *Inclusion of poset homology into Lie algebra homology.* J. Pure Appl. Algebra **111** 169-180 (1996).
23. Costas S. Iliopoulos. Worst case bounds an algorithms for computing the canonical structure of finite Abelian groups and the Hermite and Smith normal forms of an integer matrix. *SIAM J. Comput.* **18** (1989) 658-669.
24. Costas S. Iliopoulos. Worst case bounds an algorithms for computing the canonical structure of infinite Abelian groups and solving systems of linear diophantine equations. *SIAM J. Comput.* **18** (1989) 670-678.
25. Gil Kalai. Algebraic Shfiting. *Computational Commutative Algebra and Combinatorics* Advanced Studies in Pure Math., Vol. 33. Tokyo. Math. Soc. of Japan. 121-165 (2002).
26. Erich Kaltofen, Wen-Shin Lee and Austin A. Lobo. Early termination in Ben-Or/Tiwari sparse interpolation and a hybrid of Zippel's algorithm. In Carlo Traverso, editor, *Proceedings of the 2000 International Symposium on Symbolic and Algebraic Computation, Saint Andrews, Scotland,* pages 192–201. ACM Press, New York, 2000.
27. Volker Kaibel and Marc E. Pfetsch. Some algorithmic problems in polytope theory, this volume, pages 23–47.
28. Dimitris Kavvadias and Elias Stavroploulos. Evaluation of an algorithm for the transversal hypergraph problem. in: *Proc. 3rd Workshop on Algorithm Engineering (WAE'99),* Lecture Notes in Comp. Sci. 1668, Springer, 1999.
29. Anthony W. Knapp. *Lie Groups, Lie Algebras, and Cohomology,* Mathematical Notes **34**, Princeton University Press, 1988.
30. Bertram Kostant. Lie algebra cohomology and the generalized Borel-Weil theorem, *Ann. Math. (2)* **74** (1961) 329-387.
31. Robert H. Lewis. *Fermat, a computer algebra system for polynomial and matrix computations,* 1997. http://www.bway.net/~lewis.

32. James R. Munkres, *Elements of Algebraic Topology*, Addison-Wesley, Menlo Park, 1984.
33. James R. Munkres, *Topology*, 2nd Edition, Prentice Hall, 2000.
34. Sergey P. Novikov. On the algorithmic unsolvability of the word problem in group theory, *Trudy Mat. Inst.* **44** (1955) 143.
35. Sarah Rees and Leonard H. Soicher An algorithmic approach to fundamental groups and covers of combinatorial cell complexes *J. of Symb. Comp.* **29** (2000) 59-77.
36. Ernest Sibert, Harold F. Mattson and Paul Jackson. Finite Field Arithmetic Using the Connection Machine. In Richard Zippel, editor, *Proceedings of the second International Workshop on Parallel Algebraic Computation, Ithaca, USA*, volume 584 of *Lecture Notes in Computer Science*, 51-61. Springer, 1990.
37. Arne Storjohann, Near Optimal Algorithms for Computing Smith Normal Forms of Integer Matrices, Lakshman, Y. N. (ed.), *Proceedings of the 1996 international symposium on symbolic and algebraic computation, ISSAC '96*, New York, NY: ACM Press. 267-274 (1996).
38. Olga Taussky. Bounds for characteristic roots of matrices. *Duke Math. J.* **15** 1043-1044 (1948).
39. Vladimir Tuchin. Homologies of complexes of doubly connected graphs. Russian Math. Surveys **52** 426-427 (1997).
40. Richard S. Varga. *Matrix iterative analysis*. Number 27 in Springer series in Computational Mathematics. Springer, second edition, 2000.
41. Michelle Wachs. Topology of matching, chessboard and general bounded degree graph complexes. Preprint 2001.
42. Douglas H. Wiedemann. Solving sparse linear equations over finite fields. *IEEE Trans. Inf. Theory* **32** 54-62 (1986).

The Geometry of $\mathbb{C}^n$ is Important for the Algebra of Elementary Functions

James H. Davenport*

Department of Computer Science, University of Bath, Bath BA2 7AY, England
J.H.Davenport@bath.ac.uk

Abstract. On the one hand, we all "know" that $\sqrt{z^2} = z$, but on the other hand we know that this is false when $z = -1$. We all know that $\ln e^x = x$, and we all know that this is false when $x = 2\pi i$. How do we imbue a computer algebra system with this sort of "knowledge"? Why is it that $\sqrt{x}\sqrt{y} = \sqrt{xy}$ is false in general ($x = y = -1$), but $\sqrt{1-z}\sqrt{1+z} = \sqrt{1-z^2}$ is true everywhere? The root cause of this, of course, is that functions such as $\sqrt{}$ and log are intrinsically multi-valued from their algebraic definition.

It is the contention of this paper that, only by considering the geometry of $\mathbb{C}$ (or $\mathbb{C}^n$ if there are n variables) induced by the various branch cuts can we hope to answer these questions even semi-algorithmically (i.e. in a yes/no/fail way). This poses questions for geometry, and calls out for a more efficient formulation of cylindrical algebraic decomposition, as well as for cylindrical non-algebraic decomposition. It is an open question as to how far this problem can be rendered fully automatic.

1   Introduction

It is possible to do algebra with the elementary functions of analysis, using only the tools of algebra, especially differential algebra, *if* one is content to regard these functions as abstract solutions of differential equations, so that

$$(\log f)' = \frac{f'}{f}; \qquad (\exp f)' = f' \exp f \tag{1}$$

[15], but of course these differential equations only define the functions "up to a constant". Furthermore, in this context, "constant" means something that differentiates to zero. So $\log \frac{1}{z} + \log z$ is a "constant", since $\left(\log \frac{1}{z} + \log z\right)' = \frac{\frac{-1}{z^2}}{\frac{1}{z}} + \frac{1}{z} = 0$, but in fact this "constant" is 0 off the branch cut for log, but $2\pi i$ on the branch cut[1].

---

* The author was partially supported by the European OpenMath Thematic Network. He is grateful to the referees for their comments, and to Drs Bradford, Corless, Jeffrey and Watt for many useful discussions.

[1] For the standard [1,6] branch cut: $-\pi < \arg \log z \le \pi$. If we take the branch cut along the positive real axis ($0 \le \arg \log z < 2\pi$), the constant is $2\pi i$ off the cut, and 0 on it. If we were to choose the branch cut along the negative imaginary

It is the thesis of this paper that, if instead one wishes to reason with these functions as genuine one-valued functions $\mathbb{C} \to \mathbb{C}$, or even $\mathbb{R} \to \mathbb{R}$, the geometry of $\mathbb{C}^n$ (or $\mathbb{R}^n$: n being the number of independent variables) that the branch cuts of these functions induce has to be taken into account. We claim also that this is true even for functions like

$$\sqrt{z} = \exp\left(\frac{1}{2}\log z\right), \tag{2}$$

since it is conventional to align the branch cuts so that equation (2) is true for these genuine one-valued functions, as well as being algebraically true *in abstracto*. The overall structure of the various approaches proposed in this paper is:

1. see if the simplification is true algebraically;
2. analyse $\mathbb{C}^n$ to see what different regions need to be checked for the values of the implicit "constant";
3. check a sample point in each region;
4. produce an analytically correct simplification from these regions.

In other words, this approach separates the algebraic simplification (step 1) from the analytic verification.

For the purposes of this paper, the precise definitions of log, arctan etc. as functions $\mathbb{C} \to \mathbb{C}$ (or $\mathbb{R} \to \mathbb{R}$) is that given in [6], which repeats, with more justification, the definitions of [1]. In particular, the branch cut of log (and $\sqrt{}$) is along the negative real axis, with the negative axis itself being continuous with the upper half-plane, so that $\lim_{\epsilon\to0+}\log(-1+\epsilon i) = i\pi = \log(-1)$, but $\lim_{\epsilon\to0-}\log(-1+\epsilon i) = -i\pi \neq \log(-1)$.

We will consider, as motivating examples, a variety of potential equations involving elementary functions.

$$\sqrt{z^2} \stackrel{?}{=} z \tag{3}$$

is one: it is patently false, even over $\mathbb{R}$, with $z = -1$ as a counterexample. Another pseudo-equation is

$$\log \overline{z} \stackrel{?}{=} \overline{\log z} : \tag{4}$$

this is false on the branch cut for log, and instead $\log \overline{z} = \overline{\log z} + 2\pi i$ on the cut. Similarly,

$$\log(\frac{1}{z}) \stackrel{?}{=} -\log z \tag{5}$$

---

axis, so that $-\pi/2 < \arg \log z \leq 3\pi/2$, then the branch cuts for the expression would lie along the whole of the imaginary axis, and the constant would be $2\pi i$ on the negative half-plane, including the axis, and 0 on the positive half-plane.

is false on the branch cut: instead $\log(\frac{1}{z}) = -\log z + 2\pi i$ on the cut.

Another pseudo-equation that causes problems over the reals as well as over the complexes is

$$\arctan(x) + \arctan(y) \overset{?}{=} \arctan\left(\frac{x+y}{1-xy}\right). \tag{6}$$

We should note that it follows from the work of Richardson [14] that we cannot hope to produce a complete algorithm, since the problem is undecidable. However, we hope to show that many cases can be handled at least semi-algorithmically. More precisely, we outline algorithms that, where applicable (i.e. not returning "fail") will correctly answer these questions. The question of when these algorithms can be guaranteed not to return "fail" is an interesting one — [3] gives some, overly strong, sufficient criteria for the multi-valued/cylindrical decomposition algorithm.

2   How to Handle Multi-valued Functions

[7] and [4] discuss four treatments of the inherently multi-valued functions such as $\sqrt{}$, log and arctan. These four approaches are given in the following sub-sections, along with one which has emerged subsequently.

2.1   Signed Zeros

This goes back to work of Kahan [12,13], and distinguishes 0^+ from 0^-. This technique can resolve the issues raised by pseudo-equations (4) and (5), provided that we state that $\log(-1 + 0^+i) = i\pi$ but $\log(-1 + 0^-i) = -i\pi$. This method is intrinsically no help in those situations where the pseudo-equation is invalid on a set of measure greater than 0, such as (3) and (6), and does not easily help with situations such as $\log\overline{z+i} \overset{?}{=} \overline{\log(z+i)}$, where the branch cut is $\Re(z) \le 0 \wedge \Im(z) = -1$, so we would really need a "signed -1".

2.2   Genuinely Exact Equations

[9] claims that the most fundamental failure is $z \overset{?}{=} \log\exp z$. They therefore introduce the unwinding number, $\mathcal{K}$, defined by

$$\mathcal{K}(z) = \frac{z - \log\exp z}{2\pi i} = \left\lceil \frac{\Im z - \pi}{2\pi} \right\rceil \in \mathbb{Z}, \tag{7}$$

so that

$$z = \log\exp z + 2\pi i \mathcal{K}(z). \tag{8}$$

Equation (5) is then rescued as:

$$\log \frac{1}{z} = -\log z - 2\pi i \mathcal{K}(-\log(z)).$$
(9)

They would write the equation (6) as:

$$\arctan(x) + \arctan(y) = \arctan\left(\frac{x+y}{1-xy}\right) + \pi \mathcal{K}\left(2i(\arctan(x) + \arctan(y))\right).$$
(10)

2.3 Multi-valued Functions

Though not explicitly advocated in the computational community, mathematical texts often urge us to treat these functions as multi-valued, defining, say, $\mathrm{Arctan}(x) = \{y \mid \tan x = y\}$ (the notational convention of using capital letters for these set-valued functions seems helpful). In this interpretation, equation (6), considered as

$$\mathrm{Arctan}(x) + \mathrm{Arctan}(y) \overset{?}{=} \mathrm{Arctan}\left(\frac{x+y}{1-xy}\right),$$

with the $+$ on the left-hand side representing element-by-element addition of sets, is valid.

Most of the "algebraic" rules of simplification are valid in this context, e.g. with $\mathrm{Sqrt}(x) = \{y \mid y^2 = x\} = \{\pm\sqrt{x}\}$, it is the case that

$$\mathrm{Sqrt}(x)\,\mathrm{Sqrt}(y) = \mathrm{Sqrt}(xy),$$
(11)

whereas

$$\sqrt{x}\sqrt{y} \overset{?}{=} \sqrt{xy}$$
(12)

is not true: consider $x = y = -1$.

There are a few caveats that must be mentioned, though.

1. Cancellation is no longer trivial, since in principle $\mathrm{Arctan}(x) - \mathrm{Arctan}(x) = \{n\pi \mid n \in \mathbb{Z}\}$, rather than being zero.
2. Not all such multivalued functions have such simple expressions as

$$\mathrm{Arctan}(x) = \{\arctan(x) + n\pi \mid n \in \mathbb{Z}\}.$$

For example, $\mathrm{Arcsin}(x) = \{\arcsin(x) + 2n\pi \mid n \in \mathbb{Z}\} \cup \{\pi - \arcsin(x) + 2n\pi \mid n \in \mathbb{Z}\}$. This problem combines with the previous one, so that, if $A = \arcsin(x)$,

$$A - A = \{2n\pi \mid n \in \mathbb{Z}\} \cup \{2\arcsin(x) - \pi + 2n\pi \mid n \in \mathbb{Z}\}$$
$$\cup\{\pi - 2\arcsin(x) + 2n\pi \mid n \in \mathbb{Z}\}$$
$$= \{2n\pi \mid n \in \mathbb{Z}\} + \{0, 2\arcsin(x) - \pi, \pi - 2\arcsin(x)\}.$$

Note that this still depends on x, unlike the case of $\mathrm{Arctan}(x) - \mathrm{Arctan}(x)$. In the case of non-elementary functions, as pointed out in [8] in the case of the Lambert W function, there may be no simple relationship between different branches.

3. However, some equations take on somewhat surprising forms in this context, e.g. the incorrect

$$\arcsin z \overset{?}{=} \arctan \frac{z}{\sqrt{1 - z^2}} \tag{13}$$

simply becomes the correct

$$\mathrm{Arcsin}(z) \subset \mathrm{Arctan} \left(\frac{z}{\mathrm{Sqrt}(1 - z^2)} \right), \tag{14}$$

and if we want an equality of sets, we have

$$\mathrm{Arcsin}(z) \cup \mathrm{Arcsin}(-z) = \mathrm{Arctan} \left(\frac{z}{\mathrm{Sqrt}(1 - z^2)} \right), \tag{15}$$

in which both sides take on four values in each 2π period. It is an open question to produce an alternative characterisation of just $\mathrm{Arcsin}(z)$.

4. The equation

$$\mathrm{Log}(z^2) = 2\,\mathrm{Log}(z) \tag{16}$$

is not valid if we interpret $2\,\mathrm{Log}(z)$ as $\{2y \mid \exp(y) = z\}$, since this has an indeterminacy of $4\pi i$, and the left-hand side has an indeterminacy of $2\pi i$. Instead we need to interpret $2\,\mathrm{Log}(z)$ as $\mathrm{Log}(z) + \mathrm{Log}(z)$, and under this interpretation, equation (16) is true, as a specialisation of the correct equation

$$\mathrm{Log}(z_1 z_2) = \mathrm{Log}(z_1) + \mathrm{Log}(z_2). \tag{17}$$

5. Pseudo-equations in which ambiguity disappears, such as $\sqrt{z^2} \overset{?}{=} z$, cannot be encoded as $\mathrm{Sqrt}(z^2) = z$, since this is trying to equate a set and a number, but rather have to be encoded as $z \in \mathrm{Sqrt}(z^2)$.

Nevertheless, a correct treatment of multi-valued functions can provide rigorous proofs, but unfortunately not of the results desired about single-valued functions.

2.4 Riemann Surfaces

It is often said that one should "consider the Riemann surface". Unfortunately, in the case of equation (6), $\arctan x$ is defined on one surface, $\arctan y$ on a second, and $\arctan \left(\frac{x+y}{1-xy} \right)$ on a third. It is not clear how "considering the Riemann surface" solves practical questions such as this identity.

2.5 Power Series

[11] contains an interesting result on zero-testing of functions defined by differential equations: if the power (in general Puiseux) series is zero for sufficiently many (defined in terms of the differential equation) terms, then it is identically zero. This is essentially zero testing in the sense of differential algebra, but the testing of the constant term also means that initial conditions are tested. However, it does not deal with branch cuts, so it will declare that equation (6), as expanded about $x = y = 0$ (in the form $\arctan(x) + \arctan(y) - \arctan\left(\frac{x+y}{1-xy}\right) \overset{?}{=} 0$) is valid, even though, as shown in section 5, it is only valid in the component containing 0. If one expands about $x = y = 2$, then indeed this method deduces that the equation is false by a correction of $-2\arctan(2) - \arctan(4/3)$ (which is $-\pi$).

3 Simplifying $\sqrt{z^2}$

It follows from equation (11) that $(\mathrm{Sqrt}(z))^2 = \mathrm{Sqrt}(z^2)$. Since $z = (\sqrt{z})^2$ (not all single-valued equations are false!), this means that $z \in \mathrm{Sqrt}(z^2)$, i.e. $z = \pm\sqrt{z^2}$. The question then resolves to "plus or minus", or, in line with the thesis of this paper, "where on $\mathbb{C}$ is it plus, and where is it minus"?

$\sqrt{z}$ has its branch cut on the negative real axis, more precisely on $(-\infty, 0]$. Therefore the branch cut relevant to $\sqrt{z^2}$ is when $z^2 \in (-\infty, 0]$, i.e. $z \in [0, i\infty)$ or $z \in (-i\infty, 0]$. Note that, as we will see later, it is not correct to write $z \in [0, i\infty) \cup (-i\infty, 0] = (-i\infty, i\infty)$. Therefore the complex plane is divided into five regions, as shown in Table 1.

This analysis of cases can be reduced[2] (though as yet there is no known algorithm for doing this) to $\sqrt{z^2} = \mathrm{csgn}(z)z$, which can also be expressed as

$$\sqrt{z^2} = (-1)^{\mathcal{K}(2\log z)}z, \tag{18}$$

---

[2] The csgn function was first defined in Maple. There is some uncertainty about csgn(0): is it 0 or 1, but for the reasons given in [4], we choose csgn(0) = 1.

Table 1. Decomposition of complex plane for $\sqrt{z^2}$

Region	$\Re(z)$	$\Im(z)$	$\sqrt{z^2}$
R_1	< 0	—	$-z$
R_2	$= 0$	< 0	$-z$
R_3	$= 0$	$= 0$	$z = -z$
R_4	$= 0$	> 0	z
R_5	> 0	—	z

which follows from re-writing $\sqrt{}$ via equation (2) and using the definition of $\mathcal{K}$. So in this example both methods give equivalent, though differently expressed, results: one via an explicit "case" construct and one via the unwinding number, which in this example can be decomposed into a "case", since $\mathcal{K}(2\log z)$ has only three possibilities.

4   Simplifying Equations (4) and (5)

These are very similar, and we will only deal with equation (5) in detail. The branch cut for $\log z$ is along the negative real axis $(\infty, 0)$, which is transformed into itself by $z \mapsto \frac{1}{z}$, and so is also the branch cut for $\log \frac{1}{z}$. This does not disconnect the complex plane, so there are only two cases: on the cut and off it. We know that $\log \frac{1}{z} \in -\mathrm{Log}(z)$, so the only corrections required to equation (5) are by adding multiples of $2\pi i$. Off the branch cut, an evaluation at $z = 1$ shows that no correction is necessary, whereas an evaluation at -1 shows that, on the branch cut, a correction of $2\pi i$ is necessary. If we wish, we can write this as

$$\log \frac{1}{z} = -\log z - 2\pi i \mathcal{K}(-\log z). \tag{19}$$

Similarly, equation (4) is rescued as

$$\log \bar{z} = \overline{\log z} - 2\pi i \mathcal{K}(\overline{\log z}). \tag{20}$$

5   Additivity of arctan

In this section, we investigate the pseudo-equation (6), and attempt an algorithmic correction to it.

If we differentiate both sides of this equation with respect to x, we get

$$\frac{1}{1+x^2} \overset{?}{=} \frac{1}{1+x^2},$$

which is true, and if we differentiate both sides with respect to y, we get

$$\frac{1}{1+y^2} \overset{?}{=} \frac{1}{1+y^2},$$

which is also true.

Hence the difference between the two sides is a local constant, in the sense that its derivatives are zero, and so, between jumps, is constant. Evaluating both sides of equation (6) at $x = y = 0$, we get 0 on both sides, so we conclude that here, the difference is indeed zero, so that equation (6) is indeed (locally) an identity. However, if we do the same at $x = 1.2$, $y = 0.9$, we get

$$1.608873152 \ldots \overset{?}{=} -1.532719501 \ldots :$$

clearly not valid, and indeed wrong by π.

5.1   Simplifying (6) over $\mathbb{R}$

It is not obvious why there is a problem here, since arctan is a continuous, differentiable function, indeed bijection, $(-\infty, \infty) \to (-\pi/2, \pi/2)$. Indeed, we can define $\arctan(-\infty) = -\pi/2$, $\arctan(\infty) = \pi/2$ to get a bijection $[-\infty, \infty] \to [-\pi/2, \pi/2]$.

Unfortunately, $[-\infty, \infty]$ is not the right domain for this sort of analysis. One way of seeing this is to observe that, although $(-\infty, \infty) = \mathbb{R} \subset \mathbb{C}$, the analytic completion of $\mathbb{C}$ is the one-point completion $\mathbb{C} \cup \{\infty\}$, and $[-\infty, \infty] \not\subset \mathbb{C} \cup \{\infty\}$.

In fact, arctan has a branch cut[3] at infinity. When $x = \infty$ (or $y = \infty$), this might seem to cause a problem, since $\frac{x+y}{1-xy}$ tends to $\frac{-1}{y}$. However, the problem this causes is masked by the another, more serious, problem.

It is possible for $\frac{x+y}{1-xy}$ to pass through ∞ even when both x and y are finite, namely when $xy = 1$. If $\frac{x+y}{1-xy}$ goes from "large and positive" to "large and negative", then π is subtracted from the value of its arctan, and *vice versa*. We need to add a correction term, which can be done in various ways, e.g.[4]

$$\arctan(x) + \arctan(y) = \arctan\left(\frac{x+y}{1-xy}\right) + \begin{cases} \pi & x > 0, \ xy > 1 \\ 0 & x \geq 0, \ xy \leq 1 \\ 0 & x < 0, \ xy < 1 \\ -\pi & x < 0, \ xy \geq 1. \end{cases} \quad (21)$$

Equation (10) is in one sense only another way of saying the same thing, except that it is coupled more clearly to the problem of "overflow" in adding the two arctan terms, and the boundary cases are dealt with consistently.

Equation (21) divides the (x, y)-plane into three regions (the two entries with zero correction term define one region, open on the lower-left boundary and closed on the top-right boundary. If, however, we proceed algorithmically from the cut $xy = 1$, by cylindrical algebraic decomposition [5] on the equation $xy - 1$, whose y-discriminant is x, we in fact decompose the place into seven regions, as given in Table 2.

The technique of clustering [2] would reduce this, ideally to three regions as in equation (21), but certainly to five regions, which we will describe in Table 3.

One interesting question is the extent to which equation (21) can be reconstructed automatically. Evaluation at $x = y = 0$ shows that equation (6) is true there, and therefore throughout R_1. Similarly, evaluation at $x = y = 1$

---

[3] Some people refer to this as a jump discontinuity, which of course it is, but it is better to think of it as a branch cut, since it could be moved elsewhere by another choice of branch cut: indeed, as pointed out in [6], there is no agreement on where the branch cut for the closely related function arccot should go.

[4] We are assuming a single ∞, with $\arctan(\infty) = \frac{\pi}{2}$.

Table 2. Cylindrical Decomposition of (x, y)-plane

x	xy	dimension	correction
> 0	> 1	2	π
> 0	$= 1$	1	0
> 0	< 1	2	0
$= 0$	—	1	0
< 0	> 1	2	$-\pi$
< 0	$= 1$	1	$-\pi$
< 0	< 1	2	0

Table 3. 5 Regions of the (x, y)-plane

	R_1	R_2	R_3	R_4	R_5
xy	< 1	$= 1$	$= 1$	> 1	> 1
x	—	> 0	< 0	> 0	< 0

shows that it is true there, and so throughout R_2. Evaluation at $x = y = -1$ shows that equation (6) is false, and needs a correction factor of $-\pi$ at that point, and therefore throughout R_3. By hand, evaluation at $x = y = 2$ shows that equation (6) is false, and needs a correction factor of π at that point, and therefore throughout R_4. However, it seems impossible to persuade[5] Maple (release V.5) to simplify $2\arctan 2 + \arctan\frac{4}{3}$ to π. *Mutatis mutandis*, the same remarks apply to $x = y = -2$ and R_5. Of course, the author has chosen values of x and y at which the arctan function has a simple expression, and how could an algorithm do that?

The answer is that it need not. Since we know that $\arctan(x) + \arctan(y) \in$ Arctan $\left(\frac{x+y}{1-xy}\right)$, any error has to be a multiple of π. So a (sufficiently careful) numerical evaluation is all that is needed.

5.2   Simplifying (6) over $\mathbb{C}$

The definition of $\arctan x$ is [1,6]

$$\frac{1}{2i}\left(\log(1 + ix) - \log(1 - ix)\right).$$

Hence its branch cuts in the complex plane are along $1 + ix \in (-\infty, 0]$, i.e. $x \in [i, i\infty)$, and along $1 - ix \in (-\infty, 0]$, i.e. $x \in (-i\infty, -i]$. These do not disconnect the complex plane, so the only special cases are along them. The

---

[5] Of course, direct simplification of this would use equation (6), which would spoil the point, but it ought to be doable via complex logarithms.

same is true of $\arctan y$ and $\arctan\left(\frac{x+y}{1-xy}\right)$. There is still the critical locus $xy = 1$, as in the real case.

We now have to determine how these interact. Write $x = x_R + ix_I$ etc. Then we have to consider various intersections. The author has yet to produce a fully automatic solution to this problem, though. The necessary intersections are listed below, but their own intersections need to be unravelled further.

○ The intersection of the two branch cuts for $\arctan x$ and $\arctan y$, This is clearly possible, since the two constraints are independent.

○ The intersection of the branch cuts for $\arctan\left(\frac{x+y}{1-xy}\right)$ with the branch cuts for $\arctan x$. If one solves for the two (real) equations resulting from $\Re\left(\frac{x+y}{1-xy}\right) = 0$ and $x_R = 0$, one is left with, *inter alia*, the resultant

$$-3y_R y_I^2 - y_I^6 y_R - 3y_I^2 y_R^5 - 3y_I^4 y_R^3 - y_R^7$$
$$+ y_R + y_R^3 + 2y_R^3 y_I^2 - y_R^5 + 3y_I^4 y_R. \quad (22)$$

The only solutions of this have $|y_R| \leq 1$ and $|y_I| \leq 1$, and therefore are not on the branch cut for $\arctan y$, except that when $y_R = 0$, y_I is unconstrained. Hence it is in fact possible to be on all the branch cuts of $\arctan$ simultaneously, say with $x = y = 2i$, and $\frac{x+y}{1-xy} = \frac{-4}{3}i$.

○ The intersection of the branch cuts for $\frac{x+y}{1-xy}$ with the branch cuts for $\arctan x$. This is identical to the previous case.

○ The intersection of the branch cut for $\arctan x$ with the critical locus $xy = 1$. It then follows that y_R has to be zero (since $x_I y_R = 0$), so the point is also on the branch cut of $\arctan y$, and possibly on the branch cut for $\arctan\left(\frac{x+y}{1-xy}\right)$.

○ The intersection of the branch cut for $\arctan\left(\frac{x+y}{1-xy}\right)$ with the critical locus $xy = 1$ cannot happen.

In sum, the situation here is very complicated, but can be reduced to a finite number of regions which need to be tested. It turns out that the move to the complex plane does not, in fact, add any further complication.

The unwinding number approach tells us easily that the correction is

$$-\pi\mathcal{K}\bigg(\ln(1 + ix) - \ln(1 - ix) + \ln(1 + iy) - \ln(1 - iy)+$$
$$\ln\left(1 + \frac{i(x+y)}{1-xy}\right) - \ln\left(1 - \frac{i(x+y)}{1-xy}\right)\bigg) \quad (23)$$

but there seems to be no easy simplification of this.

6   Strategies for Simplifying Elementary Expressions

Traditionally, computer algebra systems, and indeed humans, are bad at simplifying expressions containing the (inverse) elementary functions. Humans fail to spot special cases, often with serious consequences [13, pp. 178–9], and computer algebra writers are torn between:

- being careful, as in Maple's `simplify(...)`, and therefore failing to make correct simplifications such as $\sqrt{1-z}\sqrt{1+z} \to \sqrt{1-z^2}$;
- using the full range of algebraic rules regardless of correctness, as in Maple's `simplify(...,symbolic)`, and therefore making incorrect simplification such as $\sqrt{z^2} \to z$.

We now describe two strategies (at the current state of development, it would not be accurate to describe them as algorithms), based on the views outlined in sections 2.2 and 2.3 and illustrated by the examples above.

6.1   The Unwinding Number Approach

This approach, outlined in [4,7], works as follows.

While a reduction exists
    Perform a reduction
        Generally introducing unwinding numbers
        Remove as many unwinding numbers as possible
If any unwinding numbers remain
    See if case-by-case analysis removes unwinding numbers

For example, the equivalent of equation (12) would be

$$\sqrt{x}\sqrt{y} = \sqrt{xy}(-1)^{K(\ln x + \ln y)}. \tag{24}$$

An illustration of this method would be the following proof from [7].

Lemma 6.1. *Whatever the value of z,*

$$\sqrt{1-z}\sqrt{1+z} = \sqrt{1-z^2}.$$

Proof. It is sufficient to show that the unwinding number term in equation (24) is zero. Whatever the value of z, $1 + z$ and $1 - z$ have imaginary parts of opposite signs. Without loss of generality[6], assume $\Im z \geq 0$. Then $0 \leq \arg(1+z) \leq \pi$ and $-\pi < \arg(1-z) \leq 0$. Therefore their sum, which is the imaginary part of $\log(1+z) + \log(1-z)$, is in $(-\pi, \pi]$. Hence the unwinding number is indeed zero.

    Note the manual reasoning required to remove the unwinding number term that equation (24) introduced.

---

[6] An automated system would probably analyse both cases.

6.2 The "Multi-valued/Regions of Validity" Approach

This approach[7], used informally in the previous sections, works as follows.

Let E' be the multi-valued version of E.
While a reduction of E' exists
        reduce E' to yield a new E'
Let E'' be the single-valued version of the reduced E'.
Analyse $\mathbb{C}^n$ to determine corrections required to $E \overset{?}{=} E''$.

This method would prove lemma 6.1 the following way
Proof. $E' := \mathrm{Sqrt}(1-z)\,\mathrm{Sqrt}(1+z)$. This reduces to $\mathrm{Sqrt}(1-z^2)$, so we know that $\sqrt{1-z}\sqrt{1+z} \in \mathrm{Sqrt}(1-z^2)$. Hence we have to analyse $\sqrt{1-z}\sqrt{1+z} \overset{?}{=} \sqrt{1-z^2}$. The relevant branch cuts are as follows.

For $\sqrt{1-z}$: $1-z \in (-\infty, 0]$, so $z \in [1, \infty)$.
For $\sqrt{1+z}$: $1+z \in (-\infty, 0]$, so $z \in (-\infty, -1]$.
For $\sqrt{1-z^2}$: $1-z^2 \in (-\infty, 0]$, so $z^2 \in [1, \infty)$, and $z \in [1, \infty)$ or $z \in (-\infty, -1]$.

These branch cuts do not disconnect the complex plane, so the three regions are $[1, \infty)$, $(-\infty, -1]$ and the rest of the plane.

$[1, \infty)$ Take $z = 3$. Then $\sqrt{1-z}\sqrt{1+z} = \sqrt{-2}\sqrt{4} = 2\sqrt{2}i$. $\sqrt{1-z^2} = \sqrt{-8} = 2\sqrt{2}i$, so no correction is needed.
$(-\infty, -1]$ Take $z = -3$. Then $\sqrt{1-z}\sqrt{1+z} = \sqrt{4}\sqrt{-2} = 2\sqrt{2}i$. $\sqrt{1-z^2} = \sqrt{-8} = 2\sqrt{2}i$, so no correction is needed.
rest of plane Take $z = i$. Then $\sqrt{1-z}\sqrt{1+z} = \sqrt{1-i}\sqrt{1+i} = \sqrt{2}$ (clearly the correct norm, and the arguments cancel). $\sqrt{1-z^2} = \sqrt{2}$, so again no correction is necessary.

Note that, while it might appear that some intelligence was used to select the sample values, any sample value s such that $\sqrt{1-s^2} \neq -\sqrt{1-s^2}$ would do, as long as we evaluated $\sqrt{1-s}\sqrt{1+s}$ to enough precision to be sure which of the two values it was taking.

7 Strategies to Algorithms?

In this section, we explore what would be necessary to convert the strategies outlined above into complete (and possibly even efficient) algorithms.

---

[7] This approach is similar to the approach "how to decide where two analytic expressions describe the same function" of [13].

7.1 The Unwinding Number Approach

A prime requirement of this method is a complete set of *correct* transformation rules. Thus, instead of the incorrect equation (12), we need the correct equation (24). While not conceptually very difficult, to my knowledge no such complete list has been built up. It would have to include rules for "incorrect"[12] applications of inverse functions, so that the incorrect $z \overset{?}{=} \log \exp z$ becomes the correct $z = \log \exp z + 2\pi i \mathcal{K}(z)$, and similarly the incorrect equation (3) becomes the correct equation (18). The often-quoted, but incorrect[8], equation

$$\arcsin z \overset{?}{=} \arctan \frac{z}{\sqrt{1-z^2}} \tag{13}$$

has to be replaced by

$$\arcsin z = \arctan \left(\frac{z}{\sqrt{1-z^2}} \right) + \pi \mathcal{K}(-\ln(1+z)) - \pi \mathcal{K}(-\ln(1-z)). \tag{25}$$

Another obstacle, as pointed out in [7], is the tendency for expressions involving $\mathcal{K}$ to proliferate: in the course of proving equation (25), they had to handle equations such as the following.

$$2i \ \arctan \left(\frac{z}{\sqrt{1-z^2}} \right) =$$

$$2i \arcsin(z) - 2\pi i \mathcal{K} \left(2\ln \left(iz + \sqrt{1-z^2} \right) \right)$$

$$+ 2\pi i \mathcal{K} \left(\ln \left(1 + i\frac{z}{\sqrt{1-z^2}} \right) - \ln \left(1 - i\frac{z}{\sqrt{1-z^2}} \right) \right).$$

The third problem is that of understanding the result, which in this case means reducing

$$- \pi \mathcal{K} \left(2\ln \left(iz + \sqrt{1-z^2} \right) \right)$$

$$+ \pi \mathcal{K} \left(\ln \left(1 + i\frac{z}{\sqrt{1-z^2}} \right) - \ln \left(1 - i\frac{z}{\sqrt{1-z^2}} \right) \right)$$

to $\pi \mathcal{K}(-\ln(1+z)) - \pi \mathcal{K}(-\ln(1-z))$. This may well result in a difficult case-by-case analysis, so there may be no useful overall picture. See formula (23) for another example.

7.2 The "Multi-valued/Regions of Validity" Approach

In this approach, the first problem is also that of finding a "correct" set of equations, in this case replacing the incorrect equation (12) by the equality

---

[8] With our definitions. Derive's definition of arctan is to make this true, but this means that it deviates from [1].

of sets in equation (24). Again, while not conceptually very difficult, to my knowledge no complete list of such equations has been built up. It would have to include rules for "incorrect"[12] applications of inverse functions, so that the incorrect $z \overset{?}{=} \log \exp z$ becomes the correct $z \in \text{Log} \exp z$, and similarly the incorrect equation (12) becomes the correct equation (11).

One problem, though, is that not all simplification rules remain set equalities, as in equation(14). Provided the strategy only expands E', then no real harm is done, but it must never contract it. There is also a problem with equations such as $z \in \text{Sqrt}(z^2)$. It does not seem to be trivial to make an algorithm out of this.

Handling the decompositions of $\mathbb{R}^{2n}$ (as we see $\mathbb{C}^n$) is also difficult. The existing technology of cylindrical algebraic decomposition [5] has the following defects when applied to the problem at hand.

○ As we saw in section 5, it does not handle directly descriptions such as the branch cut for log, i.e. $\Re(z) < 0 \wedge \Im(z) = 0$. Instead one has to introduce the two planes $\Re(z) = 0$ and $\Im(z) = 0$, even though $\Re(z) = 0$ is irrelevant off $\Im(z) = 0$.

○ As has been remarked by many authors, even for the original purposes of quantifier elimination, cylindrical algebraic decomposition is *too* powerful, since it solves not only the question at hand, but also all other questions[9] relating to the same set of equations.

○ The algorithm produces far too many regions. This can partly be solved by the technique of clustering [2], but this is a retrospective technique (produce too many, then reduce), and there is no work on how this interacts with the first problem.

○ The algorithm will not always work when we apply many–one functions to multi-valued functions. Consider

$$\sin \left(\text{Arctan}(x)^2 \right)$$

at $x = 1$. Values of this for various branches of the Arctan expression are given in Table 4.

From this we can see that the values most certainly are not monotone, and indeed the branch closest (in the table) to the principal one is the arctan -4π branch. It seems very likely that the values for the different branches are scattered throughout $(-1, 1)$, and in fact it can be shown[10] that

$$\liminf_{|n| \to \infty} \left| \sin \left(\arctan(1)^2 \right) - \sin \left((\arctan(1) + n\pi)^2 \right) \right| = 0.$$

---

[9] To be accurate, all other questions with the same order of variables in the quantifiers, though the quantifiers themselves, and the Boolean combination of the polynomials, may be different.

[10] We can make the absolute value as small as we like by choosing n such that $n\pi$ is sufficiently close to an integer multiple of 4.

Table 4. Values of $\sin\left(\text{Arctan}(1)^2\right)$

Branch	Value
arctan -4π	0.5322322646905239425510841132497
arctan -3π	-0.6884431955627695364639646140640
arctan -2π	-0.9284667030666120770272715434323
arctan $-\pi$	-0.6680110779739474530232086002687
arctan	0.5784687893545579387490182736861
arctan $+\pi$	-0.2137981114862761779134882376938
arctan $+2\pi$	-0.2961030758899680643560838126531
arctan $+3\pi$	-0.5439513643882437590931910607038
arctan $+4\pi$	0.7180884622031624733030254926312
arctan $+5\pi$	0.9601642895219452299013225178160

Hence there seems to be no way of determining numerically that one is on the correct branch.

○ The decomposition may not be semi-algebraic. If we consider a putative simplification involving $\ln\ln z$, then the branch cut is $\ln z \in (-\infty, 0]$, which is $z \in (0, 1]$. This is semi-algebraic. However, for a putative simplification involving the similar $\ln(i + \ln z)$, the branch cut is $i + \ln z \in (-\infty, 0]$, or $\ln z \in (-\infty - i, 0 - i]$. This can be expressed as $z \in \{e^t(\cos 1 + i \sin 1) \mid t \in (\infty, 0]$, which is certainly not semi-algebraic. However, in this case it is sub-Pfaffian, where algorithms exist [10].

○ The decomposition may not be finitely generated: if we consider[11] an identity involving $\log \sin x$, this has a branch cut whenever $\sin x$ is real and negative, i.e. along all intervals $((2n-1)\pi, 2n\pi) : n \in \mathbb{Z}$. We can deal with most of $\mathbb{C}$ by assuming that the whole of $\mathbb{R}$ is a branch cut, and analysing the upper and lower half-plane separately, but on $\mathbb{R}$ itself we have an infinite number of regions, and therefore need a symbolic proof that the identity is true at, say, every $(2n - \frac{1}{2})\pi$.

The last three points make it hard to know when there is a true algorithm: [3] imposes the over-strong restrictions that the only nesting of elementary functions in the formulae is of square roots inside other functions, and where inverse functions only occur in the numerators.

7.3   Power Series

As we saw, this is not of itself a solution, as it can ignore branch cuts. However, it might prove an interesting alternative approach to the "multivalued functions" idea explored in the previous subsection, though it is not clear

---

[11] I am grateful to the referee for this example.

that it can provide information on how accurately one needs to do numerical evaluation, if that is required. It is not clear at this stage whether one needs to do

1) Check that the equation is "true" by power series
2) Verify at a critical point in each region

or

1) Find a critical point in each region
2) Check that the equation is "true" by power series about each point.

This route needs further investigation.

7.4   Putting It All Together

Assuming that we eventually want an identity relating single-valued functions, what is the simplest way of expressing it? Equation (13), viz.

$$\arcsin z = \arctan\left(\frac{z}{\sqrt{1-z^2}}\right) + \pi \mathcal{K}(-\ln(1+z)) - \pi \mathcal{K}(-\ln(1-z)).$$

is hardly simple, but is it any better as

$$\arcsin z = \overline{\arctan\left(\frac{\overline{z}}{\sqrt{1-\overline{z}^2}}\right)}? \tag{26}$$

The latter equation does not easily imply that, almost everywhere, $\arcsin z = \arctan\left(\frac{z}{\sqrt{1-z^2}}\right)$, whereas, once used to reading formulae involving $\mathcal{K}$, the former does.

If we have an analysis in terms of cases, is there any algorithmic reconstruction in terms of, say, $\mathcal{K}$? None is known, and it is worth pointing out that one can also produce "nonsense" justifications in terms of $\mathcal{K}$, e.g. equations (19) and (20) can be combined to give the correct, but unhelpful

$$\log \frac{1}{z} = -\log z - 2\pi i \mathcal{K}(\overline{\log z}). \tag{27}$$

8   Conclusion

There are several forms of "identities" relating complex-valued (or even real-valued) elementary functions.

1. Those that are true, and easily provable to be so. This class has not had much mention in this paper, but, in particular, all identities involving forward functions (exp, trigonometric and hyperbolic functions, raising to integer powers) only are correct, as are those in which inverse functions are applied correctly[12].

2. Those that are true, but whose naïve proof is false. A good example of this is Lemma 6.1, where the naïve proof by multiplying out square roots needs a lot of help to make it correct. These are the ones that computer algebra systems ought to get *reliably*, but currently cannot. The "unwinding number" method should solve some of these, and the "multi-valued/regions of validity" method will get all those satisfying the restrictions in [3], and many more.

3. Those that are "almost true". Equations (5) and (4) are good examples of this. In a numerical context, these can effectively be taken as true, since the chance of landing on an exception value is very small, and an adaptive solver will increase the sampling density near the exceptional value to reduce its impact. In some numerical contexts, signed zeros (see section 2.1) will obviate the problem. In computer algebra, there is no such excuse: the answer given is wrong. However, not warning the user that there is an "almost good enough" simplification is unhelpful, to say the least. As we saw, sometimes these can be re-written by means of a "double conjugate" trick, as in equation (26). For this class one needs a "regions of validity" method, whether applied directly or to unwinding numbers.

4. Those that are only partially true, i.e. only true on some subset, of measure less than one, of the argument range. Equations (3) and (6) are good examples of this. Here the "multi-valued/regions of validity" method will say what the regions are, but not why, and an "unwinding number" method might provide more insight.

Computer algebra systems currently give no help with the third and fourth type of "identity", or in distinguishing them from the second. The techniques outlined in this paper are, it is hoped, a start in filling this gap. However, much remains to be done.

○ The algorithms need to be implemented, and tested on real examples.
○ The limitations in [3] ensure that the "multi-valued/regions of validity" method is an algorithm, but are, as we saw in the case of $\log \log z$, too strict. Is it possible to state a weaker set of restrictions which still works?
○ There is currently no complexity analysis.

---

[12] For example, tan is many–one, hence Arctan is one–many, and therefore arctan needs to be applied carefully. In particular, $\tan(\arctan x) = x$ is valid, but $\arctan(\tan x) = x$ is not.

References

1. Abramowitz,M. & Stegun,I., Handbook of Mathematical Functions with Formulas, Graphs, and Mathematical Tables. US Government Printing Office, 1964. 10th Printing December 1972.
2. Arnon,D.S. & McCallum,S., Cylindrical Algebraic Decomposition II: An Adjacency Algorithm for the Plane. *SIAM J. Comp.* **13** (1984) pp. 878–889.
3. Bradford,R.J. & Davenport,J.H., Towards Better Simplification of Elementary Functions. Proc. ISSAC 2002 (ed. T. Mora), ACM Press, New York, 2002, pp. 15–22.
4. Bradford,R.J., Corless,R.M., Davenport,J.H., Jeffrey,D.J. & Watt,S.M., Reasoning about the Elementary Functions of Complex Analysis. *Annals of Mathematics and Artificial Intelligence* **36** (2002) pp. 303–318.
5. Collins,G.E., Quantifier Elimination for Real Closed Fields by Cylindrical Algebraic Decomposition. Proc. 2nd. GI Conference Automata Theory & Formal Languages (Springer Lecture Notes in Computer Science 33) pp. 134–183.
6. Corless,R.M., Davenport,J.H., Jeffrey,D.J. & Watt,S.M., "According to Abramowitz and Stegun". *SIGSAM Bulletin* **34** (2000) 2, pp. 58–65.
7. Corless,R.M., Davenport,J.H., Jeffrey,D.J., Litt,G. & Watt,S.M., Reasoning about the Elementary Functions of Complex Analysis. Artificial Intelligence and Symbolic Computation (ed. John A. Campbell & Eugenio Roanes-Lozano), Springer Lecture Notes in Artificial Intelligence Vol. 1930, Springer-Verlag 2001, pp. 115–126.
8. Corless,R.M., Gonnet,G.H.. Hare,D.E.G., Jeffrey,D.J. & Knuth,D.E., On the Lambert W function. *Advances in Computational Mathematics* **5** (1996) pp. 329–359.
9. Corless,R.M. & Jeffrey,D.J., The Unwinding Number. *SIGSAM Bulletin* **30** (1996) 2, issue 116, pp. 28–35.
10. Gabrielov,A. & Vorobjov,N., Complexity of cylindrical decompositions of sub-Pfaffian sets. *J. Pure Appl. Algebra* **164** (2001) pp. 179–197.
11. van der Hoeven,J., A new Zero-test for Formal Power Series. Proc. ISSAC 2002 (ed. T. Mora), ACM Press, New York, 2002, pp. 117–122.
12. IEEE Standard 754 for Binary Floating-Point Arithmetic. IEEE Inc., 1985. Reprinted in SIGPLAN Notices **22** (1987) pp. 9–25.
13. Kahan,W., Branch Cuts for Complex Elementary Functions. *The State of Art in Numerical Analysis* (ed. A. Iserles & M.J.D. Powell), Clarendon Press, Oxford, 1987, pp. 165–211.
14. Richardson,D., Some Unsolvable Problems Involving Elementary Functions of a Real Variable. *Journal of Symbolic Logic* **33**(1968), pp. 514–520.
15. Risch,R.H., Algebraic Properties of the Elementary Functions of Analysis. *Amer. J. Math.* **101** (1979) pp. 743–759.

A Visual Introduction to Cubic Surfaces Using the Computer Software SPICY

Duco van Straten and Oliver Labs

University of Mainz, Algebraic Geometry Group, Mainz, Germany

Abstract. At the end of the 19th century geometers like Clebsch, Klein and Rodenberg constructed plaster models in order to get a visual impression of their surfaces, which are so beautiful from an abstract point of view. But these were static visualizations. Using the computer program SPICY[1], which was written by the second author, one can now draw algebraic curves and surfaces depending on parameters interactively.

Using this software and Coble's explicit equations for the cubic surface that arises as the blowing-up of the projective plane in six points it was possible for the first time to visualize how some of the 27 lines upon the cubic surface coalesce when the surface develops a double point.

When the user drags one of the six points, the equation and a ray-traced image of the cubic surface are computed using external programs. As the whole process takes less than half a second, one nearly gets the impression of a continuously changing surface.

1 Introduction

A cubic surface in the real projective space $\mathbf{P}^3 := \mathbf{P}^3(\mathbf{R})$ is the vanishing set of a homogeneous cubic polynomial in $\mathbf{P}^3$, i.e. it consists of all $(x : y : z : w) \in \mathbf{P}^3$, such that

$$a_0 x^3 + a_1 x^2 y + \cdots + a_{19} w^3 = 0,$$

where $a_i \in \mathbf{R}, i = 0, 1, \ldots, 19$.

The intensive study of cubic surfaces started in 1849, when the British mathematicians Salmon and Cayley published the results of their correspondence on the number of straight lines on a smooth cubic surface ([1], [19]). In a letter, Cayley told Salmon that there could only exist a finite number of such lines – and Salmon found this number to be 27 (allowing complex lines[2]).

---

[1] SPICY – Space and Plane Interactive Constructive and Algebraic Geometry – is a dynamic constructive geometry software, that uses external programs like SINGULAR ([12]) and SURF ([9]) to visualize algebraic curves and surfaces.

[2] In fact, Clebsch constructed the famous Diagonal Surface in [5] and showed that it contained 27 real lines.

Shortly after that Steiner wrote a short but fruitful article ([21]) which became the basis for a purely geometrical treatment of cubic surfaces. In that paper he formulated many theorems, but for most of them he did not even indicate the idea of the proof. Many of these proofs were supplied ten years later independently by Cremona ([7]) and Sturm ([23]).

Other important contributions were made by Cayley, Schläfli, Klein, Rodenberg and Clebsch[3] – the latter gave, e.g., the explicit equation (in terms of determinants) of a covariant of order 9, which intersects the cubic surface exactly in the 27 lines (cf. appendix). This proves both the existence and the number of lines on smooth cubic surfaces.

After all these results had been established, the mathematicians of the 19th century started to build plaster models of cubic surfaces in order to get a visual impression of these objects, which are so beautiful from an abstract point of view.

The objective of the computer program SPICY is to allow mathematicians of the 21st century not only to make static models or movies of surfaces and curves, but also to manipulate them interactively.

For this purpose, another theorem of Clebsch ([4]) is important:

Theorem 1.1. *Every smooth cubic surface can be represented in the plane using 4 plane cubic curves through six points and vice versa.*[4]

As we will not only visualize cubic surfaces, but also the straight lines on them, we need a notation for these 27 lines. Over the last 150 years, several notations have been established, but for our needs, the one introduced by Schläfli in 1858 seems to be the most convenient one. In his article on the classification of cubic surfaces with respect to the number of real lines and triple tangent planes on them ([20]), Schläfli discovered a very interesting property of the 27 lines:

Notation. On every smooth cubic surface one can always choose two sets of six lines, say $a_1, \ldots, a_6$ and $b_1, \ldots, b_6$, such that the incidence diagram is as shown in fig. 1.

Let P be the plane in $\mathbf{P}^3$ spanned by two intersecting lines a_i, b_j for some $i \neq j$. As any plane, P intersects the cubic surface in a cubic curve. This curve is reducible; it consists of the two lines a_i, b_j and a third line c_{ij}. Repeating this process for any $i \neq j$, we get the remaining $\binom{6}{2} = 15$ lines c_{ij} and have thus established a notation for all the 27 lines, which is called *Schläfli's double-six notation*. The configuration of these 27 lines was a subject of intense investigation in the 19th century. For a more modern approach we

---

[3] See [15] and [22] for a more complete list of references.

[4] For some of those surfaces, we must allow complex coordinates for some points. In this article, we will start from a representation in the real plane, so we will not be able to get all the smooth cubic surfaces, but only those on which all the lines are real.

	a_1	a_2	a_3	a_4	a_5	a_6
b_1		×	×	×	×	×
b_2	×		×	×	×	×
b_3	×	×		×	×	×
b_4	×	×	×		×	×
b_5	×	×	×	×		×
b_6	×	×	×	×	×	

Figure 1. Incidence diagram of a double-six. An × indicates, that the two lines intersect.

refer to [17, sections 6.1.2, 6.1.3] and [11], where the 27 lines are seen as lines of the unique finite generalized quadrangle with parameters $(4, 2)$.

In our application the initial choice of the $2 \cdot 6$ lines a_i, b_i, $i = 1, \ldots, 6$, will arise in a natural way in the next section – e.g. the lines $a_1, \ldots, a_6$ will correspond to the six points in the plane.

2   Blowing-Up the Plane in Six Points

We will first construct the blowing-up of the affine plane $\mathbf{A}^2 := \mathbf{A}^2(\mathbf{R})$ at the origin $O = (0, 0)$.

Definition 2.1.   *Let (x_1, x_2) be the affine coordinates of $\mathbf{A}^2$ and let $(y_1 : y_2)$ be the projective coordinates of $\mathbf{P}^1$. We define the* blowing-up *of $\mathbf{A}^2$ at the point O to be the closed subset $\widetilde{\mathbf{A}^2}$ of $\mathbf{A}^2 \times \mathbf{P}^1$ defined by the equation $x_1 y_2 = x_2 y_1$. We have a natural morphism $\pi : \widetilde{\mathbf{A}^2} \longrightarrow \mathbf{A}^2$, obtained by restricting*

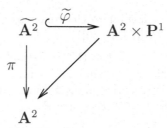

Figure 2. Blowing-up the affine plane $\mathbf{A}^2$ in a point.

the projection of $\mathbf{A}^2 \times \mathbf{P}^1$ to the first factor.

As we want to understand the blowing-up visually, we give the following properties (cf. [14, p. 28]):

Fact 1. With the notations of the previous definition, we have:

○ The restriction of π to the set $\widetilde{\mathbf{A}^2}\backslash\pi^{-1}(O)$ is bijective.

○ $\pi^{-1}(O) \cong \mathbf{P}^1$.

○ The points of $\pi^{-1}(O)$ are in $1-1$ correspondence with the set of lines through O in $\mathbf{A}^2$ (fig. 3).

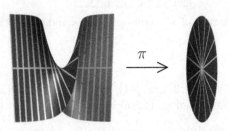

Figure 3. Blowing-up the plane in one point. This was visualized using SPICY, but we will not explain that here; the construction is similar to the one presented in section 3.3.

Definition 2.2. *If $C \subset \mathbf{A}^2$ is a curve in $\mathbf{A}^2$, we call*

$$\widetilde{C} := \overline{\pi|_{\mathbf{A}^2\backslash\{O\}}^{-1}(C\backslash\{O\})}$$

the strict transform *of C.*

All the notions introduced in this section can be generalized (cf. [14, 163-171] or [8]), so that we can talk of the blowing-up of the projective plane $\mathbf{P}^2 := \mathbf{P}^2(\mathbf{R})$ in six points and the strict transform of a curve under this blowing-up.

This enables us to formulate the following well-known facts (for proofs, cf. [14, p. 400, 401] and [15, p. 30]):

Fact 2. Let $S := \{P_1,\ldots,P_6\} \subset \mathbf{P}^2$ be a set of six points in the plane, such that no three are collinear and not all six are on a common conic. Then the blowing-up $\widetilde{\mathbf{P}^2}$ of the projective plane $\mathbf{P}^2$ in S can be embedded as a smooth cubic surface in projective three-space $\mathbf{P}^3$.

Denoting by $\widetilde{\varphi} : \widetilde{\mathbf{P}^2} \hookrightarrow \mathbf{P}^3$ this embedding and by $\pi : \widetilde{\mathbf{P}^2} \hookrightarrow \mathbf{P}^2$ the projection as in fig. 2, we have:

Fact 3. Let $S := \{P_1,\ldots,P_6\} \subset \mathbf{P}^2$ be a set of six points in the plane, such that no three are collinear and not all six are on a common conic. Let $Q_i \subset \mathbf{P}^2$ denote the unique conic through the five points $\{P_1,\ldots,P_6\}\backslash\{P_i\}$ for $i = 1,2,\ldots,6$ and let $l_{ij} \subset \mathbf{P}^2$ denote the straight line through the points P_i and P_j for $i,j = 1,2,\ldots,6$, $i \neq j$. Then the 27 lines lying on the cubic surface are as follows:

○ The 6 exceptional lines over the 6 base points P_i:

$$a_i := \widetilde{\varphi}(\pi^{-1}(P_i)) \subset \mathbf{P}^3, \ i = 1, 2, \ldots, 6.$$

○ The 6 strict transforms of the 6 plane conics Q_i:

$$b_i := \widetilde{\varphi}(\widetilde{Q}_i) \subset \mathbf{P}^3, \ i = 1, 2, \ldots, 6.$$

○ The 15 strict transforms of the $15 = \binom{6}{2}$ lines l_{ij} joining the P_i:

$$c_{ij} := \widetilde{\varphi}(\widetilde{l}_{ij}) \subset \mathbf{P}^3, \ i, j = 1, 2, \ldots, 6, \ i \neq j.$$

Furthermore, the lines a_i and b_i, $i = 1, \ldots, 6$, intersect according to the diagram in fig. 1 (i.e. they form a double six) and the c_{ij}, $i \neq j$, are the remaining 15 lines in Schläfli's notation.

3   Visualizing Cubic Surfaces Using SPICY

Using the results presented in the previous section, we can study some special point configurations and some properties of the corresponding cubic surfaces and the 27 lines on them.

First, we will see how the general situation of blowing-up the plane in six points can be visualized using SPICY and then we will focus on some interesting cases, where the use of this interactive software enlightens the situation.

3.1   SPICY

The core of the computer software SPICY ([16]) is a constructive geometry program designed both for visualizing geometrical facts interactively on a computer and for including them in publications. Its main features are:

○ Connection to external software like the computer algebra system SIN-GULAR ([12]) and the visualization software SURF ([9]), which enables the user to include algebraic curves and surfaces in dynamic constructions.
○ Comfortable graphical user-interface (cf. fig. 4) for interactive constructions using the computer-mouse including macro-recording, animation, etc.
○ High quality export to .fig-format (and in combination with external software like XFIG or FIG2DEV export to many other formats, like .eps, .pstex, etc.).

3.2 Eckardt Points

The first situation we discuss concerns a very special kind of smooth points on a cubic surface:

Definition 3.1. *An* Eckardt Point *on a cubic surface is a smooth point, where three of its straight lines meet.*

On a general cubic surface there is no Eckardt Point, but when starting from the plane representation, the following is easy to see:

Proposition 3.2. *Let $S := \{P_1, \ldots, P_6\} \subset \mathbf{P}^2$ be a set of six points, such that no three are collinear and not all six are on a common conic. Denote by l_{ij} the 15 lines through P_i and P_j for $i \neq j$ and by $Q_i \subset \mathbf{P}^2$ the unique conic through the five points $\{P_1, \ldots, P_6\} \backslash \{P_i\}$ for $i = 1, 2, \ldots, 6$. Then:*

- *If three of the lines l_{ij} meet in a point $P \in \mathbf{P}^2 \backslash \{P_1, \ldots, P_6\}$, then the corresponding lines $\widetilde{l_{ij}}$ on the cubic surface meet in an Eckardt Point (fig. 4).*
- *If l_{ij} touches the conic Q_i in P_j for some $i, j \in \{1, \ldots, 6\}$, $i \neq j$, then the corresponding lines $a_j := (\pi^{-1} \circ \widetilde{\varphi})(P_j)$, $b_i := \widetilde{Q_i}$ and $c_{ij} := \widetilde{l_{ij}}$ on the cubic surface meet in an Eckardt Point.*

Proof: Recall the properties of blowing-up given in fact 1. Then both parts are easy to see:

- The application $\pi^{-1} \circ \widetilde{\varphi}$ is bijective on $\mathbf{P}^2 \backslash S$.
- Both Q_i and l_{ij} have the same tangent direction in P_j, so the corresponding lines $b_i := \widetilde{Q_i}$ and $c_{ij} := \widetilde{l_{ij}}$ meet the line $a_j := (\pi^{-1} \circ \widetilde{\varphi})(P_j)$ in the same point. □

We will describe in the following how we can visualize the first of these situations using SPICY.

The Computer Algebra Part. First, we need to know from an abstract point of view, what we want to visualize. So we implement a function, say `cubicSurfaceFor6PointsWithLines`, for the computer algebra program SINGULAR, that takes the projective coordinates of the six points as an input and returns the equations of the surface and the 27 lines on it in a format that the visualization software SURF understands. This implies, in particular, that we have to choose a plane at infinity in order to get an equation in affine three-space. This function is part of the SINGULAR library `spicy.lib`, which can be downloaded from [16] or [22].

Example 3.3. The output for the Clebsch Diagonal Surface (without showing the lines on it) could be, e.g. `x^3+y^3+z^3+1-(x+y+z+1)^3`.

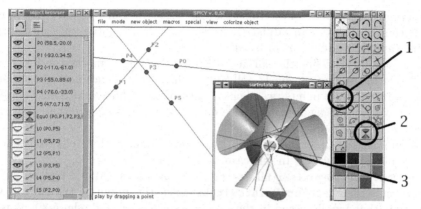

Figure 4. A screen shot of the SPICY user interface showing three lines, that meet in a point and the corresponding cubic surface, which contains an Eckardt Point (3). Buttons 1 and 2 are used to draw the lines and the surface, respectively.

In fact, the method we use there does not calculate the blowing-up by means of computer algebra. Coble explains in [6] how the equations can be calculated from the six points simply by computing some products and sums of 3×3 determinants (cf. appendix). The detailed study of this (cf. [15]) enables us to calculate the equations defining each of the 27 lines a_i, b_i, c_{ij} separately[5]. Thus, using SPICY we can not only visualize all the 27 lines at a time, but also just some selected ones.

The SPICY Part. We start the software by typing `spicy` on the command line[6]. The working area will show a visualization of Poncelet's theorem, which was historically the first problem constructed using SPICY. So, to clean up, we select `file -> new construction` from the menu.

To satisfy the condition of the first case of proposition 3.2, we need three lines that meet in a point (in fig. 4, the Eckardt Point is marked by circle 3). To construct them, we first press button 1 (marked in fig. 4) in SPICY's *tools window* to switch to the *infinite line through two points mode* and then click twice on the working area to define these two points. After that, the program switches automatically back to the *default mode*, which allows us to drag the points and thus move the lines they define.

---

[5] The SINGULAR functions related to Coble's equations are packed in the library `coble.lib`, which can also be downloaded from [16] or [22].

[6] For instructions on how to download and install the software, cf. [16]. As SPICY is implemented in Java, it runs on most systems. The external software SINGULAR ([12]) is available for most systems, too. Unfortunately, the visualization software SURF, which is also developed at the university of Mainz, is only available for Linux or Unix at the moment.

Repeating this process three times, we get three lines in the plane. By dragging the points, we can adjust them, so that all the three lines meet in a point (approximately[7]) as in fig. 4.

It remains to define the surface. We press button 2 (marked in fig. 4) to switch to the *surface mode*. To tell SPICY, that the calculation of the surface is based on the six points, we click on each of them once and then we press the mouse button again somewhere on the working area together with the `control` key held down.

A window shows up, where we can enter the command that tells SPICY how to calculate the surface. In our case, this will be

`{cubicSurfaceFor6PointsWithLines:#0.p#,#1.p#,#2.p#,#3.p#,#4.p#,#5.p#}`,

where `cubicSurfaceFor6PointsWithLines` is the name of the SINGULAR function, that we discussed in the previous subsection and where `#0.p#` is a place-holder for the projective coordinates of the 0th point that we selected earlier etc.

Once we have done this, SPICY opens a separate window that shows the cubic surface, which is the blowing-up of the six points (fig. 4). If we do not see three lines intersecting in a point yet, we can rotate the surface just by dragging the mouse on the image.

Playing with Cubic Surfaces. We could have visualized this just by using SINGULAR and SURF – without SPICY. But here, SPICY's main feature comes into play: We can drag the points and accordingly the software recomputes the image of the cubic surface. As the whole process of calculating the new equations and the new image (fig. 5) only takes less than half a second, the user nearly gets the impression of a continuously changing surface while dragging points.

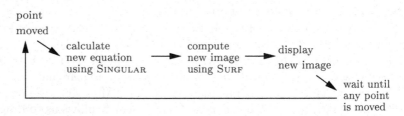

Figure 5. The whole process of recomputing the image of the cubic surface only takes less than half a second.

This could not be done using other existing dynamic geometry software like CINDERELLA or GEONEXT, because these programs can not perform

---

[7] To make them meet exactly, we may enter the coordinates of the points explicitly; cf. the tutorials on [16].

groebner basis computations or visualization of algebraic surfaces. On the other hand, standard software like MATHEMATICA and MAPLE do not have any dynamic geometry features and visualization of algebraic surfaces takes much longer and is less exact than with SURF. This is the reason why we combined our interactive geometry software SPICY with the computer algebra software SINGULAR and the visualization software SURF.

3.3 Double Points

While playing with the cubic surfaces as explained in the previous section, one could discover the following: If three of the six points are collinear or if all the six points are on a common conic, then the blowing-up is no longer smooth, but it is a cubic surface in projective three-space with singularities[8]. More exactly, using the notations of fact 3, we have in the case of six points on an irreducible conic (cf. [15] for a discussion of this and the other cases[9]):

Fact 4. Let $S := \{P_1, \ldots, P_6\} \subset \mathbf{P}^2$ be a set of six points in the projective plane, where no three are collinear. Then: If all the six points are on a conic (this conic is irreducible, because no three points are collinear), then the corresponding cubic surface contains an ordinary double point Q. More exactly, with the notations introduced in proposition 3.2, we have:

$$Q = \widetilde{Q_1} = \widetilde{Q_2} = \cdots = \widetilde{Q_6}.$$

Furthermore, if $P'_i, i \in \{1, 2, \ldots, 6\}$ is a point, such that $S'_i := S \backslash P_i \cup P'_i$ is a set of six points, where no three are collinear and not all the six points are on a common conic, denote by $\widetilde{Q'_i}$ the strict transform of Q_i under the blowing-up of the projective plane $\mathbf{P}^2$ in S'_i. Then the straight lines

$$a_i \text{ and } b'_i := \lim_{P'_i \to P_i} \widetilde{Q'_i}, \ i = 1, \ldots, 6,$$

coincide and all these six distinct lines meet in Q (fig. 6).

3.4 Cutting the Surface by a Plane

Now we describe how we can use SPICY to learn more about this surface. We cut the surface by a plane in three-space and show the resulting curve in the plane (fig. 7).

---

[8] Already Clebsch knew that the projective plane blown up in six points, which lie on a common plane conic, is a cubic surface with an ordinary double point ([4]).

[9] If three points are collinear, the corresponding cubic surface has an A_1 singularity and if six points are on a common reducible conic, the corresponding cubic surface has an A_2 singularity.

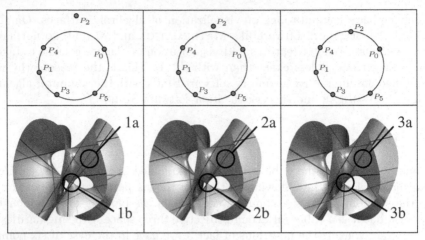

Figure 6. The blowing-up of the projective plane in six points, such that all six are on a common conic, is a cubic surface with an ordinary double point. Note the changing of the lines, when we drag the point P_2. When P_2 lies on the conic through the other five points, $2 \cdot 6$ lines meet in the double point (1b – 3b) and six pairs of two lines coincide (1a – 3a).

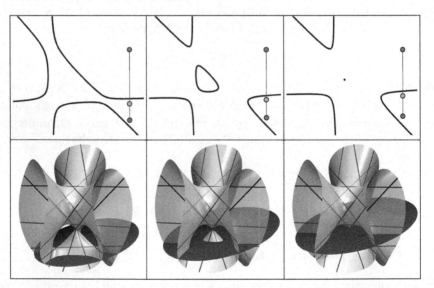

Figure 7. Cutting the surface by a plane. The equation of the plane is just $x + (8 \cdot r - 4)$, where r is defined as the ratio of the length of the upper part of the segment to the length of the whole segment. So, while dragging the point on the segment, the plane passes through the surface and SPICY shows both the plane and the surface in the 3d-view and the curve cut out by the plane in the 2d-view.

To do so, we must define a plane P in three-space and the curve in the plane $\mathbf{P}^2$ cut out of the cubic surface by P. Given the equation of a surface and a plane, it is easy to write a SINGULAR script, that computes their intersection. For the equation of the plane, we choose $x + (8 \cdot r - 4)$, where $r \in [0,1]$ is a parameter that we would like to change by dragging some point of our construction. We can do this as follows: We draw a segment and a point on this segment (fig. 7, upper part). Then we define r as the ratio of the length of the upper part of the segment and the length of the whole segment. We hide the six points and the lines between them in the SPICY construction using SPICY's so-called *object browser* (the left window in fig. 4), so that we only see the curve and the parameter segment. This visualization can now be exported in high quality to .fig format in order to use it for a publication.

References

1. Cayley, A. (1849) On the triple tangent planes to a surface of the third order. Camb. and Dublin Math. Journal, IV: p. 118–132.
2. Cayley, A. (1869) A Memoir on Cubic Surfaces. Philos. Trans. Royal Soc. CLIX: p. 231–326.
3. Clebsch, A. (1861) Zur Theorie der algebraischen Flächen. Crelles Journal, LIIX: p. 93–108.
4. Clebsch, A. (1866) Die Geometrie auf den Flächen dritter Ordnung. Crelles Journal, LXV: p. 359–380.
5. Clebsch, A. (1871) Ueber die Anwendung der quadratischen Substitution auf die Gleichungen 5ten Grades und die geometrische Theorie des ebenen Fünfseits. Math. Ann. IV: p. 284–345.
6. Coble, A.B. (1915) Point sets and allied Cremona Groups. Trans. of the Am. Math. Soc. XVI: p. 155–198.
7. Cremona. (1868) Mémoire de géométrie pure sur les surfaces du troisième ordre. Crelles Journal, LXVIII: p. 1–133.
8. Eisenbud, D. and Harris, J. (2000) The Geometry of Schemes. Springer. New York.
9. Endraß, S. et al. (2001) Surf, Version 1.03. Visualizing Real Algebraic Geometry. University of Mainz. http://surf.sourceforge.net.
10. Fischer, G. (1986) Mathematische Modelle / Mathematical Models. Vieweg.
11. Freudenthal, H. (1975) Une étude de quelques quadrangles généralisés. Ann. Mat. Pura Appl. (4), vol. 102, p. 109–133.
12. Greuel, G.-M., Pfister, G., and Schönemann, H. (2001) SINGULAR 2.0. A Computer Algebra System for Polynomial Computations. Centre for Computer Algebra. University of Kaiserslautern. http://www.singular.uni-kl.de.
13. Griffith, P. and Harris, J. (1994) Principles of Algebraic Geometry. Wiley Classics Library.
14. Hartshorne, R. (1977) Algebraic Geometry. Springer. New York.
15. Labs, O. (2000) Kubische Flächen und die Coblesche Hexaederform. University of Mainz. http://www.oliverlabs.net/diploma/.

16. Labs, O. (2001) SPICY – Space and Plane Interactive Constructive and Algebraic Geometry. University of Mainz. http://www.oliverlabs.net/spicy/.
17. Payne, S.E. and Thas, J.A. (1984) Finite Generalized Quadrangles. Pitman (Advanced Publishing Program), Boston, MA.
18. Rodenberg, C. (1879) Zur Classification der Flächen dritter Ordnung. Math. Ann. XIV: p. 46–110.
19. Salmon, G. (1849) On the triple tangent planes to a surface of the third order. Camb. and Dublin Math. Journal, IV: p. 252–260.
20. Schläfli, L. (1858) An attempt to determine the twenty-seven Lines upon a Surface of the third Order, and to divide such Surfaces into Species in Reference to the Reality of the Lines upon the Surface. Quarterly Journal for pure and applied Mathematics, II: p. 55–66, 110–220.
21. Steiner, J. (1857) Über die Flächen dritten Grades. Crelles Journal, LIII: p. 133–141.
22. van Straten, D. and Labs, O. (2001) The Cubic Surface Homepage. University of Mainz. http://www.cubicsurface.net.
23. Sturm, R. (1867) Synthetische Untersuchungen über Flächen dritter Ordnung. B.G. Teubner. Leipzig.

A   Clebsch's Explicit Equation for the Covariant of Order 9 that Meets the Cubic Surface in the 27 Lines

In [3] Clebsch gives an explicit equation for the covariant F of order 9 that meets a smooth cubic surface $u(x_1, x_2, x_3, x_4) = 0$ in projective three-space in its 27 lines in terms of three determinants. Its existence proves both the existence and the number of lines on a smooth cubic surface:

$$F = \Theta - 4\Delta T,$$

where Θ, Δ and T are defined as follows.

○ $\Delta := \det(u_{ij}) := \det\left(\frac{\partial^2 u}{\partial x_i \partial x_j}\right)$, the *hessian* of u.

For the other two determinants we write $\Delta_p := \frac{\partial \Delta}{\partial x_p}$ and accordingly $\Delta_{pq} := \frac{\partial^2 \Delta}{\partial x_p \partial x_q}$. Furthermore, we denote by U_{ij} the entries of the *Cramer Matrix* of Δ, i.e. $U_{ij} := (-1)^{i+j} u^{ij}$, where we denote by u^{ij} the determinant of the submatrix of Δ, where the ith row and the jth column are removed. With this we can define:

○ $\Theta := \sum U_{pq} \Delta_p \Delta_q := \sum_p \sum_q U_{pq} \Delta_p \Delta_q,$

○ $T := \sum U_{pq} \Delta_{pq} := \sum_p \sum_q U_{pq} \Delta_{pq}.$

B Coble's Explicit Parametrization for the Cubic Surface and the 27 Lines on it.

Let $S := \{P_1, \ldots, P_6\} \subset \mathbf{P}^2$ be a set of six points in the plane, such that no three are collinear and not all the six points are on a common conic. For such a configuration of six points, Coble ([6]) gives explicit equations for the blowing-up of the projective plane $\mathbf{P}^2$ in S and the 45 so-called *triple tangent planes* cutting out the 27 straight lines of this cubic surface. They can be calculated as follows.

Denoting by $(P_{ix} : P_{iy} : P_{iz})$ the coordinates of $P_i \in \mathbf{P}^2$, $i = 1, 2, \ldots, 6$, we write for $i, j, k, l, m, n \in \{1, 2, \ldots, 6\}$:

$$(ijk) := \det \begin{pmatrix} P_{ix} & P_{jx} & P_{kx} \\ P_{iy} & P_{jy} & P_{ky} \\ P_{iz} & P_{jz} & P_{kz} \end{pmatrix}$$

and

$$(ij, kl, mn) := (ijm)(kln) - (ijn)(klm).$$

We can now define six coefficients $\overline{x_0}, \overline{x_1}, \ldots, \overline{x_5} \in \mathbf{R}$:

$$6\,\overline{x_0} = \qquad\qquad +(15, 24, 36) + (14, 35, 26) + (12, 43, 56) + (23, 45, 16) + (13, 52, 46),$$

$$6\,\overline{x_1} = -(15, 24, 36) \qquad\qquad +(25, 34, 16) + (13, 54, 26) + (12, 35, 46) + (14, 23, 56),$$

$$6\,\overline{x_2} = -(14, 35, 26) - (25, 34, 16) \qquad\qquad +(15, 32, 46) + (13, 24, 56) + (12, 45, 36),$$

$$6\,\overline{x_3} = -(12, 43, 56) - (13, 54, 26) - (15, 32, 46) \qquad\qquad +(14, 52, 36) + (24, 35, 16),$$

$$6\,\overline{x_4} = -(23, 45, 16) - (12, 35, 46) - (13, 24, 56) - (14, 52, 36) \qquad\qquad +(15, 34, 26),$$

$$6\,\overline{x_5} = -(13, 52, 46) - (14, 23, 56) - (12, 45, 36) - (24, 35, 16) - (15, 34, 26).$$

With these notations, Coble ([6]) gives the explicit equation of the blowing-up of the projective plane in the six points $P_1, P_2, \ldots, P_6$ as an equation in projective five-space $\mathbf{P}^5$:

$$x_0^3 + x_1^3 + x_2^3 + x_3^3 + x_4^3 + x_5^3 = 0\,, \text{ where}$$

$$x_0 + x_1 + x_2 + x_3 + x_4 + x_5 = 0\,, \text{ and}$$

$$\overline{x_0}x_0 + \overline{x_1}x_1 + \overline{x_2}x_2 + \overline{x_3}x_3 + \overline{x_4}x_4 + \overline{x_5}x_5 = 0\,.$$

Via the two linear equations this surface can be embedded in projective three-space $\mathbf{P}^3$, because for our set of six points S (no three points are collinear, not all six are on a conic) there are always two indices $i \neq j$, such that $\overline{x_i} \neq \overline{x_j}$.

Furthermore, the 15 triple tangent planes which meet the cubic surface in three lines $c_{ab}, c_{cd}, c_{ef}, \{a, b, c, d, e, f\} = \{1, 2, \ldots, 6\}$ each are:

$$E_{ij} := x_i + x_j = 0, \ i, j = 1, 2, \ldots, 6, \ i \neq j.$$

There are 30 other triple tangent planes. These planes E^{ij} meet the cubic surface in the three lines a_i, b_j and $c_{ij}, \ i, j \in \{1, 2, \ldots, 6\}, \ i \neq j$. They are linear combinations of the 15 planes given above, e.g.

$$E^{12} = (531)(461)(x_0 + x_3) - (341)(561)(x_1 + x_4) = 0.$$

[6] discusses this in more detail and [15] lists all the 45 triple tangent planes and 27 lines. Here we only give the set of 9 triple tangent planes, that we used to cut out all the 27 lines for the visualizations in this article:

$$E_{01}, E_{02}, E_{05}, E^{12}, E^{23}, E^{34}, E^{45}, E^{56}, E^{16}.$$

A Client-Server System for the Visualisation of Algebraic Surfaces on the Web

Richard Morris

Department of Statistics, University of Leeds, Leeds LS2 9JT, UK

Abstract. Algebraic surfaces, defined as the zero set of a polynomial function in three variables, present a particular problem for visualising, especially if the surface contains singularities. Most algorithms for constructing a polygonization of the surface will miss the singular points. We present an algorithm for polygonizing such surfaces which attempts to get accurate representations of the singular points. A client-server approach, with a Java applet and a C program as back end, is used to enable the visualisation of the polygonal mesh in a web browser. This system allows algebraic surfaces to be viewed in any web browser and on any platform.

1 Introduction

Algebraic surfaces, defined as the zero set of a polynomial function in three variables, have a long history in mathematics. There are many famous surfaces such as Steiner's Roman Surface, Fig. 1(a), an immersion of the real projective plane, which is represented as the algebraic surface $x^2y^2 + y^2z^2 + z^2x^2 = 2xyz$.

Algebraic surfaces often contain singular points, where all three partial derivatives vanish. For example the double cone, $x^2 - y^2 - z^2 = 0$, has an A_1 singularity or node at the origin, Fig. 1(b). There are other more complicated isolates singularities such as: $x^2y - y^3 - z^2 = 0$ which has a D_4 singularity, Fig. 1(c). Other surfaces are more complicated and can contain self-intersections, $xy = 0$, and degenerate lines, $x^2 + y^2 = 0$. The cross-cap or Whitney umbrella, $x^2z + y^2 = 0$ contains a self intersection along $x = z = 0$, $y > 0$ and a degenerate line along $x = z = 0$, $y < 0$. The line forms the 'handle' of the umbrella, Fig. 1(d). The swallowtail surface, $-4z^3y^2 - 27y^4 + 16xz^4 - 128x^2z^2 + 144xy^2z + 256x^3 = 0$, is even more complicated and contains a cuspidal edge, Fig. 1(e). These and other examples of algebraic surfaces will be further examined in section 4.

These singularities cause particular problems for constructing computer models of the surfaces. Many algorithms will simply ignore the singular points [2,10]. However if information about singularities is included from the ground up it is possible to construct an algorithm, described here, which can produce good 3D models of most algebraic surfaces.

The surfaces are displayed in a web-page using a Java applet which uses the JavaView library [12,13] to allow rotation of the surface. This applet

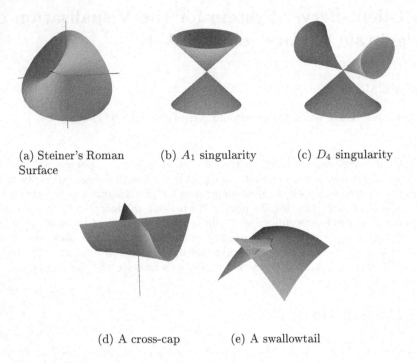

(a) Steiner's Roman   (b) A_1 singularity   (c) D_4 singularity
Surface

(d) A cross-cap   (e) A swallowtail

Figure 1. Some algebraic surfaces

connects to a server on the Internet which actually calculates polygonization of the surface.

The program described here has been adapted from an earlier program [8,9] which ran as a standalone application on SGI machines and used the Geomview [11] program for visualisation. The principal improvements have been the Java interface and an improved method of finding the polygonization 3.5.

2   The Client Applet

The client side of the system is fairly straightforward. It consists of a Java applet written using the JavaView library. It has two panels, one of which displays the surface and allows the surface to be rotated and scaled using the mouse. The other panel contains an area to input the equation of the surface as well as controls for selecting the region space in which the surface will be calculated. Several predefined equations are provided. These include many well known algebraic surfaces. The syntax of equations is standard TeX style notation and allows sub-equations to be defined as well as allowing a symbolic differentiation operator and vector operations. The user interface is shown in figure 2.

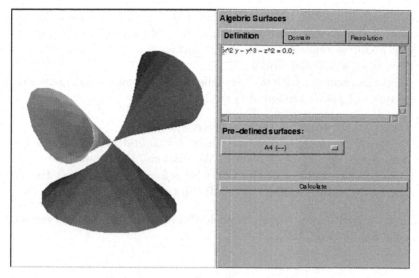

Figure 2. The user interface for the program

A button press causes the surface is to be calculated. A CGI-POST request, which encodes the defining equation and options, is sent to the server which then calculates a polygonization of the surface. This is then returned to the client in JavaView's JVX format. If the defining equation is very degenerate, say a reducible equation like $x^2 = 0$, then the server can take a long time to calculate the surface. To prevent this happening a maximum calculation time is specified by the user. If this time limit is exceeded then the calculation of the surface will end prematurely. Ideally an interrupt button could be provided to halt the calculation of the surface, but this cannot be achieved using the CGI protocol.

Due to Java security restrictions the Java applet can only connect to servers which lie on the same Internet host. This makes it difficult for users to modify the applet or include it in their own software. This could be overcome by signing the Java code.

3   The Server

The server takes the defining equation, $f(x, y, z) = 0$, of an algebraic surface and produces a polygonization of the surface inside a rectangular box specified by the user.

The basic algorithm starts with a rectangular box. Recursive sub-division is used to split that box into 8 smaller boxes, the edge-lengths of which are half those of the original box. Inside each of the smaller boxes a test based on Bernstein polynomials (Sec. 3.1) is used to determine whether the box might contain part of the surface. In such case the recursion continues breaking the box into eight more boxes. We found that three levels of recursion, giving boxes whose edge lengths are an eighth of those of the original box, are enough to give a course representation of the surface and four levels of recursion produce quite a detailed model.

After this recursion each of the smaller boxes is examined in greater detail. Three types of points are found (Fig. 3):

1. Points on the edges on the box where $f = 0$.
2. Points on the faces of the box where $f = 0$ and at least one of partial derivatives, $\frac{\partial f}{\partial x}$, $\frac{\partial f}{\partial y}$ or $\frac{\partial f}{\partial z}$ vanish. We shall call these *2-nodes*.
3. Points in the interior of the box where $f = 0$ and at least two of the partial derivatives vanish. We shall call these *3-nodes*.

Recursive algorithms are used for each of these steps which are described in sections 3.2, 3.3 and 3.4.

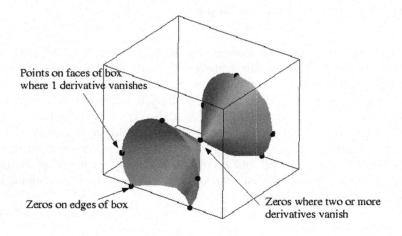

Points on faces of box where 1 derivative vanishes

Zeros on edges of box

Zeros where two or more derivatives vanish

Figure 3. The types of solutions found in a box

Finally the points found are connected together to give a polygonization of the surface which is returned to the client (Sec. 3.5).

A few assumptions about the surface are necessary to avoid degenerate cases: that the surface does not intersect the corners of the box; that none of the partial derivatives vanish at the solutions on the edges of the box; and that the 2-nodes on the faces of the box are isolated. Provided that the polynomial is not reducible, i.e. not of the form $h(x, y, z)(g(x, y, z))^2 = 0$, then all these assumptions can be satisfied by putting the surface in general position. This can always be achieved by slightly changing the bounds of the box. Typically the domain needs to be constructed with unequal bounds so that the origin, which is often a singular point does not lie at a corner of a box.

3.1   Bernstein Polynomials

The computations involved in the program are made much simpler by the use of Bernstein polynomials. These offer a quick test to see if a polynomial might have a zero inside a domain. All the results in this section are well known and the algorithms have been taken from a method for drawing algebraic curves in 2D, described by A. Geisow [6] and details of the implementation can be found in [8].

A 1D **Bernstein polynomial** $B(x)$ of degree n is written as

$$B(x) = \sum_{i=0}^{n} b_i \binom{n}{i} (1-x)^i x^{n-i}.$$

The b_i's are the **Bernstein coefficients**. We are only interested in Bernstein polynomials which are defined over the range $[0, 1]$. In three dimensions the Bernstein representation of a polynomial of degrees l, m and n in x, y, and z is

$$\sum_{i=1}^{l} \sum_{j=1}^{m} \sum_{k=1}^{n} \binom{l}{i} \binom{m}{j} \binom{n}{k} (1-x)^i x^{l-i} (1-y)^j y^{m-j} (1-z)^k z^{n-k}.$$

A Test for Zeros. If all the Bernstein coefficients of a 1D Bernstein polynomial have the same sign, all strictly positive or all strictly negative, then the polynomial has no zeros between 0 and 1. A similar result happens in the 3D case. This is easily proved by noting that $(1-x)^i x^{n-i}$ is non-negative for $x \in [0, 1]$ and $0 \leq i \leq n$. Note the converse does not always hold and it is possible to construct a Bernstein polynomial which has coefficients of different signs but no zeros on $[0, 1]$.

Other Algorithms. Several other routines are necessary for the operation of the program:

- constructing a Bernstein polynomial from a standard polynomial, this involves rescaling the domain so that it fits $[0, 1]$,
- evaluating a Bernstein polynomial at a specific point,
- calculating the derivative of a Bernstein polynomial,
- splitting the domain into two equal halves and constructing Bernstein polynomials for each half.

The last of these algorithms is necessary for the recursion steps, where a box is split into 8 smaller boxes.

3.2 Finding Solutions on Edges

A simple 1-dimensional sub-division algorithm is used to find the solutions on an edge of the box. A 1-dimensional Bernstein polynomial is constructed by restricting the function to the edge. If all the coefficients of the Bernstein polynomial are of the same sign then there is no solution on the edge. Otherwise Bernstein polynomials are constructed for each of the partial derivatives. If the Bernstein coefficients for any of the partial derivatives are not all of the same sign then there may be more than one solution on the edge. In such cases the edge is split into two and the process repeated for each sub-edge. Otherwise the signs at the end points are examined to determine whether there is a solution on the edge. If so, the solution is found by repeatedly sub-dividing the edge and looking for a change of sign. The sub-division is carried out until sub-pixel level is reached.

3.3 Finding Nodes on Faces

Another recursive procedure is needed to find solutions on the faces of the box where one or more partial derivatives vanish. This routine is also used to find lines connecting solutions on the face and its edges.

For a given face the 2-dimensional Bernstein polynomial b is constructed. Bernstein polynomials are also constructed for the three partial derivative functions. There are a number of case shown in Fig. 4.

- ○ If the coefficients of b are all of the same sign then the surface does not intersect the face and the face is ignored, Fig. 4(a).
- ○ If the coefficients of b are not all of the same sign and the coefficients of each of the partial derivative are all of the same sign, then there are exactly two solutions on the edges of the face. These solutions are connected by a line on the face and the recursion end, Fig. 4(b).
- ○ If the coefficients of any one of the derivatives fail to be all of the same sign then the face is divide into four smaller faces. Each of these face, and its edges, is then recursively tested, Fig. 4(b).

This process is carried out recursively until a pre-defined depth, typically pixel level, is reached.

When the bottom level of recursion is reached the face may contain a node and further processing is needed to deduce the geometry. If only one derivative vanishes then there may be a turning point, where $f = \frac{\partial f}{\partial x} = 0$ say. Typically there will be one of the situations shown in Fig. 5. These can be distinguished by counting the number of solutions on the edges of the face and examining their derivatives.

- ○ If there are no solutions then the face is ignored, Fig. 5(a).
- ○ If there are two solutions and the signs of the derivatives match then they are linked by a line, Fig. 5(d).

(b) face where
  no derivatives
  vanish

(c) face where
  a derivative vanishes
  which is sub–divided

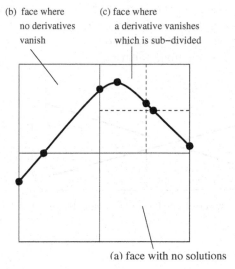

(a) face with no solutions

Figure 4. Sub-dividing a face

- If there are four solutions then they are the tested for signs of their derivatives pair-wise. Pairs with matching derivatives are linked by lines. Fig. 5(c).
- If there are two solutions which have different signs for the partial derivative then a 2-node is constructed in the centre of the face and this is linked to each of the solutions, Fig. 5(b).

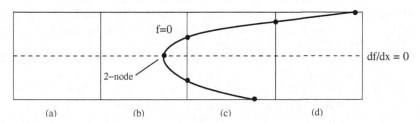

$f=0$

df/dx = 0

2–node

(a)   (b)   (c)   (d)

Figure 5. Faces where one derivative vanishes

Consider the case shown in figure 6. Here two partial derivatives vanish in both faces, yet only one contains a 2-node. To distinguish between the two cases observe that the two curves $\frac{\partial f}{\partial x} = 0$ and $\frac{\partial f}{\partial y} = 0$ only cross in the face which contains the 2-node. In this face a 2-node is created in the centre of the face and linked to the each of the solutions on the edges. In the other face the solutions on the edge are linked pair-wise. This situation typically occurs when a self-intersection of the surface crosses a face, in which case all three derivatives will vanish. A similar situation occurs when a degenerate line passes through the face: the zero sets of all three derivatives will intersect in a single point. This example illustrates the limits of using recursion, finer levels of recursion would not help resolve this case as the geometry looks similar even under greater magnification.

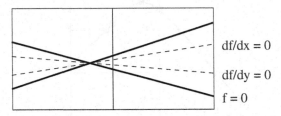

df/dx = 0

df/dy = 0

f = 0

Figure 6. Two faces where two derivatives vanish, only the left-hand one contains a node

3.4   Finding Singularities Inside a Box

A recursive procedure is used to find the 3-nodes inside a box where two or more derivatives vanish. These can either be singularities where all three derivatives vanish or points like the north-pole of the sphere, $x^2 + y^2 + z^2 = 1$, where two derivatives vanish. Including the latter type of point helps produce better polygonization as it does not truncate the surface.

This recursion splits each box into eight sub-boxes and the Bernstein test is used to tell whether the function f or its derivatives vanish.

- o If f does not vanish then the box is ignored.
- o If none of the derivatives vanish then the box is ignored.
- o If only one derivative vanishes then the number of 2-nodes on the faces of the box is found and the signs of their derivatives is examined.
  - – If there are no 2-nodes the box is ignored.
  - – If there are two 2-nodes and the signs of their derivatives match then the 2-nodes are linked by a line and the recursion ends.
  - – If there are four 2-nodes then the signs are compared pair-wise to see how they link together. Matching pairs are linked by lines.
- o Otherwise, when two or more derivatives vanish, the geometry can not easily be established and the recursion continues.

When the bottom level of recursion is reached it is assumed that the box contains a singularity (or point like the north pole of a sphere). A 3-node is constructed in the centre of the box and linked to each of the 2-nodes on the faces of the box. It may also be an isolated point where all three derivatives vanish but there are no 2-nodes on the faces.

The test for 3-nodes is too strong and it is possible that some points are incorrectly marked as singularities. This is illustrated by the swallowtail surface where several incorrect isolated points are found near the cuspidal edge, Fig. 7.

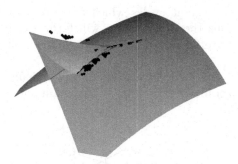

Figure 7. Incorrect isolated points found near the cuspidal edge of a swallowtail surface

3.5   Constructing a Polygonization

The final stage in the algorithm is to construct a set of polygons which approximate the surface. This is carried for each box found in the first stage of the recursion. At this stage there is a set of points linked by lines. Some of the points lie on the edges and faces of the box and others (3-nodes) may lie in the interior. However there is no information about which of the lines form the boundaries of which polygons. It would be possible to gather such information while finding the 3-nodes inside the boxes. However, this would require many more sub-boxes to be examined which would slow down the algorithm. Instead a more ad hoc algorithm is adopted, for most surfaces this will give a reasonable polygonization of the surface and there are only a few cases where it does not produce a correct polygonization. These cases typically occur when more singularities than really exist have been identified in Sec. 3.4.

The basic idea behind the algorithm is to construct polygons whose edges just consist of the lines on the faces of the box and then modify the polygon so that they include the internal lines. As a precursor to the main algorithm two sets of lines are found:

- *Cycles:* closed loops of lines which lie on the faces of the box.
- *Chains:* connected sets of lines joining 3-nodes in the interior of the box and 2-nodes on its faces. The end-points of each chain will be 2-nodes on the faces of the box.

For many simple cases where there are no internal points there will be no chains and the cycles will form the boundaries of the polygons. For other cases some refinement is necessary:

- If the cycle forms a figure of eight shape, the cycle is split into two cycles which contain no self intersections. This situation occurs when the surface has a self-intersection.

o If there are two disjoint cycles which are linked by two or more non-intersecting chains then the surface will form a cylinder. In such cases two new cycles are formed. Each form half the cylinder split along the chains Fig. 8.

o If two points on a cycle are linked together by a chain then two new cycles are formed which include the lines in the chain and some of the edges of the original cycle.

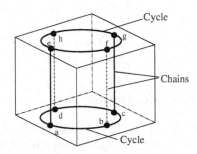

Figure 8. Constructing a cylinder. The cycles a-b-c-d and e-f-g-h and the chains a-e, c-g are linked to form two cycles a-b-c-g-f-e and c-d-a-e-h-g

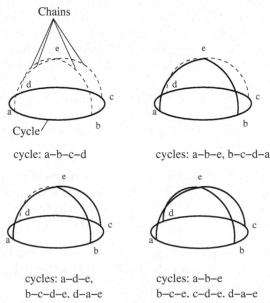

cycle: a–b–c–d

cycles: a–b–e, b–c–d–a

cycles: a–d–e,
b–c–d–e. d–a–e

cycles: a–b–e
b–c–e. c–d–e. d–a–e

Figure 9. Steps in the process of creating a polygonization of the top half of a sphere. Stating with a cycle and four chains the lines in the chains are progressively added to create four cycles used for the polygonization

These refinements happen until no more refinement are possible. The cycles then form the boundaries of the polygons. An example of this process is shown in Fig. 9 where three steps are needed to produce the final set of cycles. In practice the geometry is often more complicated than this example, Fig. 10 shows the polygonization for four boxes near a D_4 singularity. Note that several 3-nodes have been found near the singularity and the topology of the object is not quite correct.

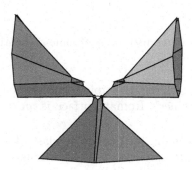

Figure 10. A close-up of the D_4 singularity showing the polygons found. Note how too many 3-nodes have been found leading to a topologically incorrect polygonization

4   Examples of Algebraic Surfaces

One area of study involving algebraic surfaces is singularity theory [3]. An important theorem of V. I. Arnold, [1, pp. 158–166] classifies the types of simple singularities which occur for functions $\mathbb{R}^n \to \mathbb{R}$. These consist of two infinite sequences: $A_k, k \geq 1$, $D_k, k \geq 4$ and three other singularities E_6, E_7 and E_8. The normal forms of these for functions $\mathbb{R}^3 \to \mathbb{R}$ are

- A_k: $\pm x^{k+1} \pm y^2 \pm z^2$, where $k \geq 1$,
- D_k: $\pm x^{k-1} + xy^2 \pm z^2$, where $k \geq 4$,
- E_6: $\pm x^4 + y^3 \pm z^2$,
- E_7: $x^3 y + y^3 \pm z^2$,
- E_8: $x^5 + y^3 \pm z^2$.

Further singularities exist which are not technically simple, however these have higher co-dimensions and are less frequently encountered. The zero sets of some of these normal forms are shown in Figures 1 and 11.

The singularities mentioned above are the only ones which occur in generic families of functions $\mathbb{R}^n \to \mathbb{R}$. In particular the singularities are always isolated. However, many of the well known functions are decidedly non-generic

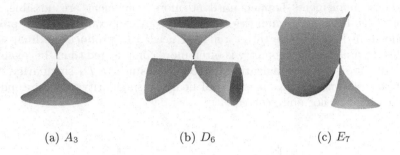

(a) A_3 (b) D_6 (c) E_7

Figure 11. Zero sets of the normal forms of some singularities

and can contain self intersections, triple points, degenerate lines, cross-caps and cuspidal edges. Steiner's Roman surface is an example which contains six cross caps.

Discriminant surfaces are an important class of surfaces which are not generic when viewed as functions. Consider the family of quartic polynomials $f(t) = t^4 + zt^2 + yt + x$, which will have a repeated root whenever $f(t) = 0$ and $\frac{df}{dt} = 4t^3 + 2zt + y = 0$. Solving these equations for t gives the swallowtail surface $-4z^3y^2 - 27y^4 + 16xz^4 - 128x^2z^2 + 144xy^2z + 256x^3 = 0$ (Fig. 12). Points of this surface will give values of x, y, z for which $f(t)$ will have a repeated root. Furthermore, if the point lies on the cuspidal edge then $\frac{d^2f}{dt^2} = 0$ and $f(t)$ has a triple root. The self-intersection of the surface gives those polynomials where $f(t)$ has two repeated roots. There is also a tail

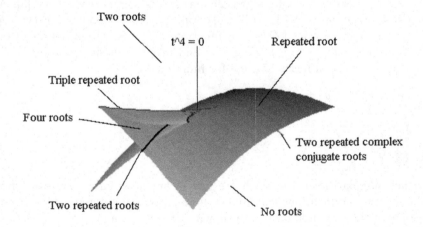

Figure 12. The discriminant surface for $t^4 + zt^2 + yt + x$ showing the types of roots which can occur.

which gives polynomials which have two complex conjugate repeated roots. Finally the swallowtail point $x = y = z = 0$ corresponds to the polynomial $t^4 = 0$.

An interesting area of study is to find low degree algebraic surfaces which contain many nodes [4]. Some examples of these include:

- Cayley's cubic, a cubic surface with the maximum of four nodes, $4(x^2 + y^2 + z^2) + 16xyz = 1$, Fig. 13(a).
- Kummer surfaces, a family of quartic surfaces some of which have 16 nodes $(3 - v^2)(x^2 + y^2 + z^2 - v^2)^2 - (3v^2 - 1)(1 - z - x\sqrt{2})(1 - z + x\sqrt{2})(1 + z + y\sqrt{2})(1 + z - y\sqrt{2}) = 0$, Fig. 13(b).
- Barth's sextic with 65 nodes $4(\tau^2 x^2 - y^2)(\tau^2 y^2 - z^2)(\tau^2 z^2 - x^2) - (1 + 2\tau)(x^2 + y^2 + z^2 - 1)^2 = 0$ where $\tau = (1 + \sqrt{5})/2$, Fig. 13(c).

These prove to be good test cases for the algorithm, which produces good but not perfect representations of the surface.

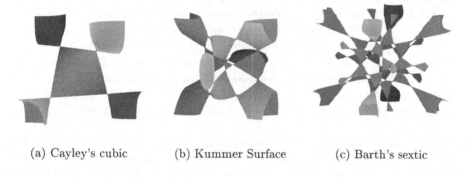

(a) Cayley's cubic   (b) Kummer Surface   (c) Barth's sextic

Figure 13. Algebraic surfaces with many nodes

5   Conclusion

This program can produce good models of many algebraic surfaces including those that contain singularities and it can even find the handle on a cross-cap. It offers considerable advantages over many other algorithms which often miss the singular points completely. The polygonization step has considerable improved its performance over previous versions.

The adaptation of the Java interface has considerably improved the usability of the software and it can now display the surfaces on most platforms without any special hardware or software requirements. It can also be used as a stand-alone program on windows machines without needing an open Internet connection. This greatly opens up the potential of the program as it could easily be used as an educational tool in schools and colleges.

The adoption of the client-server system was primarily motivated by ease of porting. The original code was written in C and it would have been a considerable task to convert this to Java. Adapting the code to run as a server and produce JVX format models was relatively straightforward. This system does have the advantage of making installation trivial. The system does also establish the program as a mathematical server which could potentially be used by other applications and incorporated into larger programs.

Whilst the models are visually good they are not always topologically correct. There are inherent problems with the algorithm as the Bernstein test for zeros of polynomials is too week and can incorrectly identify regions as containing zeros. This is particularly evident in the detection of singular points around the more complicated singularities such as the swallowtail surface. Some improvement could be made by paying more attention to the behaviour of derivatives around the singular points. One possible path for improvement is to use a particle based approach [14,15] where a set of particles surrounding the surface is allowed to converge to the surface. Another method might be to try to determine the type of singularity and use that information to inform the polygonization.

The approach we have taken here can be contrasted with ray-tracing approaches [5,7]. They produce single high quality images from a single direction. The image quality of such algorithms is better than those produce by our algorithm. However producing a 3D model which can be rotated and scaled can give a better feel for the surface and can allow particular points to be inspected.

Oliver Labs has integrated our program with the surf ray-tracer [5]: First the 3D model is calculated and rotated to produce a good view of the surface. The viewing parameters are then passed to surf to generate a high quality image from that direction.

There are several extensions to the package and the algorithm has been adapted to produce algebraic curves in 2D and 3D. The applet can be found online at http://www.comp.leeds.ac.uk/pfaf/lsmp/SingSurf.html and http://www.javaview.de/services/algebraic/.

References

1. Arnold, V. I., 1981: Singularity Theory. *London Mathematical Society Lecture Notes* **53**
2. Bloomenthal, J., Implicit Surfaces Bibliography, http://implicit.eecs.wsu. edu/biblio.html
3. Bruce, J. W., Giblin, P. J., Curves and Singularities (second edition), Cambridge, 1992
4. Endrass, S., Surface with Many Nodes, http://enriques.mathematik. uni-mainz.de/kon/docs/Eflaechen.shtml
5. Endrass, S., Huelf, H., Oertel, R., Schneider, K., Schmitt, R., Beigel, J., Surf home page, http://surf.sourceforge.net
6. Geisow, A., (1982), Surface Interrogation, Ph.D. Thesis, University of East Anglia
7. Kalra, D., Barr, A. H., Guaranteed Ray Intersection with Implicit Surfaces. *Computer Graphics*, **23(3)** 1989, 297–304.
8. Morris, R. J., A New Method for Drawing Algebraic Surfaces. In Fisher, R. B. (Ed.), Design and Application of Curves and Surfaces, Oxford University Press, 1994, 31-48
9. Morris, R. J., The Use of Computer Graphics for Solving Problems in Singularity Theory, In Hege, H.-C., Polthier, K. (Eds.), Visualization and Mathematics, Springer Verlag, July 1997. 53-66
10. Ning, P., Bloomenthal, J., An Evaluation of Implicit Surface Tilers, IEEE Computer Graphics and Applications, **13(6)**, IEEE Comput. Soc. Press, Los Alamitos CA, Nov. 1993, pp. 33-41
11. Phillips, M., Geomview Manual, Version 1.4. The Geometry Center, University of Minnesota, Minneapolis, 1993.
12. Polthier, K., Khadem, S., Preuss, E., Reitebuch, U., Publication of Interactive Visualizations with JavaView. In Borwein, J., Morales, M., Polthier, K., Rodrigues, J. F. (Eds.) Multimedia Tools for Communicating Mathematics, Springer Verlag, 2001
13. Polthier, K., Khadem, S., Preuss, E., Reitebuch, U., JavaView home page, http://www.javaview.de/
14. Saupe, D., Ruhl, M., Animation of Algebraic Surfaces. In Hege, H. C., Polthier, K. (Eds.), Visualization and Mathematics, Springer Verlag, July 1997.
15. Witkin, A., Heckbert, P., Using particles to sample and control implicit surfaces, SIGGRAPH'94, Comp. Graph. Ann. Conf. Ser. (1994), 269-277.

Visualizing Maple Plots with JavaViewLib

Steven Peter Dugaro[1] and Konrad Polthier[2]

[1] Center for Experimental and Constructive Mathematics,
Simon Fraser University, Canada
[2] Institute of Mathematics, Technical University Berlin, Germany

Abstract. JavaViewLib is a new Maple package combined with the JavaView visualization toolkit that adds new interactivity to Maple plots in both web pages and worksheets. It provides a superior viewing environment to enhance plots in Maple by adding several features to plots' interactivity, such as mouse-controlled scaling, translation, rotation in 2d, 3d, and 4d, auto-view modes, animation, picking, material colors, texture and transparency. The arc-ball rotation makes geometry viewing smoother and less directionally constrained than in Maple. Furthermore, it offers geometric modeling features that allow plots to be manipulated and imported into a worksheet. Several commands are available to export Maple plots to interactive web pages while keeping interactivity. JavaViewLib is available as an official Maple Powertool.

1 Introduction

Application connectivity refers to one programs' ability to link to other programs. In the past, application connectivity has typically been of secondary importance to the Mathematics community. Software is commonly developed from the ground up to realize the research goal, not the potential for integration with other mathematics applications. However, some applications do make provisions and have great connectivity. JavaView[5][6] provides an api for 3rd party development in addition to great import and export utility for the exchange of geometric data. Mathematica [7] has provided a very thorough interface known as MathLink on top of which a Java version known as J/Link allows Java programs to be controlled from within Mathematica, and the Mathematica kernel to be controlled from within any Java program. In fact, the ease of application connectivity provided by these two applications has already allowed for their quick and seamless integration[1].

This paper documents the authors' efforts on JavaViewLib to establish connectivity between JavaView and Maple [3], unite their strengths, and extend their functionality. JavaView and Maple vary in scope, but it is clear that one application's strengths overlaps the other's shortcomings. While Maple is a powerful tool for algebraically obtaining and generating visualization data, JavaView is a superior geometry viewer and modeling package capable of displaying geometries dynamically in HTML pages.

---

[1] http://www.javaview.de

The JavaViewLib (JVL) is an amalgam of JavaView with a Maple library; an interface between them. It makes use of the strong aspects of both to facilitate the exchange of geometries between the two, quickly build web pages with those geometries, and enhance the experience one has with Maple plots. JVL is typically used to preserve the dynamic qualities of a Maple plot upon export to the web; static plot snapshots in Maple HTML exports can now be replaced with dynamic plots in a JavaView Applet. Via JavaView, geometries can be exported from Maple to a variety of modeling packages and vice versa. JVL can quickly build geometry galleries with little effort, and these geometries can be exported and displayed in a legible XML format to ease further development. JavaViewLib is available as an official Maple Powertool [4].

2   Visualization in Maple and JavaView

2.1   Graphics in Maple

Maple is a comprehensive computer algebra system. It offers packages from various branches of advanced mathematics to aid in solution and visualization. It also offers word processing facilities with typeset mathematics and export capabilities to aid in the composition of mathematical papers. Furthermore, it provides a unique environment for the rapid development of mathematical programs using its own programming language and function libraries. These features are encapsulated in a single Maple Worksheet (mws), which can be transported and rendered between Maple applications.

Maple's mathematics engine is among the industry's best, but there is room for improvement in certain areas of the application overall. While Maple supplies numerous methods for generating graphics from mathematical expressions, the viewer itself has a primitive feature set. Once the graphic – or plot in Maple terminology – has been rendered, the viewer only provides

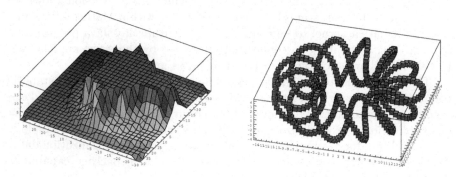

Figure 1. JavaView parses the color of meshes and viewing options like axis and frames of Maple plots.

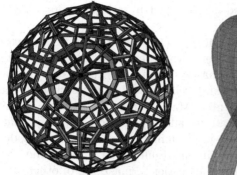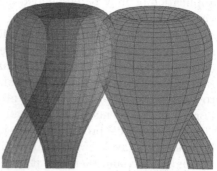

Figure 2. Maple Plots with transparency visualized and colored in JavaView, for example, to show the hidden inner tube of the Klein bottle (right).

control for color, line styles, perspective, axes, and polar coordinate rotation; sufficient control over the appearance of the visualization, but not much in the way of actually viewing it. Maple worksheets can be exported to a variety of document formats including HTML, rich text, and latex. The plots therein are exported via Graphics Interchange Format – although plots may be individually exported to various standard image formats such as eps, jpeg, and bmp. Certainly static images suffice for hardcopies, but with multimedia rich documents, such as mws and HTML, the dynamic qualities of visualizations should be preserved to convey the greatest amount of information.

Maples' programming environment is the Maple worksheet. With immediate access to the extensive mathematical function library, code is written, executed, and debugged inline. This makes coding small mathematical procedures relatively painless. However, on a larger scale such as package development, programming, compiling and debugging can quickly become cumbersome. While there are a few new developments that allow more control over the Maple interface, the absence of a software development kit (sdk), or application programmers' interface (api) makes it impossible to develop genuine, transparent plug-ins.

2.2   Interactive Visualization Online with JavaView

JavaView is a sophisticated visualization appletcation and geometric software library. It offers a superior viewing environment that employs scaling, translation, quaternion rotation, customizable axes, colouring, materials and textures, transparency, anti-aliasing, depth cueing, animation, and camera control among other features. It is capable of importing and exporting 3d geometries in a wide variety of formats, and can perform modeling upon these geometries. It also offers an API, allowing custom plug-ins and visualizations to be developed for it. Furthermore, JavaView can be used as a standalone application or may be embedded as an applet in web pages for remote use via web browsers.

JavaView is a mature and portable geometry viewer. Numerous researchers and educators have utilized it for their own experiments and as a result have developed many general JavaView based visualization tools. However, a certain amount of Java programming and web development expertise is required to employ its full functionality. Unlike Maple, where an exhaustive function library and mathematics engine can be called upon to quickly input and execute mathematical programs, more programmatic care and custom code may be needed in JavaView to achieve similar results. The scope of the application is smaller, and therefore it may be easier to make use of other software applications to generate the visualization data or geometric models for the JavaView environment.

3   JavaViewLib - A New Maple Powertool

Maple makes little provision for application connectivity. However, it does allow for packages written in the Maple programming environment to be loaded into a Maple session. This development constraint combined with lack of control over the interface makes it impossible to create a plug-in in the true sense of the word. In the absence of an api, JVL enables the two applications to exchange data through flat files. Two file formats are utilized to allow geometric information to pass between the applications; one is an .mpl file that is nearly identical to the Maple plot data structure, the other, .jvx, is JavaView's native XML based file format for the storage of geometry models, animations and scenes. JVL parses and prepares Maple plot data in one of these two formats for import into JavaView, and as these files are the means of connectivity, JVL provides the mechanism by which the data in these files can be rendered in Maple. JVL also builds the necessary HTML code that browsers require to render these geometries in an embedded JavaView applet.

3.1   Maple Plot Data Structures

The maple plotting functions produce PLOT and PLOT3D data structures describing how Maple graphics are rendered in Maple's 2D or 3D viewer respectively. The two data structures produced are themselves internal Maple functions of the form PLOT(...) or PLOT3D(...). These structures can be viewed in a maple worksheet by assigning a plot function to a variable. Executing this variable on a Maple command line will invoke the functional data structure and render the image. Normally, the structures consist of the geometric data organized in labeled point lists followed by some stylistic information. For some parts, JVL currently ignores stylistic information and uses JavaView's default options.

The PLOT data structure embeds four types of objects: points, curves, polygons, and text. In addition to these four types, the PLOT3D data structure embeds the mesh, grid and isosurface objects. A geometric point is represented by a list of floating point values; pi := [xi,yi] for the PLOT data structure and pi := [xi,yi,zi] for the PLOT3D data structure. Connected points are maintained in a geometric point list. A plot object is represented by a set of geometric points or geometric point lists wrapped in the corresponding label. For instance, the points object takes the form: POINTS(p1, p2,...,pn), the curves object takes the form: CURVES([p1,p2,...,pn], [q1,q2,...,qn], ..., [r1,r2,...,rn]), and the polygons object takes a form identical to the curves object, except the POLYGONS label indicates that the last point in each point list is to be connected to the first. The text object is nothing more than a label, a point and a string to display at that point: TEXT(p1, "string"). The mesh, and grid objects are slightly more interesting.

Typically, plot functions involving parametric equations generate the mesh object. The mesh object maintains a matrix of geometric points based on a u-parameter resolution and a v-parameter resolution. This resolution is determined by specifying the amount by which to partition the u and v parameter ranges. The matrix, a list of geometric point lists, connects the geometric points sequentially by rows and by columns. For example, the following Maple plot function encapsulates the parametric representation for a sphere, and requires a discrete u,v-domain with which to compute the mapping. This is specified with a u-parameter range, a v-parameter range and the amount to partition by in the u and v directions with the grid=[numU,numV] argument. Storing the plot object reveals its data structure, and by examining the matrix of vertices embedded within the mesh object, we are able to determine the order in which the geometric points are connected.

```
[> s:=plots[sphereplot](1,u=0..2*Pi,v=0..Pi, grid=[4,5]);

s := PLOT3D(MESH(Array(1..4, 1..5, 1..3, [
[[0.,0.,1.],[.7,0.,.7],[1.,0.,0],[.7,0.,-.7],[0.,0.,-1.]],
[[0.,0.,1.],[-.35,.61,.7],[-.49,.86,0],[-.35,.61,-.7],[0.,0.,-1.]],
[[0.,0.,1.],[-.35,-.61,.7],[-.5,-.86,0.],[-.35,-.61,-.7],[0.,0.,-1.]],
[[0.,0.,1.],[.7,0.,.7],[1.,0.,0.],[.7,0.,-.7],[0.,0.,-1.]]])))
```

Most other plot functions in Maple generate the grid object. The grid object maintains a matrix of z-coordinate values based on an implicitly derived, discretely resolved Cartesian domain. The parameter ranges are also stored in the grid object, and are used with the row and column dimensions of the z-coordinate matrix to compute the corresponding x and y coordinate of the geometric point to be rendered. The geometric points are computed and connected sequentially by rows and by columns. For instance, the following Maple plot function specifies a Cartesian grid resolution with the grid=[numX,numY] argument and the x=x1..x2 and y=y1..y2 parameter ranges. The grid argument specifies the dimensions of the z-coordinate matrix, whose values

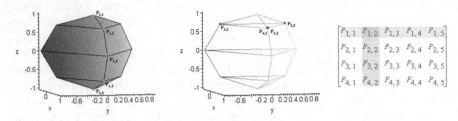

Figure 3. Polyhedral meshes in the Maple Plot data structure.

are incrementally computed with the given formula f. The first geometric point is resolved by P1,1 = [x1,y1,f(x1,y1)], and subsequent points are connected horizontally and vertically by Pi+1, j+1 [xi+(x2 - x1)/numX, yj+(y2 - y1)/numY, f(xi+(x2 - x1)/numX, yj+(y2 - y1)/numY)].

Finally, Maple achieves animation via the animate plot object. It contains a sequence of plot objects wrapped in a list where each list defines one frame in the animation. It takes the form: ANIMATE([Plot Object frame 1], [Plot Object frame 2], ..., [Plot Object frame n]). For instance, the Maple command that follows animates a sequence of mesh objects. It uses the parameter t to describe how the geometry changes from frame to frame, and builds the set of mesh objects accordingly.

```
[> catHelAnim:=animate3d([cos(t)*cos(x)*cosh(y)+sin(t)*sin(x)*sinh(y),
   -cos(t)*sin(x)*cosh(y)+sin(t)*cos(x)*sinh(y),cos(t)*y+sin(u)*x],
   x=-Pi..Pi,y=-2..2,t=0..Pi,scaling=constrained);
catHelAnim = PLOT3D(ANIMATE([MESH(...)],[MESH(...)],...,[MESH(...)]),
   AXESLABELS(x,y,""),AXESSTYLE(FRAME),SCALING(CONSTRAINED))

[> runJavaView(catHelAnim);
```

3.2   Usage of JVL

The library functions available in JVL can be viewed after successfully loading the library into a maple session:

```
[> with(JavaViewLib);

[exportHTM, exportHTMLite, exportJVX, exportMPL, genTag, genTagLite,
 getInfo, import, runApplet, runAppletLite, runJavaView, runMarkupTree,
 set, viewGallery]
```

These functions employ JVL to set configuration parameters and build web pages, allow 3rd party geometries to be imported via JavaView, and enable maple plots to be exported in a variety of ways that interface with JavaView. Notice that some function names have the 'Lite' suffix. These functions make use of an alternative version of JavaView – optimized for size and speed – intended for use as a geometry viewer only. More details on the JVL commands are described in subsequent sections of this document.

Figure 4. Maple animations are shown as dynamic geometry in JavaView, optionally with a smooth morphing between keyframes. In this figure, the surface coloring of the helicoid and catenoid, and the display of grids of the in-betweenings was fine-tuned in JavaView.

3.3   Basic Commands

The simplest way to use JVL is to wrap a 'run' function around a plot command. By default, a file called JVLExport.mpl will be created in the mpl folder of the installation directory. A call to `runJavaView` launches the JavaView standalone with the geometry contained in JVLExport.mpl. Once in JavaView, several operations can be applied to the geometry. For instance, when exporting a surface from Maple then adjacent polygons are not connected, that means, the common vertices appear multiple time. JavaView is able to identify these vertices and merge them to create a single seamless surface.

```
[> runJavaView(plot3d([4+x*cos(y),2*y,x*sin(y)],
   x=-Pi..Pi,y=-0..Pi,coords=cylindrical,grid=[2,40]));
```

The `runApplet` command is used to create an HTML file called JVLExport.htm containing the necessary applet tag and then to launch the defined browser to view it. Adding Maple plots to a JavaView enhanced web page allows geometries to be viewed remotely over the Internet. The following example also illustrates the additional flexibility of the JavaView system to analyse individual geometries of a complex scene, see Figure 5. Additional JVL commands will be discussed in the following sections.

```
[> runApplet(plots[coordplot3d](sixsphere));
```

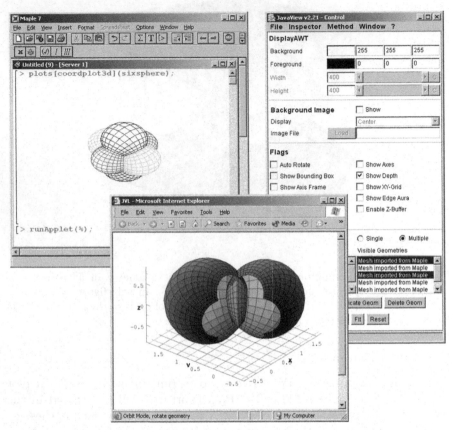

Figure 5. The JVL command `runApplet` creates an interactive web page of any Maple plot.

3.4 Development of JVL

Maple's provision for application development comes in the way of packages. Packages are implemented with Maple's module construct, and are simply a collection of data and procedures, some of which are made public to the user. A number of Maple modules can be stored together in library, which is made up of three files: maple.lib maple.rep and maple.ind. The following partial code listing outlines JVL's basic module definition and the required calls to create the library. The module definition specifies the procedures to make public using the export identifier. Private procedures and data must be declared using the local identifier. Once these are initialized, they cannot be modified. Procedures and data declared with the global modifier are not publicly exposed, but are publicly accessible and may be modified. The option identifier, among other things, specifies the function to call when the package is loaded into a maple session. After the module is defined, the march (maple archive) command creates the library files in the specified directory. Once the library files are created the savelib command adds the Maple expression defined as 'JavaViewLib' to the archive.

```
[> JavaViewLib := module()
 option package, load = setup, 'Copyright I Steven Dugaro 2001':
 export import, exportHTM, runApplet, runJavaView, ..., set:
 global 'type/JVLObject':
 local  setup, import, ..., w1:

 # muted module constants; unmodifiable strings
  w1:="Unable to find the specified file.":
  ...
 # private function definitions
  setup := proc()
   interface(verboseproc=0):
   'type/JVLObject' := proc(x) ... end proc:
   setOS():
   buildIT():
  end proc;
  ...
 # public function definitions
  import := proc() ... end proc:
  ...

 end module:

[> march('create', '/JavaViewLib', 100);
[> savelib('JavaViewLib');
```

Using Maple's plot data structures and package mechanism it is straightforward to build application connectivity via parsers, file I/O and system calls. Two parsers were written – one within JavaView and one within JVL. The JVL parser extracts the geometric information from the plot data structures, and reproduces the information in JavaViews' jvx file format. JVL also outputs the plot structures into .mpl files from which the JavaView parser extracts the necessary geometric and stylistic information. These geometry files serve as the basic means of information exchange between JavaView and Maple. The jvx file format is intended for the export of stylistic free geometries in a human readable file format. The .mpl file format should be used to export stylistic geometries in a compact file format. These files are passed to JavaView via a command line argument when it is invoked as a standalone, or via an applet tag parameter when it is rendered in a browser. JVL also allows plot data to be included as an applet parameter in the applet tag so that all the necessary information may be contained in a single HTML file. Using simple file I/O, JVL creates the geometry files, the HTML files or both then employs system calls to launch JavaView or the user's browser from within Maple. The following code snippet shows a brief example of a JVL library procedure that performs plot parsing and JVX file generation. Note, saving Maple plots as JVX file will allow JVL to include additional rendering commands like setting transparency of a geometry. However, exporting Maple plots to JavaView via MPL files will work fine, too.

```
[> exportJVX := proc()
# PUBLIC: save a Maple plot in JavaView's JVX file format

 if nargs = 1 then
  if type(args[1], JVLObject) then
   oargs:= getIOstring("JVLExport",jvx),args[1]:fi:
  else
  oargs:=check2args(args):
  oargs:=getIOstring(oargs[1],"jvx"),oargs[2]:fi:

 try
  fd:= fopen(getIOstring(oargs[1],jvx), WRITE):
  fprintf(fd,"%s", JVXHEADER):
  fprintf(fd," <jvx-model>\n\t<geometries>\n"):
  for l from 1 to nops(oargs[2]) do
   a:=op(l,oargs[2]):
   if op(0,a) = MESH then
   ...
   elif op(0,a) = GRID then
   ...
    elif op(0,a) = POLYGONS then
   ...
   elif op(0,a) = CURVES then
   ...
   elif op(0,a) = POINTS then
   ...
   else # handle other Maple objects including styles.
   ...
   fi:
  od:
  fprintf(fd,"\n\t</geometries>\n</jvx-model>"):
  fclose(fd):
  return oargs[1]:
 catch:
  error("Could not write to file ", oargs[1], lastexception):
 end try:
end proc:
```

4   Importing and Exporting Geometries

JVL has extended the capabilities of Maple to make geometric information highly portable. For the first time it is possible to export Maple plots to a variety of formats and import geometries from a variety of formats into Maple. One is finally able to export Maple worksheets into an HTML file where the dynamic qualities of the plot is preserved.

As it is typically encouraged to keep similar file types grouped together when developing web pages, JVL maintains a working directory to organize its exports. Its working directory contains four subfolders: 1) an ./mpl subfolder for mpl files, 2) a ./jvx subfolder for jvx files 3) an ./htm subfolder for HTML files, and 4) a ./jars subfolder for the JavaView applet. The working directory defaults to the JVL installation directory, so it is recommended that a working directory be set before exporting.

4.1   Import of Geometry Files into Maple

JVL does not implement its own file format parsers; these parsers are already implemented in JavaView. These parsers cannot be accessed from within JVL since application connectivity is weak. As a result, importing 3rd party geometries into Maple requires an intermediate step. After the geometry has been loaded into JavaView, it must be saved down as an mpl file. At this point JVL's import command can be called to pull the geometry into Maple. The following example and Figure 6 illustrate this.

```
[> runJavaView("/temp/hand.obj");
[> import("/temp/hand.mpl");
```

4.2   Exporting Maple Plots to Web Pages

The three fundamental JVL export functions are exportHTM, exportMPL, and exportJVX. They provide three different contexts in which to export Maple plot data. The former is used to generate HTML pages that either link to or contain the plot information and embed the JavaView applet. The latter two, are for generating the respective geometric files only. These functions require 1 argument – the plot object. A second optional argument may be given to specify the filename and path to which the file is to be exported. Maples plots commands can be wrapped in the export functions or defined in a variable to be passed by reference. Here we define a simple cube in the box variable:

```
[> box:=plots[polyhedraplot]([0,0,0],polytype=hexahedron):
```

The exportMPL interface is used to export a stylistic, compact representation of the Maple plot. JavaView interprets most of the plot attributes, and selectively discards others. The following command exports the geometry into a file called mplBox.mpl into the mpl folder of the current working directory.

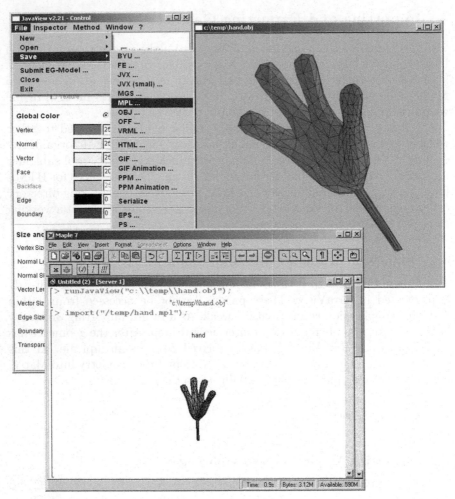

Figure 6. Import a wealth of 3D file formats into Maple via JavaView's geometry loaders. For example, OBJ is the standard file format of Java3D and Wavefront, and accepted by rendering software like Maya.

```
[> exportMPL(box, mplBox):
```

The exportJVX interface is used to export a minimal, legible representation of the Maple plot. However, the display can be embellished with the addition of attributes and other jvx tags to the jvx file. The following command exports the geometry to a file called jvxBox.jvx in the temp folder of the root directory.

```
[> exportJVX("/temp/jvxBox", box):
```

The `exportHTM` interface is used to generate and couple HTML pages with exported Maple plots. This can be done in one of three ways: 1) embed the data within the HTML page itself, 2) generate and link to a geometry file, or 3) create the HTML page for an existing geometry file. Appending a filename extension qualifies the method of plot export. Embedding the data within the HTML file is the default method for this function and so no extension is needed, however appending '.jvx' or '.mpl' will export the plot to a separate file and creates the HTML page that links to it. JVL normally exports with respect to the working directory so that all files can be relatively referenced, nevertheless, exporting to an arbitrary path will copy all the relevant files to that directory. The following commands demonstrate these four possible usage scenarios.

```
[> exportHTM(box,"box");       # export the plot into an HTML file
[> exportHTM(box,"box.mpl");   # generate box.mpl, box.htm & link
[> exportHTM(box,"jvxBox.jvx"); # generate jvxBox.jvx and jvxBox.htm
[> exportHTM(box,"/temp/");     # copy jars, JVLExport.htm to temp
```

4.3   More on the run* Commands

The JVL `run*` commands are basically wrappers around the above export functions. They do little more than specify the export method and launch the appropriate application. Consequently, the argument guidelines are the same as the export functions. A filename and/or path is optional, defaulting to `JVLExport` in the current working directory, a qualifying file extension specifies the method of export, and arbitrary paths copy the necessary files to the specified location. However, unlike `exportHTM`, existing HTML files are opened for viewing and not overwritten.

The JavaView standalone application is interfaced with Maple using the `runJavaView` command. It contains the most complete compilation of the JavaView modules as web considerations need not be taken into account. This interface is provided for the use of JavaView on your local machine, and automatically launches JavaView from within Maple. Once a model has been loaded into JavaView, Javaview's geometry tools may be utilized. This includes materials features such as texture mapping, modeling features such as triangulation, and effects features such as explosions. The following function calls illustrate how to typically make use of this interface function, see Figure 7.

```
[> runJavaView():    # launch the JavaView application
[> runJavaview(box): # launch JavaView with a Maple plot
[> runJavaview(box,myBox.mpl): # launch and save a Maple plot
[> runJavaView("models/hand.obj"): # load a 3rd party geometry
```

As an applet, JavaView can be used interactively over the Internet. The `runApplet` function is able to expedite this process by exporting maple plots to a 'skeletal web page', which can then be fleshed out into a final HTML

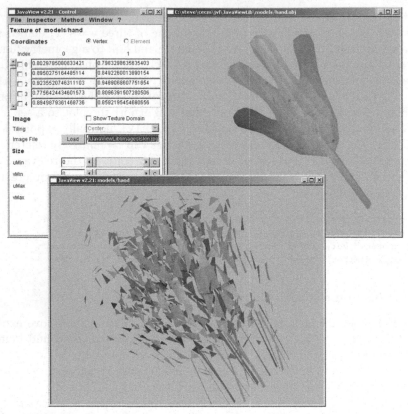

Figure 7. JavaView's advanced modeling tools allow for the fine tuning of geometric shapes.

document. This 'skeletal web page' simply contains the applet tag that embeds the JavaView applet. Tags in general are directory structure dependent as they point to the files with which the browser should render the page. Therefore it is important to keep the structure maintained by JVL in your working path - relocating files would require you to manually adjust the tag's definition. After building and exporting the necessary files, `runApplet` will automatically launch the defined browser from within Maple for viewing. The following examples demonstrate some typical uses for this function.

```
[> url:= "http://www.cecm.sfu.ca/news/coolstuff/JVL/htm/webdemo.htm":
[> runApplet(url): # open a web page from within Maple
[> runApplet(box): # launch a browser with a Maple plot
[> runApplet(jvxBox.jvx): # launch and build page for existing file
[> runApplet(box,myBox.mpl): # launch, build page, and save a plot
[> runApplet("/models/hand.obj"): # launch a 3rd party geometry
[> runApplet(box,"/temp/box.htm"): # launch, build, copy to path
```

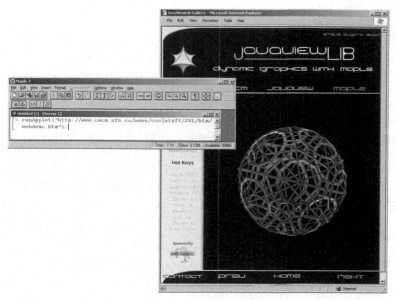

Figure 8. The original homepage of JavaViewLib launched from inside Maple.

5   Additional Features

5.1   JVL State Information

JVL maintains a small amount of state information to assist in the configuration of the JavaView viewer. For the most part, these states specify how the JavaView applet is to be rendered in a browser. These states can be set to specify the size of the JavaView viewport, its background image, the current working path as well as some viewing initializations such as auto-rotation, axes, and depth cueing. The list of state information can be obtained with the following function:

```
[> getInfo();

JavaViewLib State Information
-----------------------+------
[W ] Applet Width       | 400
[H ] Applet Height      | 400
[A ] Applet Alignment   | Center
[R ] AutoRotate         | 1. 1. 1.
[X ] Axes               | Hide
[BC] Background Colour   | 255 255 255
[BI] Background Image    | images/jvl.jpg
[B ] Border             | Hide
[BB] Bounding Box       | Show
[BR] Browser            | iexplore
[V ] Camera Direction   | 1. 2. 3.
```

270   Steven P. Dugaro and Konrad Polthier

```
[DC] Depth Cueing      | Hide
[EA] Edge Aura         | Show
[WK] Working Path      | C:\Program Files\Maple 6\JavaViewLib\
     Installation Path | C:\Program Files\Maple 6\JavaViewLib\
     Operating System  | Windows NT/2000
```

JVL states can be configured with the set command by specifying a list of attribute = value pairs. Most binary states can be toggled by either assigning a show/hide or on/off value, or by simply including its handle in the list. Here we set the working path for the export project and specify that all subsequent tags are to be rendered with a left aligned 200 pixel by 200 pixel viewport, in auto-rotate model with axes, and the currently defined background image.

```
[> set(wp="c:\\temp\\myGeoms\\", width=200, height=200, axes,
     autorotate, bg=image, align=left):
[> runApplet(plots[polyhedraplot]([0,0,0], polytype=hexahedron));
```

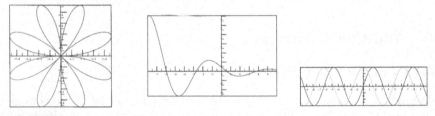

Figure 9. 2D graphs exported to a JavaView applet will keep axis and projection settings.

5.2   Markup Tree of the JVX Geometry

The runMarkupTree command simply exports a plot in the jvx format, and launches a browser to view it. The markup tree is an XML representation of the geometric data contained in a plot, and consists of tags that represent points, faces, and geometries. The listing in Figure 11 is the markup tree for a tetrahedron. Notice that the Maple plot, and therefore the .mpl format, contains many redundant points. These points can be merged with JavaView's 'Identify Points' modeling command and preserved by saving in the .jvx format.

```
[> runMarkupTree(plots[polyhedraplot]([0,0,0], polytype=tetrahedron));
```

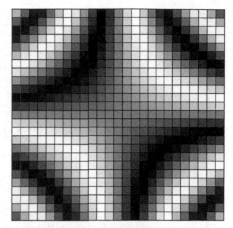

Figure 10. JavaView recognizes a large variety of different Maple plots including contour and density plots.

5.3 Create and Configure an Applet Tag

On occasion, it may be quicker to simply generate the applet tags for a series of plots, instead of exporting an HTML page for each of them. The genTag function returns the applet tag as it would be rendered in the HTML document if exported otherwise. These tags can then be cut and paste into a single HTML document kept relative to the working directory – i.e. in the htm subfolder. The following example quickly generates 3 orthogonal views and one perspective view with autorotation using the lite version of the JavaView applet.

```
[> b := plot3d((1.3)^x * sin(y),x=-1..2*Pi,y=0..Pi,
               coords=spherical,style=patch):
[> set(reset):
[> set(viewDir="0 -1 0", bg="200 200 200", border="on", axes="off",
      width=300, height=300):
[> exportHTM(b,bounty.mpl):
[> set(viewDir="0 1 0"):
[> genTagLite(bounty.mpl);
 <APPLET CODE='jvLite.class' CODEBASE='../' ARCHIVE='jars/jv_lite.zip'
  WIDTH='300' HEIGHT='300' ID='JVLAPPLET' ALT='JVL - MAPLE Export'>
  <PARAM NAME='Background' VALUE='200 200 200'>
  <PARAM NAME='Border' VALUE='Show'>
  <PARAM NAME='ViewDir' VALUE='0 1 0'>
  <PARAM NAME='Model' VALUE='mpl/bounty.mpl'>
 </APPLET>

[> set(viewDir="0 0 1"):
[> genTagLite(bounty.mpl);
 <APPLET CODE='jvLite.class' CODEBASE='../' ARCHIVE='jars/jv_lite.zip'
  WIDTH='300' HEIGHT='300' ID='JVLAPPLET' ALT='JVL - MAPLE Export'>
  <PARAM NAME='Background' VALUE='200 200 200'>
```

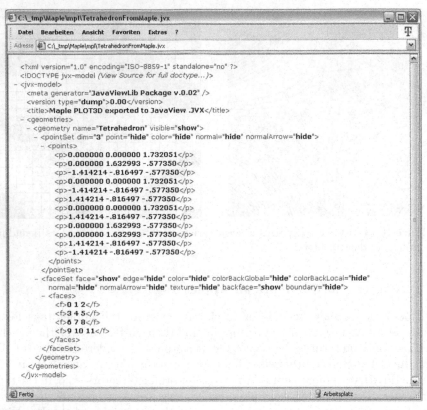

Figure 11. View a Maple plot as formatted XML document, and collaps and expand individual parts of the markup tree. The JVL command runMarkupTree makes use of JavaView's JVX file format and the formatting capabilities of the Internet Explorer. Here the vertices and faces of a tetrahedron are listed.

```
  <PARAM NAME='Border' VALUE='Show'>
  <PARAM NAME='ViewDir' VALUE='0 0 1'>
  <PARAM NAME='Model' VALUE='mpl/bounty.mpl'>
</APPLET>

[> set(viewDir="1 1 1", axes="boundingbox", rotate):
[> genTagLite(b,bounty.jvx);
<APPLET CODE='jvLite.class' CODEBASE='../' ARCHIVE='jars/jv_lite.zip'
  WIDTH='300' HEIGHT='300' ID='JVLAPPLET' ALT='JVL - MAPLE Export'>
  <PARAM NAME='Axes' VALUE='Show'>
  <PARAM NAME='BoundingBox' VALUE='Show'>
  <PARAM NAME='Background' VALUE='200 200 200'>
  <PARAM NAME='Border' VALUE='Show'>
  <PARAM NAME='AutoRotate' VALUE='Show'>
  <PARAM NAME='ViewDir' VALUE='1 1 1'>
  <PARAM NAME='Model' VALUE='jvx/bounty.jvx'>
</APPLET>
```

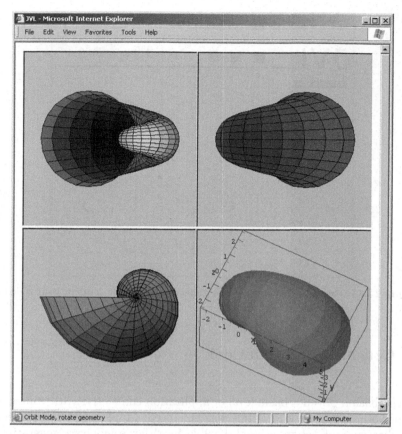

Figure 12. The JVL commands `genTag()` create an applet tag of a Maple plot. Here `genTag()` was used to show different projections of the same geometry.

5.4   Creating a Web Gallery of Maple Plots

The quickest way to make exported geometries web ready is to let JVL do it. The `viewGallery` command builds a frame based geometry gallery with a table of contents that links to all exports in the htm folder of the current working directory. The following commands export a 2D geometry to the htm folder, and build the necessary HTML files for the gallery at the top level of the current working directory. This provides a quick way to publish visualization projects on the Internet.

```
[> exportHTM(plot([sin(4*x),x,x=0..2*Pi],coords=polar):
[> viewGallery();
```

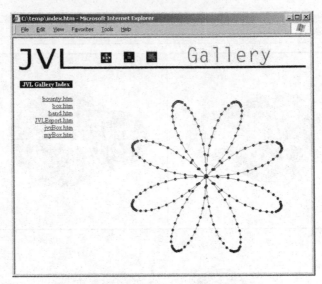

Figure 13. Automatically create a web gallery of all Maple plots in your repository with the JVL command `viewGallery()`.

6   Conclusion and Outlook

By establishing even a basic level of application connectivity, the functionality of both JavaView and Maple was enhanced. Maple was afforded greater visualization and web functionality while JavaView's geometry generation capabilities were extended to match that of Maple. The JavaViewLib has made its mark in the sand as a proponent of application connectivity. By bridging these applications, the JavaViewLib broadens the toolset available for the research and teaching of mathematics.

At the time of its development, the JavaViewLib made the best possible use of the connectivity resources made available by Maple and JavaView. However, in the year since its release, there is now room for improvement; both Maple and JavaView appear to be making considerable advancements in the provision for third party control. JavaView has introduced a secondary XML file format for the initialization and preservation of display and camera properties. Through this format, known as jvd, precise control over camera, lighting, and viewport properties can be specified and launched with JavaView. This is a step in the right direction, but covers only a small portion of JavaView's rich feature set. Ideally, this file format will mature into a full featured scripting language that allows the broad range of JavaView operations to be applied to the geometry or geometries loaded into the viewer. Two new technologies introduced with Maple 8, known as Maplets and MapleNet appear to be moving in the direction of third party integration. However, the two fall just short of providing a transparent look and feel for third party plugins.

7   Downloading JavaViewLib

The JavaViewLib has become an official Maple Powertool, and may be obtained from the MapleSoft website [3] at

http://www.mapleapps.com/powertools/javalib/javalib.shtml.

The original website of JVL [1][2] as well as new releases and updates reside at the CECM website under

http://www.cecm.sfu.ca/news/coolstuff/JVL/htm/webdemo.htm.

The JavaView [5] visualization environment, which also includes the parser for Maple plots, is contained in the JVL download but may be upgraded independently by replacing the JavaView directory with newer versions from the JavaView homepage

http://www.javaview.de.

Package downloads include a tutorial Maple worksheet *gettingStarted.mws* and a *readme.txt* file for installation instructions.

References

1. S. Dugaro. JavaViewLib homepage. www.cecm.sfu.ca/news/coolstuff/JVL/htm/webdemo.htm.
2. S. Dugaro. JavaViewLib - a visualization powertool. In *Proc. of the Maple Summer Workshop*. Waterloo Maple Inc., 2002.
3. Maple Waterloo. Homepage. http://www.maplesoft.com.
4. Maple Waterloo. Powertools homepage. http://www.mapleapps.com/powertools/.
5. K. Polthier. JavaView homepage, 1998–2002. http://www.javaview.de/.
6. K. Polthier, S. Khadem-Al-Charieh, E. Preuß, and U. Reitebuch. Publication of interactive visualizations with JavaView. In J. Borwein, M. H. Morales, K. Polthier, and J. F. Rodrigues, editors, *Multimedia Tools for Communicating Mathematics*. Springer Verlag, 2002. http://www.javaview.de.
7. Wolfram Research. Homepage. http://www.wolfram.com.

Automated Generation of Diagrams with Maple and Java

Dongming Wang

Laboratoire d'Informatique de Paris 6, Université Pierre et Marie Curie – CNRS,
4 place Jussieu, F-75252 Paris Cedex 05, France

Abstract. This note shows how to draw diagrams automatically from the predicate specification of a given set of geometric relations among a set of points in the plane. It is done first in Maple by translating the geometric relations into polynomial equations, decomposing the obtained system of polynomials into irreducible representative triangular sets, and finding an adequate numerical solution from each triangular set. A Java class coding the solution and the polynomials in each triangular set is generated, compiled, and then executed with the main Java programs to draw a diagram. The whole process combining symbolic elimination in Maple with numerical computation, graphic drawing, and letter labeling in Java is fully automatic. The drawn diagrams may be animated and fine-tuned by mouse click and dragging and saved as PostScript files. We present the drawing method, discuss some techniques of implementation, and give several sample diagrams drawn by our program.

1   Introduction and Motivation

The main motivation for this work is automated theorem proving in geometry [7], where diagrams are needed and useful for the explanation of theorems and problems in question. The author has implemented an environment GEOTHER [5] for the manipulation and proving of theorems in elementary (differential) geometry. It is natural to include a module to draw diagrams automatically for geometric theorems given in predicate specification. In order to demonstrate the validity of a theorem, it is also desirable that the drawn diagram can be animated. The construction of diagrams from arbitrary geometric constraints is a general problem that has been studied extensively in the area of geometric constraint solving [1,3]. The diagram generator described in this note works for a particular and reasonably large class of geometric constraint problems, i.e., theorems in plane Euclidean geometry.

Problem D. Given n points $P_1, \ldots, P_n$ in the plane and a set of geometric relations $\mathcal{R}_i$ among these points, draw a diagram of $P_1, \ldots, P_n$ *automatically* that satisfies the relations $\mathcal{R}_i$.

Consider, for example, the following specification of a diagram as input

```
arbitrary(A,B,C)
oncircle(A,B,C,D)
perpfoot(D,P,A,B,P)
perpfoot(D,Q,A,C,Q)
perpfoot(D,R,B,C,R)
collinear(P,Q,R)
```

where A, B, C are three arbitrary points in the plane and the predicates `oncircle`, `perpfoot`, etc. specify some geometric relations among the points involved. As output, we want the program to draw a diagram like the one in Figure 1 automatically. Note that, in general, the diagram is not uniquely determined: some points may be *free* (two degrees of freedom) or *semi-free* (one degree of freedom). Therefore, the diagram drawn is actually a (randomly yet adequately chosen) instance.

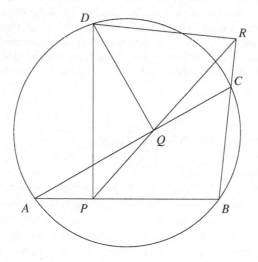

Figure 1.

Several software systems, for example, Cinderella [4] and Geometry Expert [2], allow the user to draw geometric diagrams interactively. Our emphasis here is the combination of symbolic and numerical computations performed in two environments (Maple and Java) and the automation of the whole drawing process. Although the class of problems considered by us is somewhat restricted, the automated incorporation of non-trivial symbolic expressions computed in Maple into Java programs to generate animation diagrams appears to be an interesting aspect that may be further studied for geometry software development.

2   Sketch of the Method

By taking coordinates $x_1, \ldots, x_m$ for the points P_k, the geometric relations R_i may be expressed as polynomial equations

$$\begin{cases} f_1(x_1, \ldots, x_m) = 0, \\ f_2(x_1, \ldots, x_m) = 0, \\ \quad\cdots\cdots \\ f_s(x_1, \ldots, x_m) = 0 \end{cases} \tag{1}$$

in $x_1, \ldots, x_m$ with coefficients in the field $\mathbb{Q}$ of rational numbers. The first step of drawing a diagram is to find specialized and adequate real values for the x_j that satisfy the equations in (1). This is essentially a problem of solving a system of polynomial equations (of positive dimension) (see [1,3] for more details) and thus can be done by using existing methods, for instance, those based on triangular sets, Gröbner bases, and resultants presented in [6]. To fix the idea, we use the characteristic set method of Wu [7,6]. This choice is due to the fact that Wu's method was implemented first as the algebraic engine for the provers in GEOTHER.

Let $\mathbf{P} = \{f_1, \ldots, f_s\}$, and Zero$(\mathbf{P})$ denote the set of all common zeros of $f_1, \ldots, f_s$. Using Wu–Ritt's algorithm, one can compute a Wu *characteristic set* $\mathbf{C}$ of $\mathbf{P}$, which has the following triangular form:

$$\mathbf{C} = \begin{bmatrix} c_1(x_1, \ldots, x_{p_1}), \\ c_2(x_1, \ldots, x_{p_1}, \ldots, x_{p_2}), \\ \quad\cdots\cdots \\ c_r(x_1, \ldots, x_{p_1}, \ldots, x_{p_2}, \ldots, x_{p_r}) \end{bmatrix},$$

such that Zero$(\mathbf{P}/J)$ = Zero$(\mathbf{C}/J)$, where $J = I_1 \cdots I_r$, I_i is the leading coefficient of c_i with respect to its last/leading variable x_{p_i}, and Zero$(\mathbf{P}/J)$ = Zero$(\mathbf{P}) \setminus$ Zero$(\{J\})$. If $\mathbf{C}$ consists of a single non-zero constant, then the constraints $\mathcal{R}_i$ are inconsistent. $I_i = 0$ usually corresponds to a degenerate case of the geometric problem such as a triangle degenerates to a line. We assume that Zero$(\mathbf{C}/J) \neq \emptyset$.

Let $\mathbf{y} = \{x_{p_1}, \ldots, x_{p_r}\}$ and $\mathbf{u} = \{x_1, \ldots, x_m\} \setminus \mathbf{y}$. The variables in $\mathbf{u}$ and $\mathbf{y}$ are called *parametric* and *dependent* variables of $\mathbf{C}$, respectively. By means of (algebraic) factorization, one can decompose $\mathbf{C}$ into finitely many *irreducible ascending sets* $\mathbf{C}_1, \ldots, \mathbf{C}_e$ over $\mathbb{Q}(\mathbf{u})$ such that

$$\text{Zero}(\mathbf{C}/J) = \bigcup_{l=1}^{e} \text{Zero}(\mathbf{C}_l/JJ_l),$$

where each $\mathbf{C}_l$ contains exactly r polynomials and has the same form as $\mathbf{C}$, and J_l is the product of the leading coefficients of the polynomials in $\mathbf{C}_l$ with respect to their leading variables. Let us call $\mathbf{C}_1, \ldots, \mathbf{C}_e$ a *representative characteristic series* of **P**. Under the condition $J \neq 0$, each $\mathbf{C}_l$ represents one irreducible component of the algebraic variety Zero(**P**). From each $\mathbf{C}_l$, a numerical element of Zero($\mathbf{C}_l / J J_l$) may be determined by taking values for the parametric variables in **u** and then finding the corresponding values for the dependent variables in **y** successively from the triangularized polynomials in $\mathbf{C}_l$.

Our drawing method for Problem D may be sketched as follows.

D1. Assign coordinates x_j to the points P_k involved (manually or automatically).

D2. Translate the geometric relations $\mathcal{R}_i$ (automatically) into algebraic equations (1). Fix a variable ordering, say $x_1 \prec \cdots \prec x_m$, which is either given or chosen heuristically by the program.

D3. Compute a representative characteristic series $\mathbf{C}_1, \ldots, \mathbf{C}_e$ of $\{f_1, \ldots, f_s\}$. If the series is empty, then the $\mathcal{R}_i$ are inconsistent, and the procedure terminates.

Let **y** and **u** be the sets of dependent and parametric variables of $\mathbf{C}_1$, respectively. If both coordinates of P_k belong to **u**, then P_k is a *free* point. If one and only one of them belongs to **u**, then P_k is a *semi-free* point. Otherwise, P_k is a *dependent* point.

For each $\mathbf{C}_l$ do the following steps D4–D7:

D4. Choose (randomly) a set of numerical values for the coordinates in **u** and determine the corresponding real values of the other coordinates by solving the equations $c = 0$, $c \in \mathbf{C}_l$, successively.

D5. Check whether all the points (now having numerical coordinates) are within a fixed window range and no two of them are too close. If not, then go back to step D4.

D6. Draw the points P_k (with their numerical coordinates) and the lines, circles, etc. passing through them according to the information provided by the predicate specification.

D7. Label the letters around the points appropriately (see Sect. 4.2).

In the above method, the cases $I_i = 0$ are not considered. This is mainly because for geometric theorems *adequately* specified (with correct identification of parametric variables) such cases are normally the degenerate ones, and we want to draw representative diagrams rapidly without much symbolic computation. If one is also interested in such (degenerate) cases, one can compute a full decomposition of the algebraic variety Zero(**P**) into irreducible components over $\mathbb{Q}$ (using Wu's or any other method) and then draw a diagram for each component. In this case, the computation involved is rather heavy.

While the triangularization process for arbitrary geometric constraints (of higher order) may be considerably expensive, it is usually easy to triangularize the set **P** of hypothesis-polynomials for theorems from plane Euclidean geometry. The polynomial set $\mathbf{P}$ is decomposed into irreducible representative ascending sets because the diagrams for different components may differ topologically. A diagram drawn from a reducible characteristic set may jump from one component to another during animation.

3   Implementation

The drawing method explained in the preceding section is sufficiently general because the geometric constraints R_i may be of any kind and specified in any manner as long as they can be translated into polynomial equations (and inequations). The computation of representative characteristic series is a sophisticated symbolic elimination process which requires extensive polynomial operations including factorization over successive algebraic extension fields. It is not a simple project to implement this process from scratch in a low-level programming language like C or Java. That is why we need a computer algebra system in which basic polynomial operations are already implemented. Our choice of Maple, a popular system capable of efficient algebraic computation and equation-solving, is based on the fact that a library of elimination tools including a complete implementation of the characteristic set method of Wu has been developed therein. Using our library in Maple, the implementation of steps D1–D4 of the drawing method is a relatively easy task.

Maple has a variety of functions for plotting, but its capability of drawing geometric objects is quite limited. Therefore, we have to implement a drawing module in another programming language. An implementation of this module in C has already existed since 1991, first making use of Domain Graphics Primitives on Apollo workstations and then of Xlib under Unix and Linux. As software technology advances, Java has now become one of the most convenient programming languages for graphics, animation, web and interface implementation. It is rather easy to implement the drawing steps D4–D7 in Java. Moreover, with Java our desire to animate the drawn diagrams may be easily satisfied, provided that the polynomial expressions in $\mathbf{C}_l$ are coded into Java routines. With these considerations, we decided to shift from C to Java and make use of simple interactions between Maple and Java for automated diagram generation.

A diagram changes as the values of the free coordinates do. Animation of the diagram by moving a free or semi-free point continuously allows one to observe the geometric properties of the points. To animate, the triangularized equations $c_i = 0$ need be solved in real time. It is thus necessary to code the symbolic expressions c_i into Java routines. In our implementation, this is done by generating a small piece of Java code in Maple. The generated code is compiled at the running time and then executed with the main Java programs.

The following is part of the Java code generated automatically in Maple from the specification (of Simson's theorem) given in the introductory section.

```
public class Input {
public static double Ja, Jb, Jc, Jd, x2=73, x1=175, x3=122, x4=67,
    x5, x6, x7, x8, x9;
public static String[] points = {"D", "R", "A", "B", "C", "P", "Q"};
public static int[] gPoint = {4};
public static int[] yPoint = {0, 2, 3, 5};
public static int[] rPoint = {1, 6};
public static String[] freeVar = {"_x4", "_x2", "_x1", "_x3"};
public static int[] cooX = {67, 234, -175, 175, 73, 67, -63};
public static int[] cooY = {-209, -70, 0, 0, 122, 0, 55};
            .
            .
    if ( Jp == 6 & Jq == 0 ) {
        x6 = ( -Mpow(x3,2)*x1+x4*Mpow(x2,2)+2*x4*x2*x1+x4*Mpow(x1,2)
             +x3*x5*x2+x3*x5*x1 ) / ( Mpow(x3,2)+Mpow(x2,2)+2*x2*x1
             +Mpow(x1,2) ) ;
        return x6 ;
        }
    if ( Jp == 6 & Jq == 1 ) {
        x7 = ( -x6*x2-x6*x1+x4*x2+x4*x1+x3*x5 ) / ( x3 ) ;
        return x7 ;
        }
            .
            .
    return Jz ;
    }
}
```

Note that the irreducible ascending sets C_l are computed in Maple at the running time and then coded into a Java class. There is no restriction on how the geometric constraints are specified. This is a distinction from some implementations in the existing geometry software systems where the types of constraints are fixed, the symbolic expressions for these constraints are pre-computed and incorporated *a priori* into the program, and each geometric problem may have to be stated constructively step by step.

We mention a few implementation details below, and more technical issues will be discussed in the following section. Each single equation $c_i = 0$ may be solved for its leading variable explicitly in terms of radicals of c_i's (symbolic) coefficients when the degree of c_i is smaller than or equal to 4. Since the ascending set in question is irreducible, it is sufficient to take one set of solutions. However, for drawing and animating the solutions for specific real values of the parameters must also be real. It is not easy computationally to get rid of complex solutions. We have an effective technique for quadratic equations and quartic equations without odd-degree terms (see Sect. 4.1).

Solving univariate equations of degree greater than 4 with symbolic coefficients over the reals is a more difficult problem that has not been considered in our current implementation. This is why our drawing program is incomplete for high-order constraints.

We have implemented a number of predicates, and each of them contains a standard set of information entries: for example, English meaning, algebraic expression (when the points are assigned coordinates), and geometric information. Consider, for instance, the predicate `collinear(A,B,C)`. Then the corresponding entries are: **the three points A, B and C are collinear,**

$$A_x B_y + B_x C_y + C_x A_y - A_x C_y - B_x A_y - C_x B_y,$$

and `line(A,B)`, `line(B,C)`, where (A_x, A_y), (B_x, B_y), (C_x, C_y) are the coordinates of A, B, C respectively. These information entries allow the program to translate the predicate specification of a geometric problem into English statement and algebraic expressions automatically, and to draw geometric objects like lines and circles passing through certain points. Some of the information in the entries is passed on to the generated Java class.

The set of predicates is not exhaustive and may be easily enlarged; it is easy to implement a new predicate and add it to the set.

The assignment of coordinates can be done automatically by a function called `Coordinate` from the specification, but it is suggested that the user assign the coordinates her/himself using the function `Let` (see the example for Morley's theorem in Sect. 5). It is also recommended that the user identify the parametric and dependent variables from the specification. The program may work unpleasantly or fail in some special situations.

Figure 2 shows the dump of a window in which a diagram is drawn. On top of the window there are four buttons. When the button `Animate` is clicked, the free points are shown in green, the semi-free points in yellow, and the dependent points in red. One can drag any of the green or yellow points, and the diagram changes accordingly, satisfying the given geometric relations. When `Move` is clicked, the centroid of the points is shifted to the center of the window, the color of the points changes back to black, and the points are relabeled automatically according to the centroid principle (see Sect. 4.2). Then one can move a point to any desired position by click and dragging. If `Save` is clicked, a dialog box pops up for saving the diagram as it is shown in the window into an encapsulated PostScript file, whose default name is `geother.eps`. Click on `Close` causes the window to close down.

4   Special Techniques

In this section, we discuss a few techniques that have been employed in our implementation. They are simple yet effective.

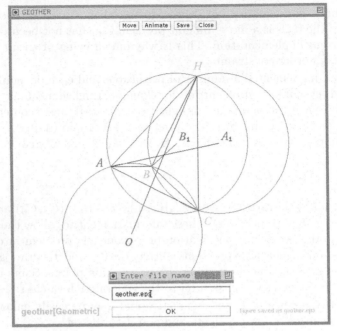

Figure 2.

4.1 Real Solving

When a free or semi-free point is dragged, its coordinates change according to the position of the cursor. For some values of the free coordinates, the coordinates of other dependent points may be complex; this has to be avoided. In other words, we need to determine the *definition domains* of the free coordinates in which the dependent coordinates always take real values. This is a difficult computational problem in general. Nevertheless, we have a simple method explained below for the quadratic case (and the case of quartic equations without odd-degree terms).

Let $\Delta_1, \ldots, \Delta_t$ be the discriminants of all the quadratic c_i, and $\Delta = \Delta_1 \cdots \Delta_t$. For the drawn diagram, we have $\Delta \geq 0$ and all the dependent coordinates are real. During dragging, the dependent coordinates may become complex only if Δ becomes negative. From non-negative to negative, Δ must change continuously in terms of the move of the cursor with the dragged point.

Our method works as follows: During the dragging of a point P, the value of Δ is calculated. If the value will change from non-negative to negative when P is moved from its current position p_i to its next position p_{i+1}, then P does not move and stays at its current position p_i.

4.2   Labeling of Letters

Where to put the letter L_i that labels point P_i? In order that the generated diagram looks friendly, it is hoped that L_i does not overlap with the lines and circles etc. in the diagram. This can be done by locating a possible region for L_i that does intersect any of the lines and circles; in this case computational overload is required. We propose a simple labeling technique according to the following principle, which needs little computation.

The *centroid principle*: Let M be the centroid of the points $P_1, \ldots, P_n$.

$$\overset{\bullet}{P_i} \quad \overset{\circ}{L_i}$$

$$\overset{\bullet}{M}$$

Label the letter L_i for point P_i at the position $\circ$ on the vector $\overrightarrow{MP_i}$ with $|L_i M| = |P_i M| + \epsilon$, where ϵ is a small positive number chosen suitably.

Labeling according to the centroid principle is ideally optimized. As the reader may see from the examples of diagram in this note, most of the labeled letters are placed in good positions. The reason is that around point P_i and towards the direction of the centroid M there are clearly more lines and circles passing through.

4.3   Generation of PostScript Files

The Java program can generate an encapsulated PostScript file directly from the current internal data used for drawing the diagram. In other words, the PostScript file saved is not converted from the graphical data shown in the window. This keeps the PostScript file small and the resolution high. One may view the PostScript file, e.g., by ghostview, instead of the Java display window. However, the diagram shown in the ghostview window cannot be modified with mouse click.

5   Some Examples

The diagrams shown in Figure 1, Figure 3(a), and Figure 4 below are as they were drawn automatically by the program without any modification. The diagram in Figure 3(b) is obtained from that in Figure 3(a) by dragging the points C and D (in the Animate mode) and by moving the labeled letters.

Figure 4 exhibits different diagrams drawn automatically for Morley's trisector theorem with the following specification:

```
Let(A=[y2,y1],B=[u1,0],C=[u2,0],P=[0,u3],Q=[y6,y5],R=[y4,y3]);
Theorem([equiangle(A,B,C,P,B,C,1,3),equiangle(A,C,B,P,C,B,1,3),
        equiangle(C,A,B,R,A,B,1,3),equiangle(A,B,R,P,B,C),
        equiangle(A,C,Q,P,C,B),equiangle(B,A,R,Q,A,C)],
        [equidistance(P,Q,P,R),equidistance(P,Q,Q,R)],
        [y1,y2,y3,y4,y5,y6])
```

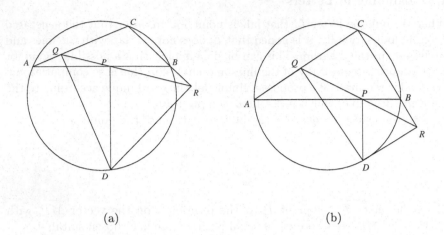

(a)   (b)

Figure 3.

We do not explain the meanings of the predicates, which the reader may easily guess. The actual input to our drawing function `Geometric` in GEOTHER is in the above form of specification, which is fixed for most of the GEOTHER functions. The lists of predicates for the hypothesis and for the conclusion of the theorem must be given, of course. A representative characteristic series Ψ is computed from the set of hypothesis-polynomials with respect to the dependent variables such that for all ascending sets in Ψ the conclusion holds true. The assignment of coordinates by `Let` and the list of dependent variables `[y1,...,y6]` are optional; the program may assign coordinates and identify the dependent variables automatically and heuristically. However, to avoid possible problems resulted from our heuristic (and incomplete) implementation of some functions, the user is advised to carry out the tasks and to specify the geometric relations in an appropriate way if she or he knows how to do them well.

The work reported in this note has been supported by the SPACES Project

http://www.spaces-soft.org

and the Chinese National 973 Project NKBRSF G19980306.

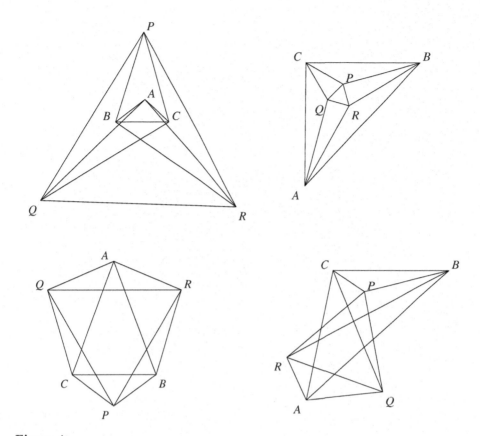

Figure 4.

References

1. Gao, X.-S. (1999): Automated Geometry Diagram Construction and Engineering Geometry. In: *Automated Deduction in Geometry* (X.-S. Gao, D. Wang, and L. Yang, eds.), LNAI 1669, pp. 232–257. Springer, Berlin Heidelberg.
2. Gao, X.-S., Zhang, J.-Z., and Chou, S.-C. (1998): *Geometry Expert* (in Chinese). Nine Chapters Publ., Taiwan (http://www.mmrc.iss.ac.cn/~xgao/gex.html).
3. Durand, C. and Hoffmann, C. M. (2000): A Systematic Framework for Solving Geometric Constraints Analytically. *J. Symbolic Computation* **30**, 493–519.
4. Richter-Gebert, J. and Kortenkamp, U. H. (1999): *The Interactive Geometry Software Cinderella.* Springer, Berlin Heidelberg (http://www.cinderella.de/).
5. Wang, D. (1996): GEOTHER: A Geometry Theorem Prover. In: *Automated Deduction* (M. A. McRobbie and J. K. Slaney, eds.), LNAI 1104, pp. 166–170. Springer, Berlin Heidelberg (http://calfor.lip6.fr/~wang/GEOTHER/).
6. Wang, D. (2001): *Elimination Methods.* Springer, Wien New York.
7. Wu, W.-t. (1994): *Mechanical Theorem Proving in Geometries: Basic Principles* (translated from the Chinese by X. Jin and D. Wang). Springer, Wien New York.

Interactive Mathematical Documents on the Web

Arjeh M. Cohen, Hans Cuypers, Ernesto Reinaldo Barreiro, and Hans Sterk

Department of Mathematics, Technische Universiteit Eindhoven, POB 513,
5600 MB Eindhoven, The Netherlands

Abstract. This paper deals with our work on interactive mathematical documents. These documents accomodate various sources, users, and mathematical services. Communication of mathematics between these entities is based on the OpenMath standard and Java technology. But, for the management of the communication, more protocols and tools are needed. We describe an architecture that serves as a framework for our work on interactive documents, and we report on what we have implemented so far.

1 Introduction

An interactive mathematical document is to be regarded as a book, an article, or announcement on computer that can be read in the way their ordinary (paper) counterparts can, but which, in addition, enables a variety of activities. Among these interactions, we count storing and communicating mathematics, presenting mathematics (e.g., in browsers) and performing mathematical operations, possibly elsewhere on the Web.

Although the notion of an interactive mathematical document has been around for several years [14] its realization is nowhere near the final stage. Recent web technological progress, for instance, has enabled a smoother communication of mathematics than ever before. The use of an interactive mathematical document can provide a window to the world of mathematical services on the internet. Moreover, a mathematical service on the internet can be created by the construction of an interactive mathematical document. In §3 of this paper we paint the contours of such a mathematical document. In particular, we describe a *mathematical document server* which takes input from various sources (the document source, mathematical services, and users), creates a highly interactive mathematical set-up, and serves it to the user. In this vein, the paper can be viewed as a sequel to [11].

We would like to stress that the technology for achieving such goals already exists, mainly (for our purposes) in the form of Java software. In §4, we describe some of the main tools we have developed for realizing the interactive mathematical documents. We envision that it will require a few more years before integrated authoring tools, as easy to use as LaTeX, will be widely available. With the work presented here, we intend to contribute to these developments.

Before going into details regarding interactive documents, in §2, we discuss the general picture of mathematics on the Web.

2   A Framework for Interactive Mathematics

In this section we describe our approach to interactive mathematics. In §2.1, we begin by overviewing OpenMath, a standard for communicating mathematics across the Web. Next, in §2.2, we discuss web services and, in particular, address two additional requirements for our purposes: query facilities and a notion of the state of the mathematics that is being communicated.

2.1   OpenMath

OpenMath A starting point for semantically rich communication across the Web is the standard for mathematical expressions OpenMath, cf. [10,25], and, for our purposes, its XML encoding. The representation of mathematics in OpenMath relies on four 'expression tree' constructors (viz., application, binding, attribution, and error), on five basic objects (byte arrays, strings, integers, IEEE floats, variables), and on a special, sixth, basic object: symbols, defined in Content Dictionaries (CDs for short). The core CDs are publicly available collections of mathematical definitions. The standard documents and the collection of public CDs for OpenMath are available in XML format from [25]. An example of the XML encoding of an OpenMath object expressing that $\cos(\pi) = -1$ is given in Figure 1.

```
<OMOBJ>
 <OMA>
  <OMS cd="relation1" name="eq"/>
  <OMA> <OMS cd="transc1" name="cos"/> <OMS cd="nums" name="pi"/> </OMA>
  <OMI>-1</OMI>
 </OMA>
</OMOBJ>
```
Figure 1. OpenMath fragment.

For a further introduction to OpenMath, see [11]. There it is also explained how OpenMath and MATHML [21] complement each other in that OpenMath objects express mathematical content whereas MATHML mainly focuses on presentation. The connection between the two rests upon

o the fact that MATHML works with a relatively small number of commonplace mathematical constructs chosen within the high school realm of applications and

○ the content symbol (`csymbol`) in MATHML for introducing a new symbol whose semantics is not one of the core content elements of MATHML. In particular, such an external definition may reside in an OpenMath Content Dictionary.

For purposes of alignment of MATHML and OpenMath, the core CDs contain symbols matching the MATHML constructs.

As it stands, the OpenMath mechanism works quite well for conveying mathematical objects: by declaring which CDs are relevant, two parties agree on a common understanding of the mathematics they communicate. The public CDs are (well-wrought) examples, but two parties may choose whichever CDs they like. They could even create CDs for the sole purpose of a brief communication. This feature ensures a great flexibility in the use of OpenMath.

In Figure 2, by way of example, we display an experimental CD for planar Euclidean geometry, which was recently constructed in joint work with Ulrich Kortenkamp for the purpose of interfacing with Cinderella, [29].

Phrasebooks provide the means to convert OpenMath objects to/from software applications. They parse OpenMath objects into an application-native language (e.g., *Mathematica*, Maple, GAP), sending the result to the application, catching the response from the application, and translating it back into OpenMath. The phrasebook communicates only OpenMath objects of which the symbols are defined in the CDs that the phrasebook recognizes. Thus, a phrasebook performs the translation back and forth as well as the communication. The actual task performed by the totality of the phrasebook actions depends on the interpretation. If the application is a computer algebra system, the interpretation is often 'evaluation' or 'simplification': when passed $2 + 3$, these applications will return 5. If the application is a proof assistant (e.g., Lego or Coq, cf. [9]), then 'verifying' or 'proving' is a more likely interpretation of what the application is supposed to do, and if the application is a browser or printing, the interpretation is to prepare the mathematical object for a presentation.

Phrasebooks providing interfaces to and from OpenMath have been built into AXIOM and GAP [2,16].

We have developed a Java library, called ROML, for building full phrasebooks outside mathematical software packages. It is described in [4] and can be found at [28]. By use of ROML, such external phrasebooks have been implemented for the proof checkers Lego and Coq, for the computer algebra packages Maple, *Mathematica*, and GAP [5,6].

2.2 Mathematical Web Services

An important mode of communicating mathematics across the Web, is by means of queries. Generally a query in Web technology refers to a request for a service, see [35]. Naturally, we would like a standard way of expressing queries, a management system for parameters accompanying the question,

```
<CD>
<CDName> plangeo1 </CDName>

 (... further data like URL, creation date, CDs on which this one depends ...)

<Description>
This CD defines symbols for planar Euclidean geometry.
</Description>

<CDDefinition>
<Name> point </Name>
<Description>
The symbol is used to indicate a point of  planar Euclidean geometry
by a variable.  The point may (but need not) be subject to constraints.
</Description>
</CDDefinition>

 (... a similar definition for 'line' ...)

<CDDefinition>
<Name> incident </Name>
<Description>
The symbol represents the logical incidence fucntion which is a
binary function taking arguments representing geometric objects
like points and lines and returning a boolean value.  It is true
if and only if the first argument is incident to the second.
</Description>

<Example> The line l through (points) A and B is given by:

<OMOBJ>
  <OMA>
    <OMS cd="plangeo1" name="line"/>
    <OMV name="l"/>
    <OMA>
      <OMS cd="plangeo1" name="incident"/>
      <OMV name="A"/>
      <OMV name="l"/>
    </OMA>
    <OMA>
      <OMS cd="plangeo1" name="incident"/>
      <OMV name="B"/>
      <OMV name="l"/>
    </OMA>
  </OMA>
</OMOBJ>
</Example>
</CDDefinition>
</CD>

 (... further definitions ...)
```

Figure 2. CD fragment.

and a reference mechanism to couple a message to the query which it answers. Such systems are under construction (e.g., work of Caprotti), often based on more general standards, such as the Web Service Description Language, WSDL. We refer to [32] for a discussion of computational and other Web services. Confidentiality and privacy are important user requirements that will have to be present in user profiles. Clearly, there is a need for the development of service management frameworks with adequate provision for resilience, persistence, security, confidentiality and end user privacy. Here, however, our focus is on the mathematical aspects. We shall depart from the OpenMath set-up for the communication of mathematical objects. A mathematical query usually refers to mathematical objects, which can be phrased in OpenMath, but often the user wants to convey more than just the mathematical objects themselves. As we have seen above, a mathematical object can often be interpreted as a query by a phrasebook (e.g., interpret $\cos(\pi)$ as 'evaluate $\cos(\pi)$' or interpret an assertion GRH as 'verify GRH'), but this is a poor way of formulating a query. A more elaborate mechanism is needed.

Regarding mathematical services across the Web, we face three issues that need further exploration:

o mathematical reliability,
o expressing mathematical queries, and
o taking into account the state, or context, in which a mathematical query takes place.

Reliability. In [9], the reliability (quality guarantee) aspects are emphasized. Up till now, complexity, the (estimated) time a computation will take, has been one of the major concerns regarding mathematical computations. Although it will remain useful for clients to be aware of feasibility, they will be more concerned with the validity of the answer. Here an interesting shift in focus from complexity to convincibility may take place. To gain experience with this issue, we work out (within our MATHBOOK technology, see §3) some concrete examples where besides the usual invocation of an algorithm, additional work is carried out to provide the users with witnesses as to the truth of the answer. In one of these examples, in response to a query for the stabilizer H of the vertex, 1 say, of a permutation group G specified by generating permutations $a_1, \ldots, a_t$, one usually expects just a set $b_1, \ldots, b_s$ of permutations fixing 1, which will generate H. Now it is a straightforward check that $b_1, \ldots, b_s$ fix 1, but it requires some work to see that these permutations actually belong to G, whence to H. Expressions of the b_j as products of $a_1, \ldots, a_t$ will solve this. Finally, a proof that each element of H lies in the subgroup of G generated by $b_1, \ldots, b_s$ requires further information, corresponding to Schreier's lemma [13]. By means of a little programming at the back engine, the additional information can be supplied, and embedded in a proof in words that supplies a substantiated answer to the query.

Queries. We have argued that, so far, the OpenMath set-up seems rather primitive in that only mathematical expressions are passed, with no indication of the required action on the object. Currently, the phrasebook makes this interpretation, and so the matter is resolved by a declaration from the phrasebook of what its action (interpretation) is, see [11]. For instance, an OpenMath object like

$$\text{Factors(Polynomial(X,X\textasciicircum2-1,Rationals))}$$

will result in a response of the form

$$\text{List(Polynomial(X,X-1,Rationals),Polynomial(X,X+1,Rationals))}$$

when sent to GAP, because its phrasebook tends to interpret the OpenMath object as an evaluation command, whereas the same expression would just be printed as something like "Factors of $X^2 - 1$" when sent to a typesetting program.

In [31], a mode of interaction is implemented where the behavior of a computer algebra system can be controlled from within a JSP page (see §4.2 below) by using a set of primitives such as assigning and retrieving OpenMath objects to CAS variables, manipulating variables using the language of the CAS. Indeed, this seems to come closer to the intended user control.

But, as hinted at in [9], there are probably better solutions from automated proof checking. A slight extension of the language in which we formulate mathematical assertions will enable us to formulate mathematical queries. Typed λ calculus expressions like $\Gamma \vdash ? : P$, where Γ represents the context (see below) and $? : P$ stands for the request for a proof of assertion P, are expected to embed into a full type checking mechanism without problems. In other words, we expect that it is possible to set up a language of well-formed query expressions, recognizable by means of a proper type inference algorithm. So, importing this language within the OpenMath framework, we expect to obtain a sound method of expressing queries by means of a CD defining the primitive symbols (corresponding to question marks used such as in $? : P$) for the most fundamental types of question asked. This approach to queries is as yet unexplored, and we intend to explore it in the near future.

Context. The problem of how to handle the state in which a mathematical query takes place has not been addressed in some of the more successful mathematical services on the internet, such as Sloane [30], Faugère's Gröbner basis service [15], Wilson's Atlas of representations of finite simple groups [34], Brouwer's coding theory data base [3], and Web*Mathematica* [33].

For example, Web*Mathematica* is a way of accessing *Mathematica* via the Web. Via browser pages users can formulate either full *Mathematica* commands or input for pre-programmed *Mathematica* commands (so that no specific knowledge of *Mathematica* is required) which will then be carried out by a *Mathematica* program run by a server accessible to the user. However, after the command is carried out, the *Mathematica* session is 'cleaned' in that the user can no longer refer to the previous command. So, it is possible to, say, compute the determinant of a matrix but the user cannot assign the matrix to a variable, say A, and change an entry of A, and/or ask for A^{-1} without re-entering the entries of A.

Clearly, as is the case for a *Mathematica* session per se, it is desirable to be able to refer to the variables at hand in a work session, to be able to ask for a second computation regarding an object passed on earlier, and so on. From a computer algebra point of view, the *context* is a list of definitions, that is, assignments to variables, of objects introduced (and computed) before (think of the assignment statements and of the In[n] and Out[n] variables for $n = 1, 2, \ldots$ in *Mathematica*). In the Javamath API, discussed in [31], there is a notion of session, in which it is possible to retain variables and their values.

However, we wish to incorporate one more feature in our notion of context. This is taken from logic, where the notion of a context is our inspiration. Indeed the symbol Γ above stands for context. Besides definitions, it contains statements which are interpreted as 'the truth' (or axioms, for that matter). This means that theorems, lemmas, conjectures, and so on, may be thrown in, and are all interpreted as 'facts'. We stress that there may very well be assumptions in the context, so that it might be possible to derive a contradiction. It would not be desirable to have the starting context of an interactive book be self-contradictory, but, in the course of a user developed proof by contradiction, there is nothing against a set of assumptions from which the user can derive $0 = 1$.

It is such a list of definitions of objects and statements, which is called context in the case of theorem provers, that gives a good starting point to what we consider to be the context of a mathematical session. It will be clear that the context is highly dynamic: for instance, if, in the example of the matrix A above, the user wants to consider the matrix over $GF(11)(x, y, z)$ rather than $\mathbb{Q}[x, y, z]$, a change of the coefficient ring should be the corresponding action on the context.

So, what is needed is a way to exploit such context data whenever a server providing a service to the user needs more knowledge. We have only made a modest beginning with the study of context, and we foresee that substantial research is needed for a successful implementation.

3   The Mathematical Document Server

In the previous section we have explained how we envision smooth communication using OpenMath with several mathematical services on the web. These services will enrich mathematical documents considerably. In this section we present a general model for a highly interactive mathematical environment embodying such features.

The heart of our architecture is a *mathematical document server*. In our approach, this server takes input from mathematical source documents, mathematical (web) services and users, and serves a view on the interactive document to the user. The document server takes care of the presentation of the document to the user, it handles the communication between user and several mathematical services. It also manages the (mathematical) context in which presentation and communication take place. Figure 3 displays the essential parts of the proposed architecture and their dependencies.

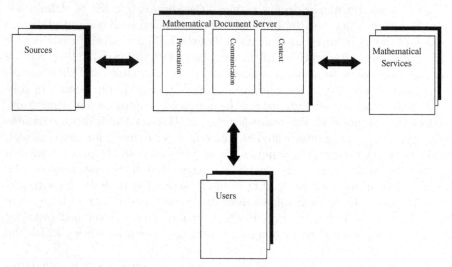

Figure 3. General architecture

The architecture we have chosen for an interactive mathematical document is based on the idea of an interactive book: the (static) mathematics is included in a source document (or *source*, for short). It is highly structured and semantically sufficiently rich to create an exact mathematical description of the content and to allow actions (see also Subsection 4.2). The source document indicates the kind of action that is supposed to be offered in the interactive environment. The actions themselves however are realized by the document server. The document source is written by an author.

The document server on the other hand, is written by tool developers. It uses the mathematical content from the sources together with input from the user and mathematical services to specify the *context* of the interactive document. The context certainly depends on user actions; a jump from one entry in the source to another may alter the context. But also results from queries to mathematical services are input to the creation of the context. Within this context a presentation of the content relevant to the mathematical setting in which the user 'resides', is realized and can be presented to the user via an interface. In line with the discussion of §2.2, the context consists of

○ assignments to variables of OpenMath objects (interpreted as definitions),

○ OpenMath objects representing mathematical assertions,

○ logistic information, for example, the user's id, mathematical background, permissions to use commercial services, etc.

At any given time, the context gives a precise description of the state the user is in by means of this data.

A simple example of this model is realized in a LATEX environment. Here the document server produces, on user's demand, a dvi or postscript file from a LATEX source and serves it using a dvi or postscript viewer to the user. Here the context is just given by the user's request to create a dvi or postscript file to view on the screen or send it to a printer, etc. In this simple example, the logistic data are relevant, but not the mathematical context.

A more advanced example can be realized in an XML-JAVA setting. Here the source consists of an XML-source. The document server creates, for example by XSL-transformations, an HTML or XML document and serves it as a web page to the user. Using a web browser as an interface, the user can view a presentation of the document. Interactivity and communication with mathematical services can be realized inside the web server using JAVA-applets or servlets. Our present approach to realizing interactive mathematical documents is based on this example and will be discussed in the next section.

Creation and bookkeeping of the context as well as presentation of the content is taken care of by the document server. This server also handles *communication* between the source, the user and mathematical services. It stores presentation information and the context as dynamic data. The context is relevant in communication with the outside world. Using the model of communication with a mathematical service, the provider of the service may be aware of the context of the user's mathematics. This can take place by means of incremental steps (loading the context at the initial stage and translating the user defined changes one by one), or by means of downloading the entire context (or relevant portions thereof) upon receipt of each new query.

Of course, our primary target is a mathematical context, where, for example, in a chapter on ring theory of an algebra book, the field of coefficients might be specified to be a finite field. By interaction of the user interface with the user, this context can be further specialized to, say, the field of order eleven GF(11).

In an example on irreducible polynomials in a polynomial ring, the document server will take care of choosing the polynomial ring $GF(11)[X]$ over GF(11). Within this context the reader can now verify that the polynomial $X^2 + 2X - 2$ is reducible, whereas the polynomial $X^2 + 2X + 2$ is irreducible.

Some variables in the context can also be of a logistic nature. For instance the user name might help to create or recognize an individual version of the context. This might be of relevance, for instance, for tracking the way a student reads an interactive text book.

4   MATHBOOK, Our Implementation

The discussion of §3 regards our conceptual framework for interactive mathematical documents. In this section we discuss our progress in implementing this architecture within a JAVA-XML set-up. Our motivating example is a forthcoming new edition of the interactive book *Algebra Interactive!* (see [7]), which is interactive course material for undergraduate algebra. We shall use the word *MathBook* for the ensemble of software tools we are building for the construction of interactive mathematical documents such as, but not limited to, 'Algebra Interactive'. The distinct components of the architecture are displayed in Figure 4. We shall deal with them separately.

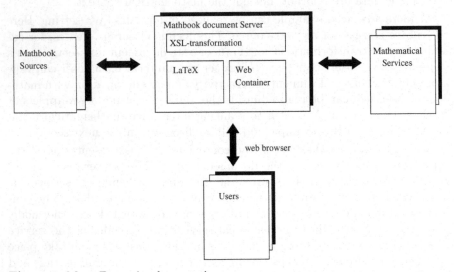

Figure 4. MATHBOOK implementation

4.1   The MATHBOOK Source

We have derived our own experimental grammar in the guise of document type definitions (DTD) for the MATHBOOK source, an XML document. As a result, there is an XML based markup language (the MATHBOOK DTD) for the creation of interactive mathematical documents.

   In creating a DTD for MATHBOOK, we have been influenced by both DocBook [8] and OMDoc [24]. The former is a fairly general standard for structuring (into chapters, sections, etc.) electronic documents, the latter is a very rich, and strongly logic-oriented standard for mathematical documents. We intend to maintain a close link with OMDoc, but found the overall machinery involved too heavy for our purposes. The connection with DocBook is of importance to us, since we expect several authoring tools for it to emerge in the coming years, tools that could be of use to us in one form or another. MATHBOOK deviates in various respects from OMDoc and DocBook and contains new features, like support for actions.

The mathematics in the source is given by means of OpenMath objects. This feature has clear advantages in terms of portability. The DocBook type grammar sees to it that there are natural scopes, where mathematical objects 'live'. For instance, when a chapter begins with "Let $\mathbb{F}$ be a field", the scope of the variable $\mathbb{F}$ is assumed to be the whole chapter (although, somewhere further down the hierarchy, say in a section of the chapter, this assignment can be overridden).

```
<OMOBJ>
 <OMA>
  <OMATTR>
   <OMATP>
    <OMS name="pres" cd="ida"/> <OMSTR "frac"/>
   </OMATP>
   <OMS name="divide" cd="arith1"/>
  </OMATTR>
  <OMA>
   <OMS name="plus" cd="arith1"/>
   <OMA> <OMS name="divide" cd="arith1"/> <OMI>3</OMI> <OMI>4</OMI> </OMA>
   <OMA> <OMS name="divide" cd="arith1"/> <OMI>2</OMI> <OMI>3</OMI> </OMA>
  </OMA>
 </OMA>
</OMOBJ>
```

Figure 5. An attributed OpenMath object.

The mathematical content is represented in OpenMath. This means that the semantics is taken care of satisfactorily, but that no attention is being paid to presentation. In general, this is in line with the idea that presentation should be taken care of by the document server rather than the source. There are however some clear exceptions. Let us give two examples. In LaTeX, for each individual fraction, the author has a choice between a slash and a fraction display. In

$$\frac{3/4 + 2/3}{5}$$

we have used both. The other example concerns the statement "$3, 4 \in \mathbb{Z}$". The corresponding OpenMath expression would be the equivalent of "$3 \in \mathbb{Z}$ and $4 \in \mathbb{Z}$", whereas the presentation in the first form is highly desirable from an esthetic point of view.

In order to have such a flexible presentation, we are using presentation annotated OpenMath. This means, that in our MATHBOOK source we allow style attributes inside OpenMath objects. Figure 5 shows an attributed expression corresponding to the L^AT_EX macro \frac, in which the fraction display is forced. It is assumed here that the default presentation is 'slash' so that the two other divides need not be attributed. Of course, the author can also force the slash presentation of these by a similar attribution. By discarding these style attributes, regular OpenMath is obtained. So, one can easily go from annotated OpenMath to 'bare' OpenMath.

Within the MATHBOOK grammar, special attention is also given to interactivity. For this purpose a whole range of tags (that is, structuring elements defined in DTDs and appearing in XML documents between pointed brackets) have been introduced. We will give two snippets of code appearing in the MATHBOOK source to illustrate some of these tags and to give the reader an idea of how an author may create interactivity.

```
<eval scope="session">
  <OMOBJ>
   <OMA>
    <OMS cd="univpoly1" name="expand"/>
    <OMA>
     <OMS cd="univpoly1" name="gcd"/>
     <getomcontent> <getvarvalue name="poly_a"/> </getomcontent>
     <getomcontent> <getvarvalue name="poly_b"/> </getomcontent>
    </OMA>
   </OMA>
  </OMOBJ>
</eval>
```

Figure 6. Some MATHBOOK tags for interactivity.

The code in Figure 6 uses the combined effects of the following tags.

o getvarvalue: Read two strings representing OpenMath objects that were previously stored in the variables poly_a and poly_b, respectively. Note that the scope is set to session. This implies that, at the time the user visits this particular part of the source, the context will have the variables poly_a and poly_b, but when the user leaves it, these will no longer stay alive. The scope is introduced by a command like
<enablescope scope="session"/>
(not displayed in the figure). The above code then creates an OpenMath object (in fact, a univariate polynomial) that is placed in the session scope.

o getomcontent: Get the content of the OpenMath object, i.e., remove the markers <OMOBJ> and </OMOBJ> at the beginning and the end of an OpenMath object.

o eval: this tag indicates that in a realization of the source as an interactive document, the constructed OpenMath object is sent to a computational backengine, like *Mathematica*, for evaluation.

```
<addtoscope name="matrixsquared" scope="session">
<OMOBJ>
 <OMA>
  <OMS cd="arith1" name="times"/>
  <getomcontent> <getfromscope name="matrix"/> </getomcontent>
  <getomcontent> <getfromscope name="matrix"/> </getomcontent>
 </OMA>
</OMOBJ>
</ida:addtoscope>
```

Figure 7. Some MATHBOOK tags for content control.

The snippet in Figure 7 shows how objects in a context can be created. Here, the OpenMath object named `matrix` is read from the session scope (by means of the `getfromscope` tag). This object is used to create a new OpenMath object that is placed in the session scope (by means of the `addtoscope` tag) with name `matrixsquared`.

4.2 The MATHBOOK Server

As mentioned in Section 3, the mathematical document server, called the MATHBOOK *server*, should cater for presentation, communication, and context in our implementation.

One part of our MATHBOOK server consists of an XSL transformer together with a set of XSL stylesheets. The transformer picks up the MATHBOOK sources and transforms them into LaTeX files for creating printouts or JAVASERVER PAGES (JSP) to create an interactive realization of the source. JSP technology [17] is designed to develop dynamic web pages easily. A JAVASERVER PAGE is a template containing standard HTML code, user defined tags, and JAVA scriptlets encoding the logic and the required behaviour of the dynamic web page.

The JAVASERVER PAGES together with our JAVA tools form a Web application residing in a Web container of a standard (JSP/Servlet) server, allowing us to extend the functionalities of any such server. Note that, as opposed to Web*Mathematica*, we are not modifying the server itself.

So far, besides ROML (discussed at the end of §2.1), our JAVA tools include phrasebooks and a (JSP based) tag mechanism for bringing to life the interaction specified in the MATHBOOK source. The phrasebooks deal with the following kinds of action.

- ○ Sending an OpenMath object to a backengine (e.g. Mathematica) for evaluation and returning an OpenMath object.
- ○ Retrieving the answer to a mathematical query from a web service and reacting on the outcome.
- ○ Transforming OpenMath into MATHML presentation.

```
<phreval id="result" name="GapPhrasebook" method="EVAL" scope="session">
 <OMOBJ>
  <OMA>
   <OMS cd="integer1" name="factorof"/>
   <OMI> <expression>b</expression> </OMI>
   <OMI> <expression>a</expression> </OMI>
  </OMA>
 </OMOBJ>
</phreval>
<if> <condition> <expression>result</expression> </condition>
 <then>
  <para> Answer:
   <expression>b</expression> divides <expression>a</expression>.
  </para>
 </then>
 <else>
  <para> Answer:
   <expression>b</expression> does not divide <expression>a</expression>.
  </para>
 </else>
</if>
```

Figure 8. Retrieving an answer.

As an example of the first two features, consider the code displayed in the beginning of Figure 8. This is part of a JSP page obtained from a MATHBOOK source. There, the OpenMath object contained in the `phreval` tags is sent to the **GAP** Phrasebook and passed to **GAP** for evaluation, and the response from **GAP** and the **GAP** Phrasebook is an OpenMath object (a Boolean) assigned to the variable `result` which lives inside the scope called `session`.

Communication is governed by JAVA servlets and phrasebooks; the actions defined within the MATHBOOK sources are mapped onto and taken care of by a JAVA tag library, called the MATHBOOK *tag library.*

The tag library is the tool that allows a practical implementation for actions in MATHBOOK grammar. As a side issue, we mention that the library can also be used independently by someone who is just interested in developing his/her own JSP pages.

In the above example, the phrasebooks are invoked by the `phreval` tag. Furthermore, the tag mechanism handles

- the flow within a document, like if/then/else (see Figure 8), for loops, etc.
- the context. For example, objects can be stored and retrieved from the context.
- Casting OpenMath objects to OpenMath objects of another (often more structured) kind. This mechanism is especially useful for software systems that do not type their objects very strongly. For instance, **GAP** does not distinguish a list of lists from a matrix, so when we expect a matrix to

be returned (a fact that is noticeable in the presence of a context), we cast the list of lists onto a matrix of the right kind. Observe that this cast indeed belongs to the MATHBOOK server and not to the source.

In particular, the tags in the library deal with two of the three features of a mathematical document server: communication, and context. The remaining feature, presentation, is dealt with by a separate Java program that translates the OpenMath objects of the source into MathML. It keeps track of the attributes for presentation such as the one for fraction discussed in §4.1. The MathML appearing in a JSP page can be rendered by a browser like Mozilla or Amaya, cf. [23,1].

Experimental MATHBOOK servers for 'Algebra Interactive', with both *Mathematica* and GAP as backengine, have been realized. We intend to experiment further with CoCoA, Maple, and formal automated proof assistants such as COQ.

At the moment we are investigating the possible extension of the tag library in such a way that we are able to interface with other more geometry or visualisation oriented packages. We have started work on interfacing to the geometry package Cinderella [29].

4.3 Examples

We make the picture described above more concrete by considering three scenarios. In each of these scenarios a suitable interactive mathematical document can offer the appropriate mathematical services via the internet. In [12] we have described a way to provide the computing facilities of various mathematical software packages via OpenMath servers to the internet community. However, in these scenarios the mathematical services cannot consist solely of web interfaces with computational backengines, but ask for more specialized activities.

1. An author of a book on mathematical analysis has used both Maple [19] and *Mathematica* [20] to write algorithms discussed in his/her book. At times, the results of one system are fed into the other. Preferably, the author would like to write the code only once.

By use of the MATHBOOK tools, the algorithms written in either system can be made to run virtually within the electronic version of the book. The MATHBOOK tag library then takes care of communication with the backengines Maple and *Mathematica* and the computer algebra code is stored in the source. An alternative to sending native code to the backengines is to work with OpenMath. If the commands are confined to standard applications such as factorization of polynomials, the EVAL interpretation of the phrasebooks suffice (currently, a Maple phrasebook based on ROML does not exist, but one based on [26] could probably be used). In this case, we can express input and output as well-understood mathematical objects (using the cast of the tag library, if necessary); moreover, we can ask for an evaluation by any third computer algebra system for which a phrasebook exists. For the author, the implementation has been reduced to writing a simple `phreval` tag. There is a third option in which more elaborate commands can be run on back engines. It uses a first version of an algorithm CD. We expect to be able to write most of the 130 gapplets (i.e., interactive examples using GAP) from the former edition of 'Algebra Interactive', in this OpenMath code, in such a way that each of the systems GAP, CoCoA and *Mathematica* will be able to run the code at the server end.

2. At a high school, students have been assigned a project on cryptography. In this context, information about prime numbers is required. They need to know the definition of a prime number, to find a few prime numbers of 200 digits and to compute related encoding and decoding keys.

There are many home pages about prime numbers, see e.g. [27] for an interesting one. Most of these sites contain a lot of static information, but lack available computation power, for instance to check primality of a given number. As part of the ESPRIT OpenMath project, we have made sample pages on prime numbers, backed up by GAP and *Mathematica*, in which the students can actually profit from the computation power of these backengines without having to know anything from the syntax of these backengines. They can retrieve primes with 200 or more digits to build a realistic and safe RSA cryptosystem, they can break such systems using too small primes, they can search for Mersenne primes, etc.

Upon request, calling a 'Pocklington' server (cf. [6]), the students can obtain a full proof (in words) of the primality of a given number. This works in much the same way as the stabilizer subgroup example discussed in §2.2.

3. Cinderella [29] is a beautiful program for exploring traditional planar geometry. Configurations can be constructed corresponding to classical theorems such as Pappus' Theorem, in which the conclusion is that three points are on a line. Once the configuration of points and lines is drawn, the fact will present itself 'automatically'. The user can drag and rescale the configuration while the three points stay on a line. An OpenMath representation of this configuration, using the OpenMath CD 'plangeo1', see Figure 2, can be translated into a formal description, a presentation in words, or to Cinderella input. Conversely, Cinderella (at least an experimental version) is able to provide such OpenMath expressions. We intend to explore these interactions further in collaboration with Ulrich Kortenkamp.

5 Conclusion

We have argued that OpenMath objects suffice to communicate mathematics in a rigorous way between software systems, but that two more features are of immediate need: query facilities and management of the context of the mathematics in which the user is immersed. A solution of the query problem seems feasible on a fundamental level within the OpenMath framework, but the context problem requires more experimentation. We expect that the MATHBOOK tag library will solve some of the most urgent matters in this respect. Authors can use it to augment their XML sources (in MATHBOOK format), so as to obtain a high degree of structured mathematical interactivity.

A major obstacle to authoring an interactive document is the inaccessibility of XML source code, the enormous number of brackets and labels, such as in the examples of code in Figure 8. The general expectation is that, once good special purpose editors have been developed, no author will need to work with the elaborate XML sources. However, currently there is no alternative at hand. There are two editors for OpenMath objects, viz. [18,22], but these do not suffice for the more elaborate source documents described in §4.1. By means of the MATHBOOK tag library, we have tried to reduce the difficulties of authoring as far as possible, but some XML editing remains necessary.

Another issue to be explored is 'searching for mathematical content'. Standard XML techniques might work on CDs. Although the interdependence of CDs is rather loosely organized (there is a CDUses field in a CD indicating on which other CDs the definition of symbols contained in it depend), standard XML tools will be able to produce the dependence trees. For example, via the CDUses construct we will be able to unravel that times in the core CD group refers to times in the core CD monoid, which in turn refers to times in arith1. Mathematical knowledge always has a hierarchical structure. It is the question whether the CDUses construct will suffice for an efficient implementation of the full hierarchy and the related searches.

References

1. Amaya, W3C's Editor/Browser, www.w3.org/Amaya.
2. Axiom interface to OpenMath. OpenMath ESPRIT Deliverable, 2000, www.nag.co.uk/projects/OpenMath/final/node10.htm.
3. A. E. Brouwer. *Coding theory server, for bounds on the minimum distance of q-ary linear codes*, $q = 2, 3, 4, 5, 7, 8, 9$, http://www.win.tue.nl/~aeb/voorlincod.html.
4. O. Caprotti, A. M. Cohen, and M. Riem. *Java Phrasebooks for Computer Algebra and Automated Deduction*. SIGSAM Bulletin, 2000. Special Issue on Open-Math.
5. O. Caprotti and A.M. Cohen. *Connecting proof checkers and computer algebra using OpenMath*, pp. 109–112 in *The 12th International Conference on Theorem Proving in Higher Order Logics* (Y. Bertot, G. Dowek, A. Hirschowitz, C. Paulin, L. Théry eds.) Nice, France, September 1999. Springer Lecture Notes in Computer Science, vol. 1690.
6. O. Caprotti and M. Oostdijk. *How to formally and efficiently prove prime(2999)*, in *Proceedings of Calculemus 2000: 8th Symposium on the Integration of Symbolic Computation and Mechanized Reasoning*, St. Andrews, Scotland, August 2000.
7. A.M. Cohen, H. Cuypers, H. Sterk. *Algebra Interactive!*, Interactive lecture notes on Algebra (paper book and CD-Rom), Springer-Verlag, Heidelberg, August 1999.
8. DocBook, http://www.docbook.org.
9. H. Barendregt and A.M. Cohen. *Electronic communication of mathematics and the interaction of computer algebra systems and proof assistants*. J. Symbolic Computation **32** (2001) 3–22.
10. O. Caprotti, A.M. Cohen, D. Carlisle, *The OpenMath Standard*, monet.nag.co.uk/cocoon/openmath/standard/.
11. O. Caprotti, A. M. Cohen, H. Cuypers, H. Sterk. *OpenMath Technology for Interactive Mathematical Documents*, pp. 51–66 in Multimedia Tools for Communicating Mathematics (Jonathan Borwein, Maria H. Morales, Konrad Polthier, and José F. Rodrigues, editors), Springer-Verlag, Berlin, Heidelberg, 2002.
12. O. Caprotti, A. M. Cohen, H. Cuypers, M. N. Riem, and H. Sterk. *Using OpenMath Servers for Distributing Mathematical Computations*, pp. 325–336 in: ATCM 2000: Proceedings of the Fifth Asian Technology Conference in Mathematics, Chiang-Mai, Thailand, Wei Chi Yang, Sung-Chi Chu, Jen-Chung Chuan (eds.), ATCM, Inc., 2000.
13. A.M. Cohen, H. Cuypers, H. Sterk (eds.). *Some Tapas of Computer Algebra*, Springer-Verlag, Heidelberg, 1999.
14. A.M. Cohen and L. Meertens. *The ACELA project: Aims and Plans*, pp. 7–23 in *Computer-Human interaction in Symbolic Computation* (ed. N. Kajler), Texts and Monographs in Symbolic Computation, Springer-Verlag, Wien, 1998.
15. J.C. Faugère's Polynomial Equations Server, www-calfor.lip6.fr/~jcf.
16. GAP interface to OpenMath. OpenMath ESPRIT Deliverable, 2000, www-groups.dcs.st-andrews.ac.uk/~gap/Info4/deposit.html.
17. JavaServer Pages, for dynamically generated Web content, java.sun.com/products/jsp/.
18. Jome: Java OpenMath editor, http://mainline.essi.fr.

19. Maple, the computer algebra system, www.maplesoft.com.
20. *Mathematica*, the computer algebra system, www.wolfram.com.
21. MATHML, Mathematical Markup Language, www.w3.org/TR/MathML2/.
22. MathWriter, Stilo's editor for rapid generation of mathematical expressions for display and processing on the web (handles MATHML and OpenMath), STILO: www.stilo.com.
23. Mozilla, a browser development project, www.mozilla.org.
24. OMDoc, a standard for open mathematical documents, www.mathweb.org/omdoc/.
25. OpenMath Society Website, www.openmath.org.
26. PolyLab Java Phrasebook for Maple, team.polylab.sfu.ca/~warp/openmath0.7.6.tar.
27. Prime Pages, http://www.utm.edu/research/primes.
28. ROML, The RIACA OpenMath Library, www.riaca.win.tue.nl.
29. J. Richter-Gebert and U. Kortenkamp. *Cinderella. The interactive geometry software* (book and CD-Rom), Springer-Verlag, Berlin, Heidelberg, 1999. See also www.cinderella.de/en/index.html.
30. N.J.A. Sloane. Online Encyclopedia of Integer Sequences, www.research.att.com/~njas/sequences.
31. A. Solomon, C.A. Struble. *JavaMath: an API for Internet accessible mathematical services,* to appear in Proceedings of the Asian Symposium on Computer Mathematics (2001), www.illywhacker.net/papers/ascm.ps.
32. A. Solomon, *Distributed Computing for Mathematical System Integration,* to appear in this volume, 2001.
33. Webmathematica, Mathematica on the Web, www.wolfram.com/products/webmathematica.
34. R. A. Wilson. Atlas of Finite Group Representations, www.mat.bham.ac.uk/atlas.
35. www.w3.org/XML/Query.

Distributed Computing for Conglomerate Mathematical Systems

Andrew Solomon

Faculty of Information Technology, University of Technology,
Sydney, NSW 2007, Australia
andrews@it.uts.edu.au

Abstract. This work is informed by two important schools of thought in distributed computing: *computational grids* and *web services*. We use the main ideas from these disciplines to give a conceptual framework for *conglomerate* mathematical systems which are distributed over the Internet.

When implemented, such a framework will support interactive mathematical web content, and it will help augment the functionality of popular computer algebra systems.

1 Introduction

The present volume responds to an economic imperative. As the demand for mathematical computation outstrips the time, money and expertise available to build systems, we must fulfill our needs by combining functionality from existing systems. This effort can be seen in two distinct but related strands – the drive toward *interoperability* of systems, and the *integration* of existing systems. According to Pollock [35]:

> *Interoperability-based approaches focus on the exchange of meaningful, context-driven data between autonomous systems. Integration approaches in contrast, typically attempt to build a monolithic view of the enterprise.*

Conglomeration lies somewhere in-between. As with integration, our goal is to build unified systems for a particular purpose. Unlike integration, the parts are separately owned and located and are available to participate in other systems. The parts of the system are interoperable, in a limited sense, as a result of having simple, well-defined interfaces. Having a framework for building conglomerate mathematical systems will confer the following mutually reinforcing benefits:

- o as components are deployed within the framework, the cost of developing a conglomerate system is reduced toward that of the top-level system;
- o a framework which supports accounting and payment for the use of components will encourage further component deployment, which will in turn stimulate the development of more top-level systems.

Two particular applications define the scope of our work and serve to illustrate many general issues in this area.

Interactive mathematical web content can be achieved in a limited way using Java applets in a web browser. This quickly becomes infeasible since computing even the most fundamental mathematical operations, such as the integral, can require a large and complicated piece of software. We therefore need to incorporate existing mathematical software into our web application to achieve the required functionality.

Most mathematicians who perform computer calculations have one or two favourite computer algebra systems (CASs) such as Maple, Mathematica or MATLAB. There are many pieces of mathematical functionality which are implemented in one but not others, as stand-alone programs, or which run best on special hardware such as a parallel machine. Whichever the case, it is desirable to be able to augment a CAS with functionality from an external system.

Our focus is on the conceptual aspects of this subject. Inevitably though, it will be necessary to make reference to more technical matters and, rather than attempting to explain them within the body of the text, the reader is referred to the Appendix for technical definitions and explanations.

2   Current Techniques for Integration of Mathematical Systems

Although we present a general framework for conglomerate mathematical systems, we focus on the two application domains mentioned above in order to illustrate the concepts and define the scope of the work.

This section reviews the techniques currently used to integrate systems in these application domains. Through this discussion, desiderata for a conglomerate architecture will be identified.

2.1   Interactive Mathematical Web Content

In the context of interactive mathematical web content, we concentrate on higher education since it is almost certainly the application which justifies the highest expenditure on system development as a result of its recent and widespread commercialization. From this point of view, higher education is the litmus test of viability for a conglomerate framework in support of interactive mathematical web content.

Landau [23] discusses web based materials for a computational physics course where students can 'experience' physical phenomena by modeling them mathematically and then creating graphical or auditory simulations based upon these models. However, on the economics of this enterprise:

> the investment in time needed to develop and maintain the materials
> is tremendous, and in our project has only been possible through the

hard work of graduate students and staff, and with the support of grants.

From the Australian Senate Inquiry into Higher Education, 2001 [3]:

> ... *there is a concern that innovative teaching materials of high quality are not being produced in sufficient quantity ... It has been estimated that in the United States the cost of developing quality on-line packages is about $25,000 per 'instructional hour', or $650,000 per semester. This is far above the cost of delivering conventional programs... High development costs incur high maintenance costs ... there remains the issue of whether the technology has reached a point where it can deliver what is required without breaking down.*

These reports reflect the fact that while interactivity which requires computation on a remote server has been achieved by ad-hoc techniques such as CGI scripting (see Section A.6 and [47]), these solutions tend to be neither robust nor scalable and have a poor cost/benefit ratio since there is little re-usability.

For another example, Cohen, Cuypers and Sterk [8] have created an interactive textbook to be read in a web-browser. It uses the free CAS GAP [13] via applets in the web-browser to demonstrate concepts in abstract algebra. Since there is no infrastructure for load and session management on a central server, it is necessary to install GAP on the student's machine via CD – a distribution channel fraught with difficulty. Firstly, automating the installation of such a large and complex piece of software is usually unreliable. More insurmountably, this method does not extend to an interactive document based on a commercial CAS – distributing a commercial CAS in this way would make the book unaffordable.

To understand precisely the source of these economic problems, we conceptualize the steps required to implement interactive mathematical web content as follows:

1. Install and maintain the component which actually performs the mathematical computation. This might be anything from a simple C program, to an entire CAS or a specially compiled program running on a parallel architecture (for example, the Tübingen Parallel Gröbner Basis system [42]);
2. Write the wrapper (see Section A.4) for this component to enable it to be called from another program – in this case the wrapper must present an API (Section A.1);
3. Write an applet or HTML form to present the interactive element together with a CGI script or Java servlet which will interface with the wrapper (Section A.6);
4. Write the actual content and embed the applet or form into the text with the appropriate parameters.

The author's JavaMath software development kit [38] greatly assists in Step (3) – the construction of and access to server-side functionality, while Cohen's group at RIACA, Eindhoven [7] is addressing the problem of creating a language richer than HTML in which to write interactive mathematical texts (4) and generate connections to interactive elements without low-level programming. Nevertheless, it is easy to see that developing online content with more than a very few interactive elements will certainly go beyond the most generous course development budget.

If the routine production of interactive content is to be feasible there needs to be a framework in which a single person or teaching unit need not bear the responsibility for all four of the tasks above. It would be more cost effective if the content developer (4) could buy or lease the interactive element (3) from an institution who had already developed it for its own course, which is in turn supported by a computational service (1 and 2) provided by an organization with expertise in that piece of software.

We anticipate that just as each teacher does not invent his or her own way of teaching calculus, online content developers will not have to devise and implement all their own interactive elements. Instead, content developers will incorporate into their own material interactive elements made available on the Internet by other developers, configuring them to suit their own pedagogic aims.

2.2 Augmenting CASs with Functionality of External Systems

There are many situations where it is desirable to add to a CAS functionality which has already been implemented efficiently elsewhere. For example:

o Vega [44] has an interface from Mathematica to McKay's C program Nauty [29] for calculating automorphism groups of graphs;
o the author [9] has augmented Maple with three dimensional graphical rendering of posets as implemented in Java by Freese [11].
o it would be useful to be able to access directly from a CAS the fast parallel Gröbner Basis computations provided on the Web by the University of Tübingen [42] or Faugère's FGb [15] which is available for non-commercial use via a web interface.

The steps required to add functionality from an external system to a CAS can be conceptualized as follows when the CAS and adjunct system both reside on the same computer:

1. Install and maintain the component which performs the computation;
2. Write the wrapper for the component with a command-line interface (Section A.1) and formal syntax (Section A.3);
3. Devise a scheme whereby the wrapper can be invoked from *within* the CAS, usually via a command shell.

Step (2) is often the most challenging because it requires disambiguating the user output of the original system into formal output syntax of the wrapper. Step (3) is more awkward than the corresponding step (3) in the previous section as most CASs do not support any programmatic interface to external programs and must therefore perform all interaction through the command shell (see Section A.1) and input/output files.

The difficulty with this picture is that the user of a CAS usually does not want the inconvenience of installing and maintaining an external system. Nor do they want to grapple with the problem of transforming the syntax from another system into that of the CAS.

We have sought to avoid this problem in our Hasse diagram renderer [9] for Maple and GAP. At a single command within the CAS, a web browser is invoked via a command shell, and loaded with an HTTP request for a service which returns a Java applet displaying the Hasse diagram. In this way it is unnecessary to install the rendering engine on the user's computer.

Unfortunately, this is not a general solution to the need to take responsibility for the adjunct system out of the CAS user's hands, since it doesn't provide a mechanism for getting the results of a remote computation back into the CAS. Two recent developments in CASs point the way forward.

Mathematica's MathLink [48] enables Mathematica to interface with external programs via an API interface rather than a command line (see Section A.1). Such an external program could send its arguments to a mathematical computation service on the Internet and return the result directly into Mathematica.

Another encouraging development is that Maple 7 has a sockets [27] library for communicating over the Internet, and a library for parsing XML [54]. These ingredients suffice to make Maple a client for a computation service available on the Internet.

Ultimately we want to achieve a single generic mechanism which could be used for all computation requests with little or no extra software needing to be loaded into the CAS to interface with each new online service.

In achieving this, the first problem is to determine how CASs will phrase requests for computation and how they will interpret the results. That is, we need to establish a standard for request-response exchanges. Part of this is a standard for the representation of the mathematical objects exchanged. OpenMath [34], MathML [53] and MP [21]Multi Protocol are well advanced on the path to solving this problem. Beyond that comparatively little progress has been made in standardizing the *form* the requests and their responses may take. For example, one needs to decide whether the response to $\texttt{Solve}(\sin(x) = 0)$ is to be an exact symbolic solution or a floating point approximation, and whether it is to be a representative subsequence or all the infinitely many solutions. A CAS must know what to expect in order to process the answer. The approaches to this problem range from specifying the form of the response in the definition of a mathematical service ([7], [25]) to permitting the requestor to specify the required form and having the service attempt to cast the result in that form ([6], [25], [1]).

The other problem is to locate the service providing a particular computation without knowing its URL beforehand. This requires a mechanism for *discovery* which is discussed in more detail below.

3   A Synopsis of Distributed Computing

This section briefly sketches two streams of technological development in distributed computing which are convergent [18] on an architecture with many of the key features required for building conglomerate systems.

3.1   Grid Computing

The *grid* paradigm defines a modular architecture for systems of distributed data and computation on the Internet [16], [17]. It is an emerging framework to enable applications to share and access data and computation transparently over the Internet in much the same way as we already obtain electricity over a power grid without having generators in our homes.

In the early 1990's as the cost of commodity hardware and operating systems plunged and high bandwidth, low latency networks became readily available, cluster computing emerged as a low cost alternative to multiprocessor, shared memory supercomputers. For the so called "embarrassingly parallel" computations which involve non-communicating processes invoked with different parameters, a shell script (a program run in the operating system's command shell) suffices to run several parts of the computation on different nodes. Standard APIs now exist for achieving more structured parallelism, such as PVM (Parallel Virtual Machine [32]) and MPI (Message Passing Interface [2]).

At the next level of aggregation, the cluster paradigm is the foundation of a number of terascale computing facilities which are the result of harnessing together clusters in different geographic locations via high speed data links. Several of the most impressive initiatives in this direction are supported and administered in the USA by the National Science Foundation's Division of Advanced Computational Infrastructure and Research [28]. This work has driven research into the problems of heterogeneity of the components and variability in the quality of the network [4]. By analogy with the electric power grid, such distributed reservoirs of computing power have come to be known as *computational grids*.

NetSolve is a computational grid which has moved beyond simply providing raw computing power in the form of CPU cycles and memory. The components of NetSolve are services which solve particular numerical linear algebra problems [31]:

NetSolve searches for computational resources on a network, chooses the best one available, and using retry for fault-tolerance solves a problem, and returns the answers to the user. A load-balancing policy is used by the NetSolve system to ensure good performance by enabling the system to use the computational resources available as efficiently as possible.

Neos [30] is a similarly structured computational grid for solving mathematical programming (optimization) problems.

In recognition of the importance of these high-level grid functions, the Globus [19] project has formed to develop technology, standards and infrastructure to accommodate and manage distributed computational resources. This project is independent of any particular problem domain and computational grid.

3.2   Web Services

Web services [22] have emerged from the technological efforts to make functions of programs running on one computer, available to be called from other computers on the Internet (see Section A.5 for background).

The idea of organizations having an Internet accessible, programmatic interface to their system enables the construction of aggregate and complex services to be integrated into a single system involving several organizations. An e-commerce example [14]: *"a Web service could provide a set of high-level travel features by orchestrating lower-level Web services for car rental, air travel, and hotels."*

While mathematical web services at this level can easily be achieved with JavaMath [38], the same article [14] lists the following unresolved technological issues which are also of concern for mathematical applications:

○ *Discovery* refers to two separate issues - service *description* and service *location*. The WSDL (Web Services Description Language) enables resources to be described in terms of their input and output types. This description is lodged with a resource broker. The program requiring the service gets the broker to select a service with a particular name (which describes what it does, e.g `PolynomialFactorization`) together with the required signature (e.g `DistributedPoly` → `List(DistributedPoly)`). The broker then returns the location of the polynomial factoring service as a URL.

○ *Reliability*: the requesting program doesn't depend on the vagaries of the services provided by particular organizations, but discovers the best service available when it is required.

○ *Scalability*: As demand for a particular type of service increases, it must be possible to deploy more components providing this service and make them available transparently. This is also possible with a good discovery mechanism where clients become aware of services as they become available.

o *Accountability* is one of the primary concerns if we are to distribute costs associated with aggregating mathematical software so that there is incentive for the providers of mathematical components to make them available to other mathematical systems. A beginning on how to structure a grid economy is made in the papers [49] and [5].

Sophisticated systems for discovery, reliability and scalability are already implemented in Grid systems for numerical computing such as NetSolve [31].

4   A Distributed Framework for Mathematical Systems

We describe an ongoing project at the University of Technology, Sydney to design and implement a framework in which to deploy mathematical web services in support of conglomerate mathematical systems.

The architecture we use is common to both the grid and web services world views and has three layers: the *application* layer, the *collective* layer and the *resource* layer. An application (such as an interactive document server or a SOAP enabled CAS) can contact a *resource broker* in the collective layer and request any services it needs – such as data or a particular computation. The broker will then look up its directory of services (or *resources*) to provide the application with the address of a resource it can use. The resource suggested depends upon availability, cost and demand on the resource. The actual location and owner of the resource is irrelevant to the application since all resources which provide a particular service have a standard interface.

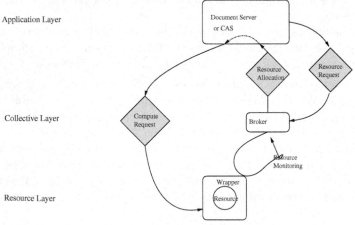

4.1   Resource Layer

JavaMath [38] provides an API and server for constructing mathematical components for the resource layer. We have developed a number of such services including an online database of group character tables. At present Java-Math provides an RMI over HTTP interface, however we anticipate making public interfaces to our services using SOAP over HTTP.

4.2   Collective Layer

This layer is a place to organize services into a coherent collection to enable discovery. Regardless of the particular registry system adopted there are two necessary conditions: there must be a standard behaviour for particular service names; and there must be a standard format in which to express the arguments and return values of the service. The standardization of service names is a matter for community agreement, so that the behaviour of $Solve(\sin(x) = 0)$, for example, is universally accepted within the user community. The specification of arguments and return values is done in the service's WSDL description document stored at the broker's service registry. The argument types will be expressed using the Simple Type System for Open-Math [10] and the actual arguments will be represented using the OpenMath standard [34].

The service description at the repository will contain not only the resource's location, signature and an (informal) description of its behaviour, but perhaps also a measure of the cost of the service in terms of both its price for usage, and its algorithmic complexity – this latter being a feature of NetSolve [31].

4.3   Application Layer

For the purposes of extending CASs with external functionality, we believe that systems such as Maple 7 are paving the way forward by implementing functionality for sockets and XML, which is enough to implement a client for the broker. We anticipate implementing our own (perhaps UDDI compliant) client in Maple 7.

The implementation of an interactive document server would be straightforward as a client for the UDDI broker - using one of the publicly available software development kits such as UDDI4J [41]. More complicated however, is the task of compiling a high level language (suitable to describing interactive documents) into Java Server Pages with embedded calls to mathematical web services. A language for describing interactive mathematical web content is under development by Cohen's group in Eindhoven [7], and we are collaborating on a compiler to interface with our distributed framework for mathematical services.

While mechanisms for accounting for usage don't appear to be settled in the case of web services, we will work toward the mechanisms being used in the Grid computing framework, or adopt them completely if possible.

5   Conclusion

Recent developments in distributed architecture and protocols such as the Grid for scientific computing, and Web services for e-business offer insights into how to achieve conglomerate mathematical systems more cheaply and effectively.

The desirable qualities of a framework to support conglomerate mathematical software are: a system for service discovery, a way of measuring service reliability, scalability and an economy to facilitate compensation for organizations providing services on the Internet. As we saw in Section 2.2, in addition to these generic issues in distributed computing, the special case of mathematical systems requires request-response standardization in a domain where there are usually many correct answers to any given question.

There is no "out-of-the-box" solution which is a complete framework, however current work on standards for mathematical object representation, and software development kits for mathematical web services (such as Java-Math [38], PolyMath [36] and ROML [37]), together with the toolkits and architectures for computational grids (e.g Globus [19]) and web services (e.g WSDL and UDDI [41]) fill in large parts of the conceptual and technical jigsaw puzzle.

Our immediate outstanding task is to implement prototypes in order to evaluate the available technologies and understand how they fit together to complete the picture.

References

1. Alessandro Armando, Michael Kohlhase and Silvio Ranise, *Communication Protocols for Mathematical Services based on KQML and OMRS*, to appear in the Proceedings, Calulemus 2000.
2. Argonne National Laboratories, *The Message Passing Interface*, http://www-unix.mcs.anl.gov/mpi/.
3. Australian Senate Inquiry into Higher Education, 2001, http://www.aph.gov.au/senate/committee/eet_ctte/public%20uni/report/contents.htm
4. Rajkumar Buyya, Hai Jin and Toni Cortes, *Cluster Computing*, Future Generation Computer Systems, **18** (2002) v–viii.
5. Rajkumar Buyya, David Abramson and Jonathan Giddy, *Grid Resource Management, Scheduling and Computational Economy*, in Proceedings, WGCC2000.
6. O. Caprotti, A. M. Cohen, H. Cuypers, M. N.Riem, H. Sterk, *Using OpenMath Servers for Distributing Mathematical Computations*, Proceedings of Asian Technology Conference on Mathematics, 2001.
7. A. M. Cohen, H. Cuypers, E. Reinaldo Barreiro and H. Sterk, *Interactive Mathematical Documents on the Web*, to appear in this volume, 2002.
8. A. M. Cohen, H. Cuypers and H. Sterk, *Algebra Interactive*, http://www.win.tue.nl/~ida/.
9. Alan Cooper, Stephen A. Linton and Andrew Solomon, *Constructing mathlets using JavaMath*, (2001) to appear in the Journal of Online Mathematics and its Applications, http://www.illywhacker.net/papers/mathlets.html.
10. James Davenport, *A small type system for OpenMath*, http://www.nag.co.uk/projects/OpenMath/omstd/sts.pdf

11. Ralph Freese, *LatDraw* http://www.math.hawaii.edu/~ralph/LatDraw/.
12. Todd Gayley, *A MathLink Tutorial*, MathSource http://www.mathsource.com.
13. GAP - Groups, Algorithms, Programs http://www.gap-system.org/.
14. Graham Glass, *Applying Web services to applications*, http://www-106.ibm. com/developerworks/webservices/library/ws-peer1.html, IBM developer-Works, November 2000.
15. J.-C. Faugère, *FGb*, http://calfor.lip6.fr/~jcf/Software/FGb/index.html
16. Ian Foster, *Internet Computing and the Emerging Grid*, http://www.nature. com/nature/webmatters/grid/grid.html, Nature, 7 December 2000.
17. I. Foster, C. Kesselman, S. Tuecke, *The Anatomy of the Grid: Enabling Scalable Virtual Organizations* http://www.globus.org/research/papers/ anatomy.pdf, to appear in Intl J. Supercomputer Applications, 2001.
18. I. Foster, C. Kesselman, J. Nick, S. Tuecke, *The Physiology of the Grid: An Open Grid Services Architecture for Distributed Systems Integration* http:// www.globus.org/research/papers.html#OGSA, (2002).
19. The Globus Project, http://www.globus.org/.
20. JavaMath, http://javamath.sourceforge.net/.
21. Simon Gray, Norbert Kajler and Paul Wang, *MP: A Protocol for Efficient Exchange of Mathematical Expressions*, Proceedings ISSAC 1994.
22. Heather Kreger, *Web Services Conceptual Architecture (WSCA 1.0)*, IBM Software Group, May 2001.
23. Rubin H. Landau, Hans Kowallik and Manuel J. Paez, *Computational Physics Education: A Course and a Web-Enhanced Book*, http://www.physics.orst. edu/~rubin/TALKS/CPtalk/.
24. Weidong Liao, Dongdai Lin and Paul S. Wang, *OMEI: an open mathematical engine interface*, Proceedings, Fifth Asian Symposium on Computer Mathematics, 2001.
25. Stephen A. Linton and Andrew Solomon, *OpenMath, IAMC and GAP*, proceedings of ISSAC/IAMC workshop (1999).
26. Masahide Maekawa, Masayuki Noro, Nobuki Takayama, Yasushi Tamura and Katsuyoshi Ohara, *The design and implementation of OpenXM-RFC 100 and 101*, Proceedings, Fifth Asian Symposium on Computer Mathematics, 2001.
27. Qusay H. Mahmoud, *Sockets programming in Java: A tutorial*, JavaWorld, (1996) http://www.javaworld.com/javaworld/jw-12-1996/jw-12-sockets. html.
28. National Science Foundation, Division of Advanced Computational Infrastructure and Research, http://www.cise.nsf.gov/acir/
29. Brendan McKay, *Nauty*, http://cs.anu.edu.au/~bdm/nauty/.
30. Neos Server for Optimization, http://www-neos.mcs.anl.gov/neos/.
31. NetSolve, http://icl.cs.utk.edu/netsolve/.
32. Oak Ridge National Laboratory, *Parallel Virtual Machine*, http://www.epm. ornl.gov/pvm/pvm_home.html.
33. Object Management Group, *CORBA*, http://www.corba.org/.
34. OpenMath Consortium, *The OpenMath Standard*, OpenMath Deliverable 1.3.4a, O. Caprotti, D. P. Carlisle and A. M. Cohen Eds. http://www.nag. co.uk/projects/OpenMath.html, February 2000.
35. Jeffrey T. Pollock, *The BIG Issue: Interoperability vs. Integration*, eAI Journal, October 2001.
36. PolyLab, *The NAOMI OpenMath Java Library*, http://pdg.cecm.sfu.ca/ openmath/#lib

37. RIACA, *ROML: The RIACA OpenMath Library*, http://crystal.win.tue.nl/download/.
38. Andrew Solomon and Craig A. Struble, *JavaMath: an API for Internet accessible mathematical services*, Proceedings, Fifth Asian Symposium on Computer Mathematics, 2001.
39. Sun Microsystems, *Java Remote Method Invocation*, http://java.sun.com/products/jdk/rmi/
40. Sun Microsystems, *Java Servlet Technology*, http://java.sun.com/products/servlet/.
41. Doug Tidwell, *UDDI4J: Matchmaking for Web Services* http://www-106.ibm.com/developerworks/opensource/library/ws-uddi4j.html?dwzone=opensource, IBM Developerworks, January 2001.
42. University of Tübingen, *Parallel Gröbner Basis Computation*, http://www-sr.informatik.uni-tuebingen.de/~www-sr/java_gui/groebner.html
43. Universal Description, Discovery and Integration Project, http://www.uddi.org/.
44. http://vega.ijp.si/Htmldoc/vinfo.htm
45. P. Wang, S. Gray, N. Kajler, D. Lin, W. Liao, X. Zou, *IAMC Architecture and Prototyping: A Progress Report*, Proceedings ISSAC 2001.
46. WebCT http://www.webct.com/
47. WIMS, Xiao Gang, University of Nice, http://wims.unice.fr/
48. Wolfram Research, MathLink http://www.wolfram.com/solutions/mathlink/.
49. Rich Wolski, James S. Plank, John Brevik and Todd Bryan, *G-commerce: Market Formulations Controlling Resource Allocation on the Computational Grid*, http://www.cs.utk.edu/~plank/plank/papers/CS-00-450.html.
50. World Wide Web Consortium, *CGI: Common Gateway Interface*, http://www.w3.org/CGI/.
51. World Wide Web Consortium, *Web Services Description Language*, http://www.w3.org/TR/wsdl.
52. World Wide Web Consortium, *Simple Object Access Protocol*, http://www.w3.org/TR/SOAP/.
53. World Wide Web Consortium, *The W3C's Math Homepage*, http://www.w3.org/Math/.
54. World Wide Web Consortium, *Extensible Markup Language*, http://www.w3.org/XML/.

A   Technical Concepts and Terminology

Throughout the text we have skipped lightly over technical detail. In this section we give a brief guide to the most important terms and technologies.

A.1   Interfaces

We distinguish three types of interface a system may have: a *user interface*, an *application programming interface* (API); and a *command line interface*.

A user interface expects user input (e.g from the keyboard) and produces human readable output, generally to the screen.

An API specifies a set of function calls or methods by which a system can provide functionality to another program. For example, Mathematica provides an API interface called MathLink [48]. A fragment of C code to add two integers using Mathematica would look like [12]:

```
MLPutFunction(lp, "Plus", 2);
MLPutInteger(lp, i);
MLPutInteger(lp, j);
MLEndPacket(lp);
```

A command line interface is the simplest of all, being simply a program to be executed in a command shell, taking its input from a file and returning its output to another file.

A.2   Sessions vs Stateless Access

A *session* with a system is a sequence of interactions where the system's state continues to evolve throughout. This is most often the case with systems with a user interface. Normally command line systems do not provide access to a session, but are instead *stateless* and process all input beginning with the same initial state.

Systems with an API can provide stateless access (like a programming library), or a session in the way MathLink provides a session with the Mathematica kernel [12].

A.3   Syntax, Semantics, and Standards

Input and output to a system may take a number of forms. Firstly, when a system has a user interface, the input and output will be human readable strings. One characteristic of user output is that the meaning of these strings on their own is often inherently ambiguous depending upon the user's knowledge of the context for their meaning.

An API can have exactly the same syntax as a user interface, simply exchanging human readable input and output strings with a system via function calls such as

```
SendUserCommand(''AutomorphismGroup(FieldExtension(Rationals(),
                                    x^2-1));'');
```

and `String RetrieveUserOutput()` taking a value such as "$\langle x \mid x^2 = 1 \rangle$" which is only meaningful due to an awareness of mathematical notation in this particular problem domain. More commonly, though, an API will have a more formal representation of the data being exchanged with the return value being, perhaps an array of the form

```
[GroupPresentation, [x], [x^2=1]]
```

Having an interface with a formal syntax such as this, together with an awareness of the semantic mapping permits an unambiguous transfer of mathematical objects between systems.

To further ease the transfer of objects between systems, a number of standards for the formal representation of mathematical objects have been proposed. Among them MP [21] and OpenMath [34] and MathML [53]. MathML is, however, limited to a fixed repertoire of object types while MP and Open-Math have the notion of a *content dictionary* which enables new encodings to be defined for each new type of object to be transferred. The advantage of a standard for object representation is that, while in principle the encoding of any particular type of object consists of the same ingredients, the difference between structures and syntax used is a routine but cumbersome extra layer of translation. If two systems recognize XML encoded OpenMath there is no need to translate

```
[GroupPresentation,[x], [x^2=1]]
```

into

```
struct
{
        name="GroupPresentation";
        varnames = ["x"];
        relations = [[[1,2],1]];
}
```

since both systems will correctly interpret

```
<OMOBJ>
  <OMA>
    <OMS cd="group_presentations" name="presentation"/>
    <OMBIND>
      <OMBVAR>
        <OMV name="x"/>
      </OMBVAR>
      <OMA>
        <OMS cd="arith1" name="equals"/>
        <OMA>
          <OMS cd="arith1" name="mult"/>
          <OMV name="x"/>
          <OMV name="x"/>
        </OMA>
        <OMS cd="group1" name="id"/>
      </OMA>
    </OMBIND>
  </OMA>
</OMOBJ>
```

A.4   Wrappers

It is often necessary to transform a component system's interface type and syntax in order for it to become part of a conglomerate system. Of particular concern to us are the following two examples.

Most CASs (excluding Mathematica) can only invoke other programs via a command line interface to the operating system shell. Therefore, if a system has a user interface it can only be used as an adjunct to a CAS if we construct a command-line interface to it. Similarly, if we wish to invoke a system from another programming language such as C, it is most convenient if it can be given an API (see Section A.1).

Another important transformation on a program's interface which makes its inclusion into a conglomerate system easier is if it is given a standard formal syntax for object transfer such as OpenMath.

To accomplish these transformations, a piece of software known as a *wrapper* is implemented. The wrapper takes the place of the component system in our application, presenting the desired interface to the operating environment and translating external interactions to and from the native interface and syntax of the component system.

The construction of wrappers is probably the most difficult part of building conglomerate systems, especially when the native interface of a system is a user interface, with all the attendant ambiguities of interpretation of output.

A.5   Web Services: RMI, CORBA, and SOAP

Object oriented systems are built from interacting components consisting of data and methods. They tend to model objects in the application domain. For example a mathematical system might have a `Field` object with methods `Field.zero()`, and `Field.add(FieldElt a, FieldElt b)`.

RMI (Remote Method Invocation) [39], CORBA (Common Object Request Broker Architecture) [33] and SOAP (Simple Object Access Protocol) [52] are all protocols for taking an object in a system running on some computer and exposing it to, and accessing it from, a remote system as though it were local to that system. RMI is native to the Java platform and CORBA is an established cross-platform standard.

SOAP is an emerging cross-platform standard based on XML [54], its advantages are that one may easily incorporate standard representations for data types such as OpenMath; and the protocol is itself plain text, allowing messages between computers to be inspected for debugging purposes. Another attribute which sets SOAP apart from the other protocols for object access is the fact that by default it communicates via Hypertext Transfer Protocol (HTTP), the ubiquitous, stateless protocol which enables the World Wide Web. SOAP over HTTP shows great promise for aggregating services over the Internet where firewalls and quality of service issues undermine persistent socket connections required by RMI and CORBA.

Associated with SOAP are WSDL (Web Services Description Language [51]) and UDDI (Universal Description, Discovery and Integration of Business for the Web [43]). WSDL is a standard way of describing the objects made available on the Internet as services available to other systems and UDDI is a framework for storing and searching these descriptions to discover a desired service and its location. WSDL and UDDI are being developed particularly for integrating business functions between organizations.

A.6   CGI and Servlets

In creating interactive web content, unless the interactive element can be implemented as a small Java program (known as an *applet*) running in the browser, it is necessary to orchestrate an interaction between the user and a program on the web server. This is generally referred to as *server-side interactivity*. There are two principal technologies for achieving this: CGI (Common Gateway Interface [50]) and Java Servlets [40].

CGI is a standard by which an external program interfaces to a web server for the purpose of generating a response to HTTP requests. A program which acts in this capacity is commonly referred to as a CGI *script*. Similar to CGI scripts, Java Servlets are programs which run within a Java enabled web server to generate responses to HTTP requests.

While the response to an HTTP request is often an HTML page, it can also be any other object such as plain text, an XML document or a serialized Java object. This last possibility is the basis of being able to use HTTP for emulating RMI over the Internet as described in the following section.

A.7   JavaMath

There are a number of toolkits for aggregating disparate mathematical systems over the Internet. OMEI (Open Mathematical Engine Interface [24]) is an API and architecture for accessing CAS sessions remotely. This may be done from within code or from a special browser (prototyped as *Dragonfly* [45]) but in either case, using the informal human readable syntax provided natively by the CAS.

At the other extreme with a formal syntax, OpenXM (Open message eXchange for Mathematics) [26] defines a stack machine API to a system and a standard instruction set and data representation for interacting with it.

Produced by the author and his collaborators, JavaMath [38] is an SDK (Software Development Kit) to enable the development of conglomerate systems in Java from existing components. It gives a template for writing wrappers and a standard API for creating and using sessions with these components. The programmer may interact with these sessions using both user (informal) and OpenMath (formal) syntax.

The JavaMath SDK provides an RMI-like interface for the client program to communicate with the server using HTTP and Servlets, ensuring reliable service over the Internet. To impart the flavour of programming with JavaMath we give a snippet of code to factor a polynomial using Maple. Typically, the code would be part of a JSP page or servlet on a web server, and it would make use of Maple running on a JavaMath server (the fictitious compserv.javamath.net in the present example).

```
// Find server
proxyCAServer =
    new ProxyCAServer("http://compserv.javamath.net/ProxyCAS");
// Set up the Maple session
proxyMapleSession = proxyCAServer.createSession("Maple");

/* <snip> Some code here which retrieves a string p
 * representing a polynomial from user input. */
proxyMapleSession.sendNativeCommand("evala(AFactor(p));");

fp = proxyMapleSession.getNativeOutput();
// fp is now the factorization
// (a string representing a list of polynomials)
```

Demonstrations of applications built with JavaMath can be found at the JavaMath home pages [20].

Index

Software Systems